Green Nanotechnology

Green Nanotechnology

Editor: Jeff Norton

R CALLISTO REFERENCE

www.callistoreference.com

Callisto Reference,
118-35 Queens Blvd., Suite 400,
Forest Hills, NY 11375, USA

Visit us on the World Wide Web at:
www.callistoreference.com

ISBN: 978-1-64116-133-6 (Hardback)

Cataloging-in-Publication Data

Green nanotechnology / edited by Jeff Norton.
      p. cm.
Includes bibliographical references and index.
ISBN 978-1-64116-133-6
1. Nanotechnology. 2. Green technology. 3. Sustainable engineering.
I. Norton, Jeff.
T174.7 .G74 2019
620.5--dc23

Table of Contents

Permissions

List of Contributors

Index

Preface

Over the recent decade, advancements and applications have progressed exponentially. This has led to the increased interest in this field and projects are being conducted to enhance knowledge. The main objective of this book is to present some of the critical challenges and provide insights into possible solutions. This book will answer the varied questions that arise in the field and also provide an increased scope for furthering studies.

Green nanotechnology is the field that is concerned with the application of nanotechnology for environmental sustainability. It aims to minimize potential environmental and human health risks associated with the development, manufacture and use of nanotechnology products through the adoption of new nano-products that are ecofriendly throughout their lifecycles. The other goal of green nanotechnology is to develop products that solve environmental problems. The applications of this field include cleaning of hazardous wastes, water treatment, environmental remediation and monitoring of environmental pollutants. Other applications include the use of nanocomposites, fuel cells and light-emitting diodes for saving fuel, reducing pollution, etc. Current research in this domain explores solar cells, nanoremediation, cleaning oil spills, removing plastics from oceans, etc. The various studies that are constantly contributing towards advancing technologies and evolution of this field are examined in detail in this book. It provides significant information of this discipline to help develop a good understanding of green nanotechnology. As this field is emerging at a rapid pace, the contents of this book will help the readers understand the modern concepts and applications of the subject.

I hope that this book, with its visionary approach, will be a valuable addition and will promote interest among readers. Each of the authors has provided their extraordinary competence in their specific fields by providing different perspectives as they come from diverse nations and regions. I thank them for their contributions.

Editor

Cd(II) Sorption on Iranian Nano Zeolites: Kinetic and Thermodynamic Studies

Taher Yousefi*, Meisam Torab-Mostaedi, Amir Charkhi, Abolfazl Aghaei

Nuclear Fuel Cycle Research School, Nuclear Science and Technology Research Institute, Tehran, Iran

ABSTRACT

An uptake of Cd(II) from aqueous solutions by ion exchange on Iranian natural zeolitic (TOSKA) has been studied. Experiments were carried out using batch method as a function of the initial concentration of metal ions, contact time, and temperature. The adsorbent is characterized using XRF, FTIR, TEM, and XRD. The TEM images showed that the zeolite particle sizes are reduced into the size range of less than 90 nm by means of ball milling. The characterization of sample indicates that the natural zeolite used in this study was classified into clinoptilolite. Equilibrium modelling data were fitted to linear Langmuir and Freundlich models. Thermodynamics parameters such as change in free energy (ΔG°), enthalpy (ΔH°) and entropy (ΔS°) were also calculated. The negative values obtained for ΔG° indicated that the sorption of Cd(II) on natural zeolite was spontaneous at all studied concentrations. These results show that natural zeolites hold great potential to remove Cd(II) from industrial wastewater.

KEYWORDS: *Cd(II); Clinoptilolite; Ion exchange; Kinetic; Sorption; Nano zeolite*

INTRODUCTION

The heavy metals are not biodegradable and their presence in streams and lakes leads to bioaccumulation in living organisms, causing health problems in animals, plants, and human beings[1]. Cadmium in particular, is a very toxic element that can be present in aqueous wastes from many industries, such as oil refineries, metal plating, mining operations, pigments and batteries. Excessive human intake of Cadmium leads to damage of kidney and renalsystem, skeletal deformation, cardiovascular diseases and hypertension. Therefore, the removal of excess heavy metal ions from wastewater is essential to protect human and environmental health. As a result, the removal of toxic heavy metal ions from sewage and from industrial and mining effluents

has been widely studied in recent years[2].

Numerous processes such as precipitation, phytoextraction, ultrafiltration, reverse osmosis, and electrodialysis exist for removing dissolved Cd(II) ions [2-5]. A major problem with this type of treatment is the disposal of the precipitated wastes. Ion exchange treatment which is the second most widely used method for metal ion removal does not present a sludge disposal problem and has the advantage of recovery of Cd (II). This method can reduce heavy metals to very low levels. However, usage of synthesized ion exchanger materials does not appear to be economical. The use of alternative low-cost materials as potential sorbents for the removal of heavy metals has been emphasized recently. Cost comparisons are difficult to make due to the scarcity of consistent

* Corresponding Author Email: taher_yosefy@yahoo.com

cost information. Although many experiments have been accomplished in the area of low-cost sorbents, a great deal of work is necessary to understand better low-cost adsorption processes and to demonstrate the technology.

Inorganic ion exchange materials have emerged as an increasingly important replacement or complement for conventional organic ion exchange resins, particularly in liquid waste treatment due to their chemical stability and greater selectivity for certain important species, such as heavy metal ions [6]. Among of inorganic ion exchange materials, natural zeolites, have been received great attention, as potential adsorbents especially for application in liquid waste treatment due to their high cation exchange capacities and low cost[7].

Natural zeolite is porous material with high cation exchange capacity, cation selectivity, higher void volume and great affinity for metal ions [8]. Zeolites posses a negative net charge compensated by the presence of exchangeable cations. A variety of cations can be adsorbed on zeolites by the cation exchange mechanism.

Therefore, natural zeolites are known as efficient adsorbents for cation water pollutants[9].

The aim of this work was to determine the capacity of Iranian natural zeolite to remove cadmium ions from aqueous solutions and to predict and compare their performances from the adjustment of experimental data to models.

EXPERIMENTAL
Materils and Characterization

The Iranian natural zeolite used as an adsorbent in this study was a commercial sample, supplied from Toska Mining Company, Miyaneh, Iran. Ball milling of zeolite was performed by mean of a planetary ball mill (PM100; Retsch Corporation). It was characterized by X-ray diffraction (XRD, Phillips PW-1800), Fourier transform infrared (FT-IR, Bruker Vector 22 FT-IR spectrometer) and X-ray fluorescence (XRF, STIDY-MP-Germany). The morphology of natural zeolite after ball milling was characterized by Transmission electron microscopy (TEM, Phillips EM 2085). Inorganic chemicals were supplied by Merck as analytical-grade reagents and deionized water was used. The stock solution of Cd(II) was prepared using $Cd(NO_3)_2.H_2O$ salts. The Cd(II) ion concentrations were determined using an inductively coupled plasma atomic emission spectroscopy machine (ICP-AES, Optima 7300 DV, Perkin Elmer Co. USA).

Adsorption Studies

The ion exchange behavior of the zeolite materials for Cd(II) ions was carried out using batch equilibrium method. The influence of parameters such as shaking time, intial concentration and temperature were tested to evaluate the zeolite material characteristics. Batch adsorption experiments were conducted using 0.1 g of adsorbent with 20 ml of solutions containing Cd(II) ions of desired concentrations at different temperatures in 50 ml plastic bottles with continuous stirring at 600 rpm. After shaking (in order to determine the amount of Cd(II) ions sorbet) the solid phase was separated from the solution by filtration and the concentration of Cd(II) was determined in the liquid phase using ICP-AES.

In order to obtain the sorption capacity, the amount of ions adsorbed per unit mass of adsorbent (q_e in milligrams of metal ions per gram of adsorbent) was evaluated using the following expression[10]:

$$q_e = \frac{(C_o - C_e)V}{M}$$

where C_o is the initial metal ion concentration (mg l^{-1}), C_e the equilibrium metal ion concentration (mg l^{-1}), V the volume of the aqueous phase (l), and M the amount of the adsorbent used (g). Distribution ratio (Kd) were calculated using the equation[11];

$$K_d = \frac{C_o - C_e}{C_e} \times \frac{V}{M}$$

where V is the volume of the solution (ml) and M is the weight of the adsorbent (g).

RESULTS AND DISCUSSION
Characterization

The XRD pattern of natural zeolites is shown in Fig. 1. The XRD pattern showed that the sample exhibited high crystallinity, with the characteristic reflection peaks at Bragg angle $(2\theta)=10°, 11.3°, 13.2°, 22.5°, 27°, 30.02°$ and $32°$, respectively. The pattern indicates that the Iranian natural zeolite used in this study was classified into clinoptilolite[12]. Clinoptilolite with the ideal formula of $(Na,K)_6$ $Si_{30}Al_{16}O_{72}.nH_2O$ is the most common natural zeolite found mainly in sedimentary rocks of volcanic origin[13].

The FT-IR spectra of the natural zeolite was investigated in the 4000–400 cm^{-1} region (Fig. 2), Peaks at 1060 cm^{-1}, 794 cm^{-1} and 609 cm^{-1} were characteristic of clinoptilolite [13-15].

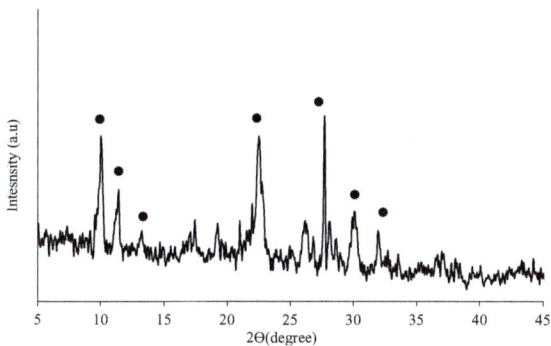

Fig. 1. XRD analysis of natural zeolite.

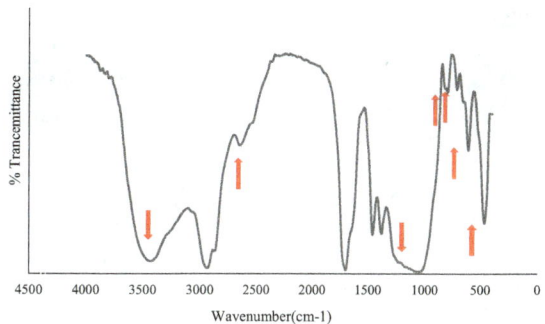

Fig. 2. FTIR spectra of natural zeolite

The strongest band observed at 1060 cm^{-1} was assigned to asymmetric stretching of the external tetrahedral linkages[13-14]. The second strongest band at 465 cm^{-1} corresponded to internal tetrahedral bending. The band observed at 609 cm^{-1} was related to the presence of double rings in the framework structure. Other bands at ca. 1208 cm^{-1}, 790 cm^{-1} and 711 cm^{-1} were assigned to the asymmetric stretching modes of internal tetrahedra, symmetric stretching of external tetrahedra and symmetric stretching of internal tetrahedra, respectively[13-15]. The 670 cm^{-1} band arises from symmetric tetrahedral stretching. In addition, the FT-IR spectra exhibited several bands at 1690 cm^{-1}, 3450 cm^{-1} and 3630 cm^{-1} arising from the deformation of water molecules, the bending vibrations of sorbed water, hydrogen-bonded –OH groups and isolated –OH groups, respectively[12-15].

The chemical composition characterization of the natural zeolite was carried out by XRF technique and is presented in Table 1. This table shows that silica and alumina are the major components of the adsorbent. Oxides of other metals are present in trace amounts [16].

The morphology of zeolite was analyzed by TEM (Fig. 3a, b and c). The TEM images show the zeolite is composed of an irregular particles with the different size range about 60–1000 nm. A irregular surface with particles size less than 100 nm as a separated particle or in the form of larger agglomerates shown in Fig.3 a and c. Moreover, it can be seen that the rods with a diameter about 60-65 nm without agglomeration are revealed by TEM analysis (Fig. 3 b). The nanostructure formation is result of zeolite's ball milling. The pervious study showed that larger than 1 mm particle size of clinoptilolite powder may mechanically be reduced into the size range of less than 100 nm by means of planetary ball milling[17].

Effect of Initial Concentration and Temperature

The effect of the initial ion concentration was performed at initial concentrations of 50, 250, 500, 750 and 1000 mg/L at different sorption temperatures of 298, 313 and 333 K for the sorption of Cd(II) ions onto the natural zeolite the results were shown in Fig. 4. It is clear that the sorption amount of Cd(II) ions increase with increasing the initial ion concentration. These results reflect the efficiency of the zeolite materials towards Cd(II) ions. Also the Fig. 4 shows the variation of the amounts of Cd(II) ions sorbed at fixed initial concentration, at different sorption temperatures of 298, 313 and 333 K. The data showed that the amount of the sorbed Cd(II) ions increases with the increase in temperature indicating an endothermic nature of the sorption processes.

Effect of Contact Time and Initial Concentration

Fig. 5 shows the effect of shaking time on the removal of Cd(II) ion using the natural zeolite. The amount of Cd(II) ions sorbet sharply increases for

Table 1. XRF pattern of natural zeolite

Analyte	Concentration	
LOI	13.15	(Wt %)
Na$_2$O	0.55	(Wt %)
MgO	0.78	(Wt %)
Al$_2$O$_3$	8.58	(Wt %)
SiO$_2$	72.59	(Wt %)
P$_2$O$_5$	0.37	(Wt %)
SO$_3$	0.18	(Wt %)
K$_2$O	1.08	(Wt %)
CaO	1.22	(Wt %)
TiO$_2$	0.19	(Wt %)
Fe$_2$O$_3$	1.13	(Wt %)
Sr	600	ppm
ZrO$_2$	300	ppm

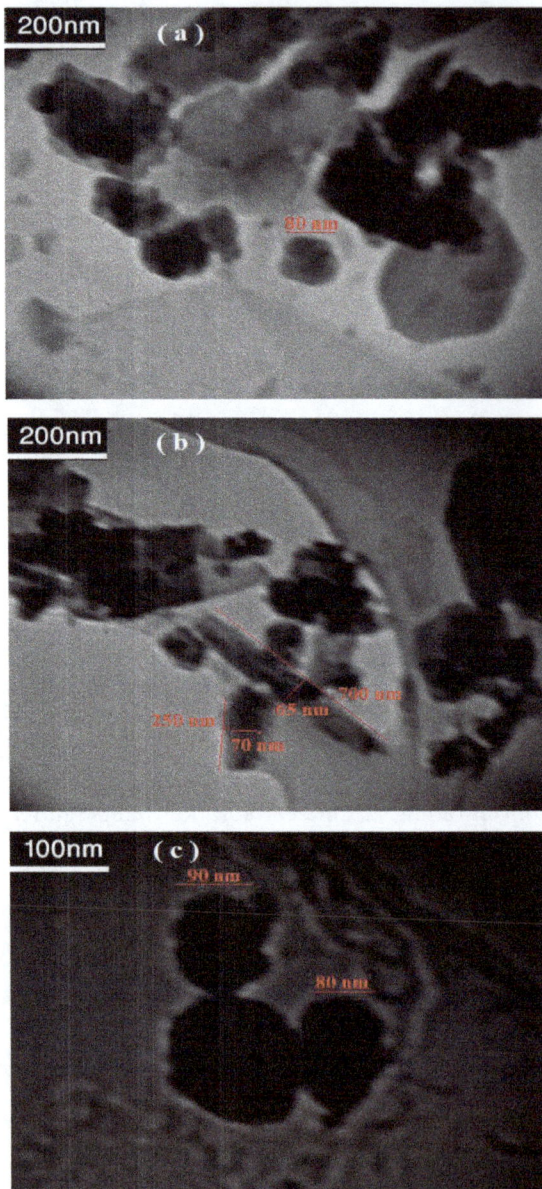

Fig. 3. TEM images of zeolite.

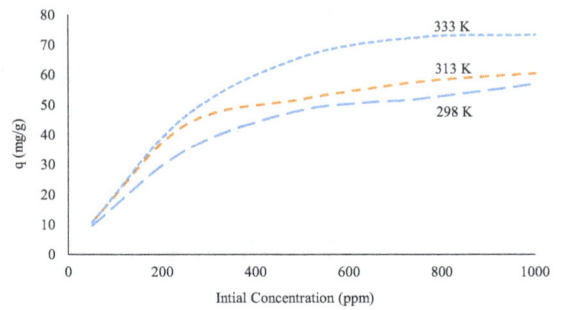

Fig. 4. Effect of initial concentration on the amount sorbet of Cd(II) ions onto natural zeolite at different temperatures

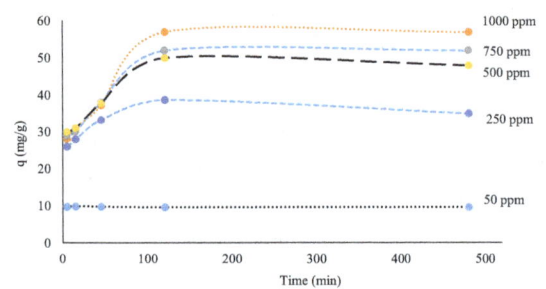

Fig. 5. Effect of initial ion concentration and contact time on the amount sorbet of Cd2+ ions onto natural zeolite.

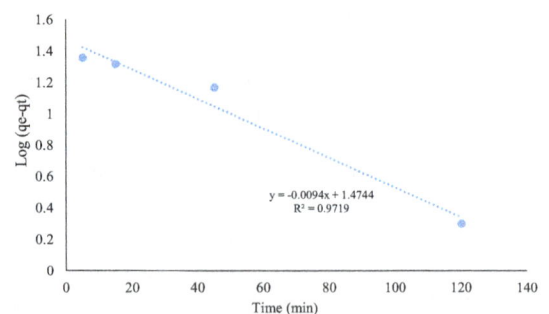

Fig. 6. Pseudo first-order kinetic model fit for Cd2+ sorption onton natural zeolite.

each adsorbent with time in the initial stage (0–100 min range). It reached up 90% within 100 min and became very slow with increasing of shaking time. This may be ascribed to the utilization of the most readily available adsorbing sites of the zeolite that leads fast diffusion and rapid equilibrium attain [18].

Kinetics Analysis

The Cd(II) sorption dependence on time was investigated at 298 K for solutions with C_0=50, 250, 500, 750 and 1000 ppm. The time dependence was followed until the sorption equilibrium has

essentially been reached. Fig. 5 shows the variation of the amounts uptake values of Cd(II) from solution at different time intervals.

It is well recognized that the characteristic of the sorbent surface is a critical factor that affect the sorption rate parameters and that diffusion resistance plays an important role in the overall transport of the ions. To describe the changes in the sorption of metal ions with time, two simple kinetic models were tested.

The data of the kinetics such as rate constants of Cd(II) ions removal from aqueous solutions by natural zeolite at different intial concentration,

Table 2. Adsorption kinetic model rate constants for Cd(II) adsorption on nnatural zeolite.

q(mg/g)	\multicolumn{6}{c}{Time (min)}					
	5	15	45	120	20	480
q_t	29	31	37	50	52	52
q_e-q_t	23	21	15	2	0	0
Log(q_e-q_t)	1.36	1.32	1.17	0.30	-	-

Metal ion	\multicolumn{3}{c}{First-order kinetic parameters}			\multicolumn{3}{c}{Second-order kinetic parameters}			q_e, exp. (mg/g)
	k_1(cm^{-1})	q_e, calc,(mg/g)	R^2	k_2(cm^{-1})	q_e, calc,(mg/g)	R^2	h(mg/g min)		
Cd^{2+}	0.0046	29.81	0.9713	0.0016	54.34	0.9983	4.72		60.5

as illustrated in Fig. 6 and 7, were analyzed using pseudo first-order and pseudo second-order. The conformity between experimental data and each model predicted values was expressed by the correlation coefficient (R^2). A relatively high R^2 values indicates that the model successfully describes the kinetics of metal ion sorption removal.

The sorption kinetics of Cd(II) ions from liquid phase to solid is considered as a reversible reaction with an equilibrium state being established between two phases. A simple pseudo first order model was, therefore, used to correlate the rate of reaction and expressed as follows:

$$\frac{dqt}{dt} = k_1(q_e-q_t) \tag{1}$$

where q_e and q_t are the concentrations of ion in the adsorbent at equilibrium and at time t, respectively, (mg/g) and k_1 is the pseudo first-order rate constant (min^{-1}). After integration and applying boundary conditions t=0 to t=t and q_t = 0 to $q_t = q_t$, the integrated form of Eq. (1) becomes:

$$\log(q_e-q_t) = \log q_e - \frac{k_1}{2.033}t \tag{2}$$

The slopes and intercept of the plots of $\log(q_e -q_t)$ versus t, as shown in Fig. 6, were used to determine the first order rate constant (k_1) and the theoretical equilibrium sorption capacities (q_e), respectively. The calculated values of k_1 and q_e with the values of the linear correlation coefficients (R^2) of plot are presented in Table 2. Approximately linear fits were observed for the adsorbent, over the entire range of shaking time explored and at all temperatures with low correlation coefficients, indicating that the pseudo first-order kinetic model is not so valid for the present systems. Another important issue that in the study of a kinetic model must be considered is the agreement of the theoretically calculated equilibrium sorption capacities, q_e

and the experimental sorption capacity value q_{ex}. As can be seen from Table 2, although the linear correlation coefficient of the plot is acceptable, the q_e (calculated) value is not in agreement with q_{ex} (experimental) for studied sorption processes. So, it could suggest that the sorption of Cd(II) ions onto zeolite is not a first-order reaction.

A pseudo second-order rate model is also used to describe the kinetics of the sorption of Cd(II) ions onto adsorbent materials. The pseudo second-order rate model is expressed as[19, 20]:

$$\frac{dqt}{dt} = k_2(q_e-q_t)^2 \tag{3}$$

For the boundary conditions t=0 to t=t and q_t=0 to $q_t=q_t$, the integrated form of Eq. (3) becomes:

$$\frac{1}{qe-qt} = \frac{1}{qt} + k_2t \tag{4}$$

$$\frac{t}{qt} = \frac{1}{k_2qe^2} + \frac{1}{qe}t \tag{5}$$

where k_2 is the rate constant of pseudo second-order equation (g/mg min).
If the initial sorption rate h (mg/L h) is:

$$h = k_2qe \tag{6}$$

Then Eqs. (10) and (11) become:

$$\frac{t}{q} = \frac{1}{h} + \frac{1}{qe}t \tag{7}$$

The kinetic plots of t/q_t versus t for Cd(II) ions sorption is presented in Fig. 7. The relation is linear, and the correlation coefficient (R^2), suggests a strong correlation between the parameters and also explains that the sorption process follows pseudo second-order kinetics.

From Table 2 it is also can be seen that theoretically calculated equilibrium sorption capacities, q_e, is very close to the experimentally

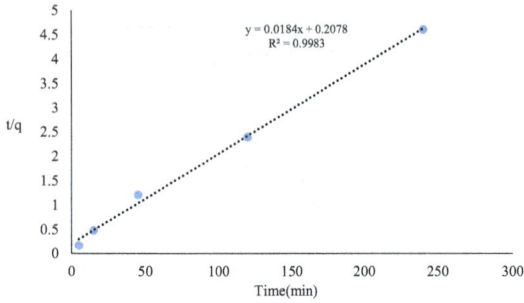

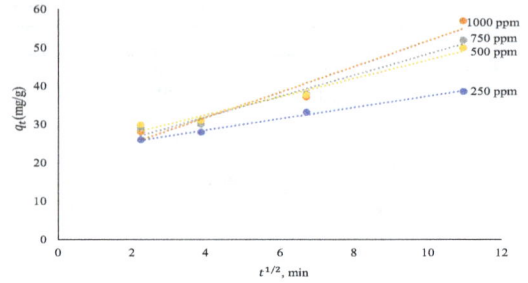

Fig. 7. Pseudo second-order kinetics of Cd2+ sorption ontonnatural zeolite.

Fig. 8. Morris–Weber kinetic plots for the sorption of the sorption of Cd2+ ions from aqueous solutions.

Table 3. Internal diffusion rate constant for the sorption Cd2+ ions onto natural zeolite at different initial concentration.

Initial concentration(ppm)	k_{ad}(mg/g min$^{1/2}$)	Intercept (b)	R^2
1000	3.33	18.5	0.956
750	2.74	20.92	0.973
500	2.39	22.96	0.975
250	1.48	22.64	0.994

measured values, q_{ex} for pseudo second-order kinetic model indicating that the pseudo-second- order model gives a better description of the Cd(II) ions sorption kinetics as judged by the R^2 values (Table 2). This agrees with general observations in heavy metal sorption studies.

Diffusion Model

Though the Cd(II) ions removal by clinoptilolite occurs via ion exchange, but the ion-exchange reaction might not alone be adequate in explaining the sorption kinetics of Cd(II); diffusional processes have also to be taken in to account. In order to examine the role of diffusion in the sorption process, the data were also analyzed by the Weber–Morris mass transfer model [21]. This model is defined by the rate equation (5):

$q_t = k_{ad} t^{1/2} + b$

Where the q_t (mg g^{-1}) is the adsorption capacity at time t (min), k_{ad} (mg g^{-1} min$^{-1/2}$) is the diffusion rate constant and, and b (mg g^{-1}) is the boundary layer diffusion. According to this model, plotting a graphic of q_t vs. $t^{1/2}$ for various initial Cd (II) concentrations, if a straight line with intercept b is obtained, it can be assumed that the involved mechanism is a diffusion of the species as shown in Fig. 8.

As can be seen in all case a liner relationship

between q_t and $t^{1/2}$ are found suggesting that the diffusion mechanism is involved. In this case the slope of the linear plot is the rate constant of intraparticle transport. The values of K_{ad} were calculated, from the slope of the linear plots obtained, and the values of b were calculated from the intercept that presented in Table 3.

Thermodynamic Study

The data obtained by sorption experiments at 298, 313 and 333K, and the initial Cd(II) concentrations of 50, 250, 500, 750 and 1000 ppm, were used for the estimation of some thermodynamic parameters. The standard free energy of sorption ($\Delta G\circ$) was calculated by Eq. 8

$$\Delta G\circ = -RT \ln K \qquad (8)$$

R is the universal gas constant and K is the equilibrium constant at the temperature T. The constant K was calculated by using Eq. (9) [20]:v

$$K = \frac{qe}{Ce} \qquad (9)$$

Ce is the equilibrium concentration of the Cd(II) ions (mg l^{-1}). The enthalpy and entropy of sorption were calculated from Eq. (10):

$$\Delta G\circ = \Delta H\circ - T\Delta S\circ \qquad (10)$$

$$\ln K = \frac{\Delta S\circ}{R} - \frac{\Delta H\circ}{RT} \qquad (11)$$

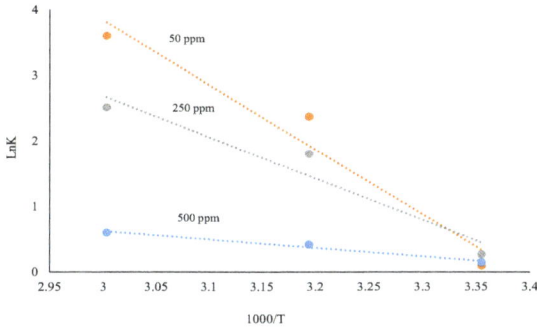

Fig. 9. Variation of the distribution coefficient (Kd) in Cd(II) ion removal using the natural zeolite as a function of temperature

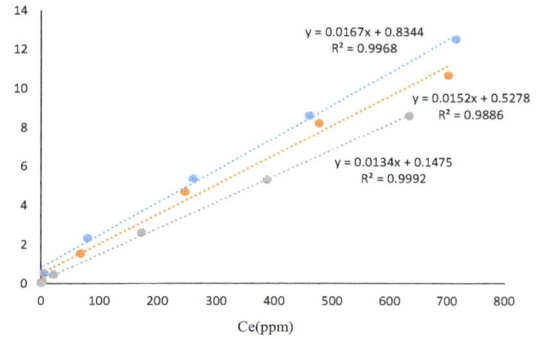

Fig 10.The linearized Langmuir isotherm for adsorption of Cd(II) by natural zeolite

Table 4. Thermodynamic parameters for zeolite adsorption of cadmium ions.

C_i (ppm)	T (K)	$\Delta G°$(kj mol^{-1})	$\Delta H°$(kj mol^{-1})	$\Delta S°$(jK^{-1})	R^2
	298	-0.28	81.72	277.10	0.952
50	313	-6.167			
	333	-10			
	298	-0.67	52.21	178	0.9352
250	313	-4.88			
	333	-7			
	298	-0.36	10.6	36.8	0.9733
500	313	-1.10			
	333	-1.7			

The plot of ln K vs. 1/T gives a straight line(Fig. 9), and the values of $\Delta S°$ and $\Delta H°$ are evaluated from its intercept and slope, respectively. The calculated thermodynamic parameters are listed in Table 4.

As can be seen from Fig. 9 and fromt he R^2 values in Table 4, the linearity of the ln K vs. 1/T plot is satisfactory for the initial concentrations of 500, and 250 ppm, and less so for 50 ppm. The $\Delta G°$ values given in Table 4 show that the sorption of Cd(II) on zeolite occurs spontaneously in the 298–333K range. The spontaneity slightly increases with temperature for all initial Cd(II) concentrations of 500 ppm. The Cd(II) sorption is endothermic ($\Delta H°>0$) and proceeds with an increase inentropy. The positive $\Delta S°$ values reflect the fact that the sorption involves the liberation of more ions when one Cd(II) ion is bound to the sorbent. It is also evident from Table 4 that the spontaneity (at a given temperature), as well as the $\Delta H°$ and $\Delta S°$ values all decreaseas the initial Cd(II) concentration increases.

Isotherms Equations

Several sorption isotherm models have been applied to describe experimental data of sorption isotherms. For the sake of convenience, explicit and simple models are preferred and commonly used; these include several two- and three-parameter isotherm models. The Langmuir and Freundlich models are the most frequently employed models.

The data obtained were applied to the Langmuir adsorption isotherm and linear expression of this model has been demonstrated as below (eq.11):

$$C_e/q_e=1/bK+C_e/b \qquad (11)$$

where q_e is the amount of metal ion sorbed per unit weight of zeolite (mg/g), Ce the equilibrium concentration of the metal ion in the equilibrium solution (mg/L), and K (L mg^{-1}) and b are the Langmuir constants related to the sorption capacity and energy of adsorption ($b \propto e^{-\Delta G/RT}$), respectively.

Langmuir, the simplest type of isotherm, is based on the view that every adsorption site is equivalent and independent; the ability of a molecule to bind is independent of whether or not neighboring sites are occupied.

Another adsorption isotherm, the Freundlich isotherm, was calculated from the adsorption data. This isotherm is more general than the Langmuir isotherm since it does not assume a homogenous surface or constant sorption potential and this

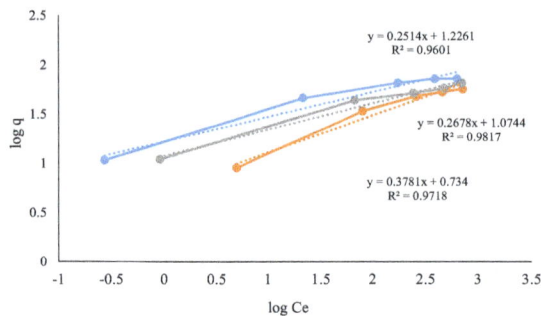

Fig. 11. The linearized Freundlich isotherm for adsorption of Cd(II) by natural zeolite

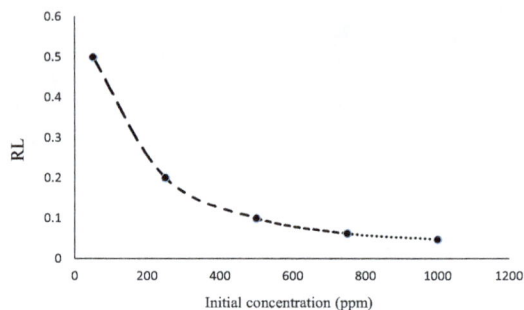

Fig. 12. Variation of separation factor (RL) as a function of initial Cd2+ ion concentration at 298 °K.

Table 5. The Langmuir and Freundlich constants and correlation coefficients of isotherm models at different temperature

Temperature (°K)	Langmuir isotherm			Freundlich isotherm		
	b	K	R^2	1/n	K_f	R^2
298	59.88	0.020	0.9968	0.3781	5.42	0.9718
313	65.78	0.028	0.9886	0.2678	11.75	0.9817
333	74.62	0.090	0.9992	0.2514	16.83	0.9601

model has a linear expression, which has been demonstrated as below (eq.12):

$$q_e=K_FC_e^{1/n} \text{ or } \log q_e=\log K_F+1/n \, LogC_e \quad (12)$$

where K_F (mg g^{-1}) is the Freundlich constant related to the sorption capacity of the sorbent, and 1/n is the Freundlich constant related to the energy heterogeneity of the system and the size of the adsorbed molecule. Other parameters (Ce, q_e) have been mentioned above.

The resulting adsorption isotherms for the Zeolite used in this study are shown in Fig. 10 and 11. The regression values and correlation coefficients (R^2) presented in Table 5 indicate that the adsorption data for Cd(II) removal best fitted the Langmuir adsorption isotherm. However, the Freundlich isotherms are important because they do not assume a homogeneous surface. The maximum adsorption capacity based on the Langmuir isotherm is 74.6 (mg g^{-1}).

The Freundlich isotherm constants K_F and n are determined from the intercept and slope of a plot of log qe versus log Ce (Fig. 11). In this study n values are greater than unity indicating chemisorption (Table 3) [23]. Isotherms with n > 1are classified as L-type isotherms reflecting a high affinity between adsorbate and adsorbent and is indicative of chemisorption [24]. The Freundlich constant, K_F, which is related to the adsorption capacity, increased with temperature, indicating that the

adsorption process is endothermic.

On the other hand, the dimensionless equilibrium parameter or separation factor, R_L, based on the further analysis of Langmuir equation can be given by (eq.13);

$$R_L=\frac{1}{1+KC0} \quad (13)$$

where C_0 (mg L^{-1}) is the initial amounts of adsorbate and k is Langmuir constant. The R_L parameter is considered as more reliable indicator of the adsorption. There are four probabilities for the R_L value; (A)for favorable adsorption, $0 < R_L < 1$, (B) for unfavorable adsorption, $R_L > 1$, (C) for linear adsorption, $R_L =1$, (D) for irreversible adsorption, $R_L =0$ [25]. Fig. 12 shows the variation of R_L with initial concentration of the Cd(II) ions. As could be seen from the curve, the R_L parameters lied between 0 and 1 represent that the removal of Cd(II) ions by natural zeolite is propitious. Fig. 12 also indicates that the R_L value approaches zero as the C_0 value is increased, and it means that the adsorption of Cd(II) ions onto natural zeolite is less favorable at high concentration of the solution.

CONCLUSIONS

Zeolites have exchange sites with different positions in the lattice and thus different bond energies. Steric hindrance and unfavorable charge distributions can affect extent and kinetics of cation exchange in zeolites. This study showed the possibility of the selected sorbents (Iranian natural zeolite, Toska) utilization for Cd(II) ions removal from aqueous solutions. The influence of parameters such as shaking time, intial concentration and temperature were tested to evaluate the zeolite material characteristics. The characterization of sample indicates that the natural zeolite used in this study was classified into clinoptilolite.

The Langmuir isotherm showed a better fit than the Freundlich isotherm, thus, indicating the applicability of monolayer coverage of cadmium on zeolite surface. Thermodynamic analysis showed that the adsorption process was endothermic and spontaneous in nature and increasing temperature improved adsorption performance for the zeolites. Results from this study suggest that Iranian natural zeolite (Toska) is a very effective adsorbent for cadmium(II) ions, as anticipated.

CONFLICT OF INTEREST

The authors declare that there are no conflicts of interest regarding the publication of this manuscript.

REFERENCES

1. Monser, Land N. Adhoum, 2002. Modified activated carbon for the removal of copper, zinc, chromium and cyanide from wastewater. Separation and Purification Technology, 26(2-3): 137–146.
2. Katsumi, Z.Z., T. Li, S. Imaizumi, X.W. Tang and T. Inui, 2010. Cd(II) adsorption on various adsorbents obtained from charred biomaterials. Journal of Hazardious Materials, 183: 410-420.
3. Sengupta A.K and D. Clifford, 1986. "Important process variables in chromate ion exchange. environmental science and technology, 20: 149-155.
4. Sharma Y.C., 2008. Thermodynamics of removal of cadmium by adsorption on an indigenous clay. Chemical Engineering Journal, 145: 64–68.
5. Applegate L.E., 1984. Membrane separation processes. Chemical Engineering, 91(12): 64-89.
6. Yousefi T., A.R. Khanchi, S.J. Ahmadi, M.K. Rofouei, R. Yavari, R. Davarkhah and B. Myanji, 2012. Cerium(III) molybdate nanoparticles: Synthesis, characterization and radionuclides adsorption studies. Journal of Hazardous Materials, 215–216: 266– 271.
7. Cortés-Martínez R., M.T. Olguín and M. Solache-Ríos, 2010. Cesium sorption by clinoptilolite-rich tuffs in batch and fixed-bed systems. Desalination, 258(1): 164–170.
8. Kusuma R.I., J.P. Hadinoto, A. Ayucitra, F.E. Soetaredjo and S. Ismadji, 2013. Natural zeolite from Pacitan Indonesia, as catalyst support for transesterification of palm oil. Applied Clay Science, 74: 121–126.
9. Martínez R.C., M.S. Ríos, V.M. Miranda and R.A. Cuevas, 2009. Sorption behavior of zeolitic rock in batch and fixed bed systems. Water Air Soil Pollution 196: 199–210.
10. Erdem E., N. Karapinar and R. Donat, 2004. The removal of heavy metal cations by natural zeolites. Journal of Colloid and Interface Science, 280(2): 309–314.
11. Yousefi.T, M.T.Mostaedi, M.A.Moosavian and H.G.Mobtaker,2015. Effective removal of Ce(III) and Pb(II) by new hybrid nano-material: $HnPMo_{12}O_{40}@Fe(III)xSn(II)ySn(IV)_{1-x-y}$. Process safety and environmental process protection, 98 : 211-220.
12. B.E.˘an Alver B.E., M. Sakizci and E.˘rul Yorukog˘ullari, 2011. Study of thermal and CH4 adsorption properties. Adsorption Science & Technology 29(4): 413-422.
13. Doula M.K., 2007. Synthesis of a clinoptilolite-Fe system with high Cu sorption capacity. Chemosphere, 67: 731.
14. Smičiklas I., S. Dimović, and I. Plećaš, 2007. Removal of Cs^{1+}, Sr^{2+} and Co^{2+} from aqueous solutions by adsorption on natural clinoptilolite. Applied Clay Science. 35 (1-2): 139-144.
15. Evangelos P. F., G. T. Constantinos, A.A. Sapalidis, G.T. Tzilantonis, S.K. Papageorgiou and A.C. Mitropoulos, 2016. Clinoptilolite, a natural zeolite material: Structural characterization and performance evaluation on its dehydration properties of hydrocarbon-based fuels. Microporous and Mesoporous Materials. 225:385-391
16. Semmens M. J. and W. P. Martn, 1988. The influence of pretreatment on the capacity and selectivity of clinoptilolite for metal ions. Water Research, 22: 537-542.
17. Charkhi A., H. Kazemian and M. Kazemeini, 2010. Experimental design optimized ball milling of natural clinoptilolite zeolite for production of nano powders. Powder Technology, 203: 389–396.
18. Karadag D., Y. Koc, M. Turan and B. Armagan, 2006. Removal of ammonium ion from aqueous solution using natural Turkish clinoptilolite. Journal of Hazardous Materials, 136: 604-609.
19. McKay G and Y.S. Ho, 2006. The sorption of lead (II) on peat. Water Research, 33(10): 585–587.
20. McKay G and Y.S. Ho, 1999. Pseudo-second order model for sorption processes. Process Biochemistry, 34: 451–460.
21. Weber Jr. W.J and J.C. Morris, 1963. Kinetics of adsorption on carbon from solution. Journal of the Sanitary Engineering Division, American Society of Civil Engineering, 89(1): 31–60.
22. Gupta S.S., K.G. Bhattacharyya, 2005. Adsorption of Ni (II) on clays. Applied Clay Science, 30(3): 199–208.
23. Jiang J.Q., C. Cooper, S. Ouki, 2002. Comparison of modified montmorillonite adsorbents. Part I: Preparation, characterization and phenol adsorption. Chemosphere, 47(7): 711–716.
24. Yousefi.T, M.T.Mostaedi, M.A.Moosavian and H.G.Mobtaker,2015, Potential application of a nanocomposite:HCNFe@polymer for effective removal of Cs (I) from nuclear waste. Progress in Nuclear Energy,85: 631-639.
25. Gupta S.S. and K.G. Bhattacharyya, 2005. Interaction of metal ions with clays. I. A case study with Pb(II), Applied Clay Science. 30: 199–208.

Silver Doped TiO$_2$ Nanoparticles: Preparation, Characterization and Efficient Degradation of 2,4-dichlorophenol Under Visible Light

Zahra Sarteep[1], Azadeh Ebrahimian Pirbazari[*,1, 2], Mohammad Ali Aroon[2]

[1]*Fouman Faculty of Engineering, College of Engineering, University of Tehran, Tehran, Iran*

[2]*Caspian Faculty of Engineering, College of Engineering, University of Tehran, Tehran, Iran*

ABSTRACT

Hydrothermally synthesized TiO$_2$ nanoparticles containing different amounts of silver were characterized by X-Ray diffraction (XRD), Fourier transform infrared (FT-IR) and scanning electron microscopy equipped with energy dispersive X-ray microanalysis (SEM/EDX) techniques. XRD results showed prepared samples include 100% anatase phase. The presence of silver in TiO$_2$ nanoparticle network was established by XRD, SEM/EDX and FT-IR techniques. The photocatalytic performance of the prepared catalysts was tested for the degradation of 2,4-dichlorophenol (2,4-DCP) under visible light.. The experiments demonstrated that 2,4-DCP was effectively degraded in the presence of Ag/TiO$_2$ samples. It was confirmed that the presence of Ag on TiO$_2$ catalysts could enhance the photocatalytic degradation of 2,4-DCP in aqueous suspension. It was found that an optimal dosage of 1.68 wt% Ag in TiO$_2$ achieved the fastest 2,4-DCP degradation (95% after 180 min irradiation) under the experimental conditions. On the basis of various characterizations of the photocatalysts, the reactions involved to explain the photocatalytic activity enhancement due to Ag doping include a better separation of photogenerated charge carriers. GC-MS analysis showed the major intermediates of 2,4-DCP degradation are simple acids like oxalic acid, acetic acid, etc. as the final products.

KEYWORDS: *2,4-dichlorophenol; Degradation; Silver; TiO$_2$*

INTRODUCTION

Recently the research based on advanced oxidation methods for water and wastewater purification are expanding progressively. Photochemical and photocatalytical methods are used for treating wastewater containing chloroaromatic compounds [1,2]. There has been a growing distress related to the environmental, health impacts and environmental damage due to chloroaromatic compounds during the last two decades [3]. The presence of chloroaromatic compounds in the aqueous environment is a result of the extensive use of chlorinated compounds in a range of industrial processes. These compounds are resistance to degradation and accumulate in the environment [4]. Recently adverse effects of chloroaromatic compounds on the human nervous system have been reported and have been linked to many health disorders [5]. Therefore, it is important to find novel and cost-effective methods for the safe and complete degradation of chloroaromatic compounds such as chlorophenols (CPs).

Heterogeneous photocatalysis technology based

* Corresponding Author Email: aebrahimian@ut.ac.ir

on semiconductors is one of the advanced oxidation processes used for degradation of organic pollutants [6]. In this process the semiconductor absorbs irradiation and generates active species such as hydroxyl radicals which leads to complete oxidation of organic components present in wastewater. Titanium dioxide (TiO_2) is a semiconductor with wide band gap (3.2eV), used as photocatalyst for destruction and removal of highly toxic and non-biodegradable pollutants. TiO_2 has its advantages over the other semiconductor photocatalysts such as ZnO, SnO_2 and etc. TiO_2 is chemically stable, environmentally friendly, non-toxic and cheaper [7].TiO_2 is synthesized using various methods such as chemical vapor deposition (CVD), plasma, hydrothermal and sol-gel [8]. Despite the many known advantages of using TiO_2, it suffers from the shortcoming of having a large band gap (~3.2 eV) which restricts its use to the ultraviolet (UV) region. Such radiation is not very abundant in the solar radiation that reaches the Earth, which limits the use of TiO_2 in solar energy utilization [9].

During two last decades, there has been a great deal of interest in the preparation of nanostructured titania [10–13] to increase the efficiency of these photochemical processes. In addition, a variety of attempts has been made to incorporate various metal species into the TiO_2 network to enhance the photocatalytic activity and/or broaden the absorption of the solar spectrum by the doped TiO_2. Methods that have been used include ion exchange, impregnation and co-precipitation [14,15]. Some interesting effects have been observed from these implantation studies. Metal-doped TiO_2 has been widely studied for improving photocatalytic performance on the degradation of various organic pollutants under visible-light [16–18]. The intrinsic properties of TiO_2, such as charge carried recombination rates, the particle size and interfacial electron transfer rates and extending photoresponse of TiO_2 into the visible range influence directly by silver doping [19–21]. The metal acts as a sink for photoinduced charge carriers and promotes interfacial charge-transfer processes.

To the best of knowledge, the application of TiO_2 nanoparticles containing different amounts of silver for heterogeneous photocatalytic degradation of 2,4-dichlorophenol (2,4-DCP) has not been reported so far. The aim of this work is,

the silver doped to TiO_2 nanoparticles (Ag/TiO_2) with photocatalytic activity were synthesized by hydothermal method. The as-prepared sample was characterized by X-ray diffraction (XRD), fourier infrared spectroscope (FTIR), scaning electron microscopy/energy dispersive X-ray (SEM/EDX). Moreover, the photocatalytic performance of the prepared samples in degrading 2,4-DCP was estimated using UV–Vis spectrophotometry

EXPERIMENTAL
Materials and Reagents

Silver nitrate ($AgNO_3$), 99.9%, was supplied by (Merck, No.101510). Tetraisopropylorthotitanat, for synthesis, 99.9%, (Merck No. 8.21895), ethanol, 96%, (Merck, No.159010), acetylacetone, 99% , (Merck No. 800023) and deionized water were used for photocatalyst synthesis. High-purity 2,4-DCP, 98%, (Merck No. 803774) was used as a probe molecule for photocatalytic tests.

Photocatalyst Preparation

Pure TiO_2 and TiO_2-based photocatalyst containing different amounts of silver were synthesized with little modification in the described procedure in ref. [22]. In a typical reaction, a mixture of 20 mL $Ti(OC_3H_7)_4$, 20 mL ethanol and 1.62 mL acetylacetone was prepared and stirred for 30 min at room temperature. Then, a new mixture containing X mg $AgNO_3$ (this amount calculated for 0.5-3 wt% Ag in final solid), 80 mL ethanol and 2 mL H_2O was added into the first solution, which led to clear solution. This solution was transferred into an autoclave, and then heated to 240 °C at a heating rate of about 2 °C min^{-1}. Finally, the temperature was kept at 240 °C for 6 h. After cooling, the obtained solid washed with ethanol and water and dried at 100 °C for 2 h in air. These photocatalysts are labelled as Ag/TiO_2 (a), where (a) is the weight percentage of Ag in the final solid that obtained by EDX analysis. A solution without $AgNO_3$ was also prepared for pure TiO_2 synthesis.

Characterization

Fourier transform infrared (FTIR) analysis was applied to determine the surface functionalgroups, using FTIR spectroscope (FTIR-2000, Bruker), where the spectra were recorded from 4000 to 400 cm^{-1}. The XRD patterns were recorded on a Siemens, D5000 (Germany). X-ray diffractometer

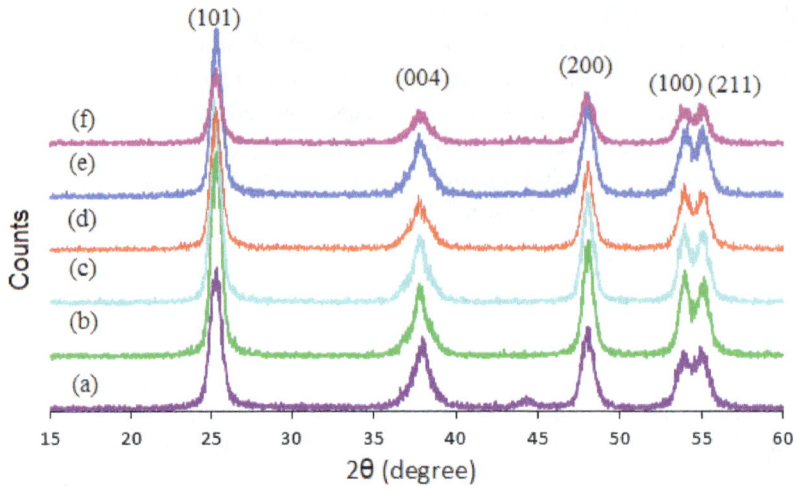

Fig. 1. XRD patterns of a) TiO$_2$, b) Ag/TiO$_2$ (0.17), c) Ag/TiO$_2$ (0.73), d) Ag/TiO$_2$ (1.68). e) Ag/TiO$_2$ (1.62) and f) Ag/TiO$_2$ (1.98)

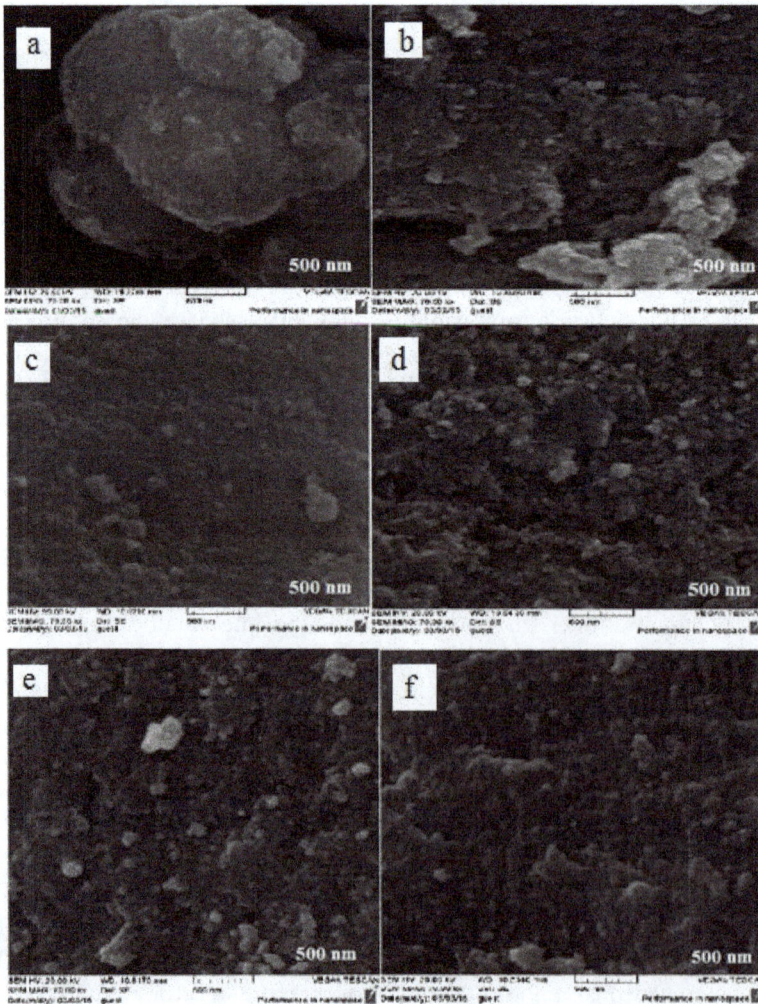

Fig. 2. SEM microghraphs of a) TiO$_2$, b) Ag/TiO$_2$ (0.17), c) Ag/TiO$_2$ (0.73), d) Ag/TiO$_2$ (1.68). e) Ag/TiO$_2$ (1.62) and f) Ag/TiO$_2$ (1.98)

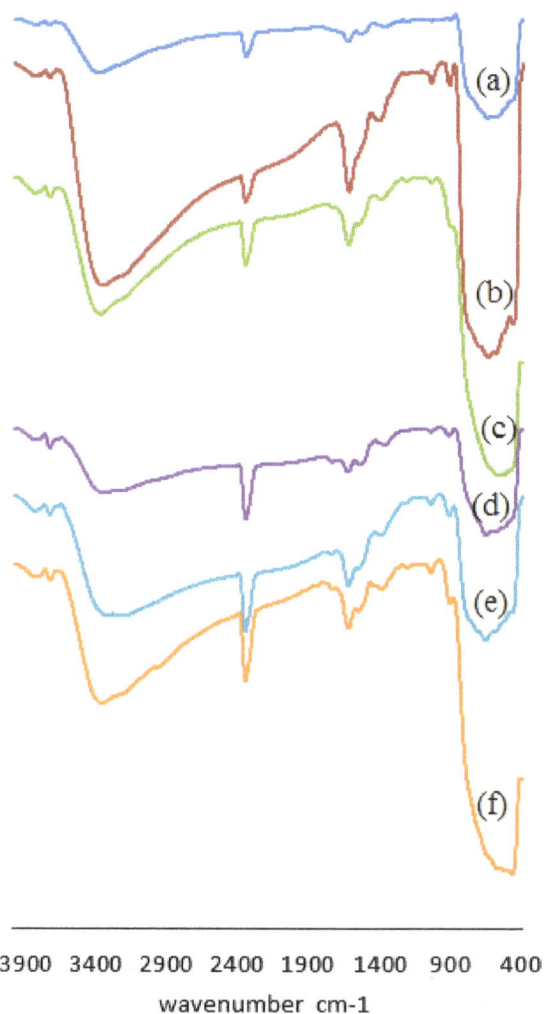

Fig. 3. FTIR spectra of a) TiO_2, b) Ag/TiO_2 (0.17), c) Ag/TiO_2 (0.73), d) Ag/TiO_2 (1.68). e) Ag/TiO_2 (1.62) and f) Ag/TiO_2 (1.98)

using Cu K_α radiation as the X-ray source. The diffractograms were recorded in the 2θ range of 15-60°. The morphology of nanoparticles were characterized using scanning electron microscope (SEM) (Vegall-Tescan Company) equipped with an energy dispersive X-ray (EDX).

Photocatalytic Degradation Monitoring

In a typical run, the suspension containing catalyst and 50 mL aqueous solution of 2,4-DCP (40 mg/L) was stirred first in the dark for 30 min to establish adsorption/desorption equilibrium. Irradiation experiments were carried out in a self-built reactor. A visible (Halogen, ECO OSRAM, 500W) lamp was used as irradiation source (its emitting wavelength ranges from 350 nm to 800 nm with the predominant peak at 575 nm). At certain intervals, small aliquots (2 mL) were withdrawn and filtered to remove the catalyst particles. These aliquots were used for monitoring the degradation progress, with Rayleigh UV-2601 UV/VIS spectrophotometer.

Statistical Analysis

All experiments were performed in triplicate and the average values were presented. The data were analyzed by one-way analysis of variance (ANOVA) using SPSS 11.5 for Windows. The data was considered statistically different from control at $P < 0.05$.

Identification of Intermediates

The reaction intermediates were identified by GC–MS in an Agilent 190915-433 instrument equipped with a HP-5MS capillary column (30 m × 0.25 mm). The column temperature was programmed at 50 °C for 2 min, and from 50 to 250 °C at a rate of 10 °C min^{-1}. The samples used for GC–MS analysis were prepared according to the following procedure: The obtained degradation product was acidified to pH 1 and subsequently extracted with dichloro-methane. After dichloromethane was evaporated to dryness under vacuum, 10 mL methanol was added to dissolve the residue. Then, 1 mL concentrated sulfuric acid was added and the combined solution was refluxed for about 3 h. The solution was further extracted with dichloromethane followed by concentrating to about 1 mL under reduced pressure. The released chloride ions originating from the degradation of 2,4-DCP were identified and determined by the $AgNO_3$ method.

RESULT AND DISCUSSION

X-ray Diffraction Analysis

The x-ray diffraction patterns of the pure TiO_2 and silver doped TiO_2 nanoparticles are shown in Fig. 1. The nanocrystalline anatase structure was confirmed by (101), (004), (20 0), (105) and (211) diffraction peaks [23]. The XRD patterns of anatase have a main diffractions at $2\theta = 25.2°$ corresponding to the 101 plane (JCPDS 21-1272) while the main diffractions of rutile and brookite phases are at $2\theta = 27.4°$ (110 plane) and $2\theta = 30.8°$ (121 plane), respectively. Therefore, rutile and brookite phases have not been detected [24,25].

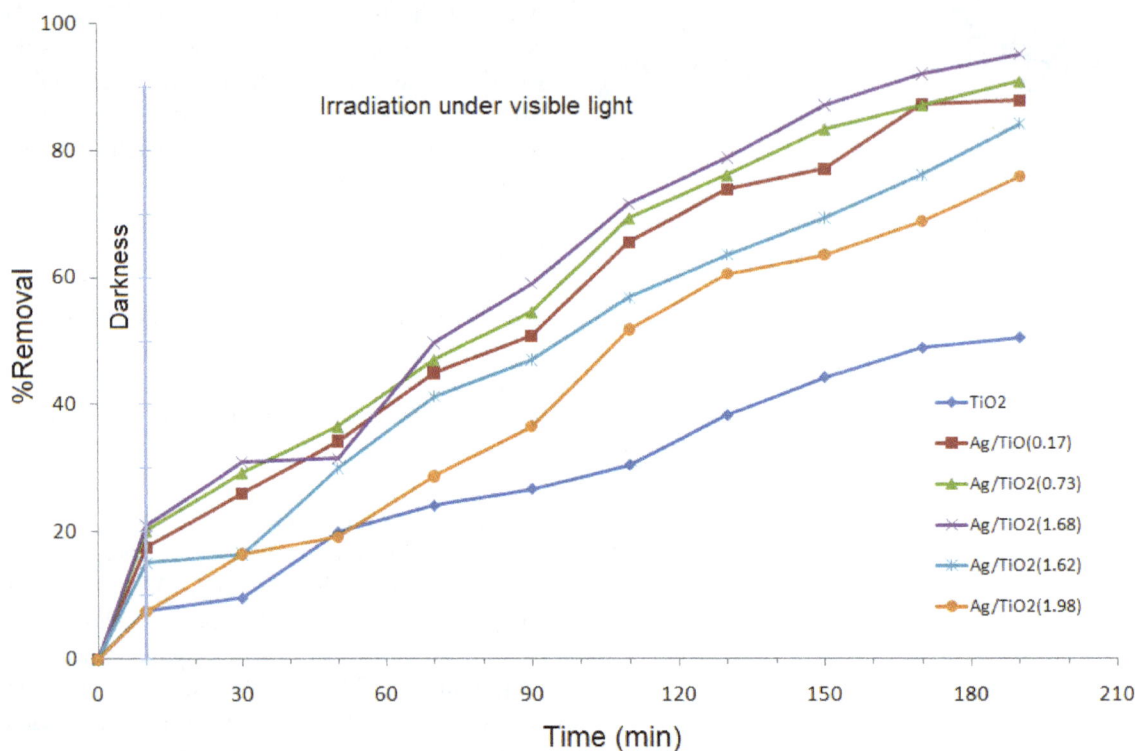

Fig. 4. Photocatalytic degradation of 2,4-DCP in the presence of prepared samples under visible light. (Initial concentration of 2,4-DCP, 40 mg /L; volume, 100 mL; catalyst dosage. 50 mg)

The x-ray diffraction patterns of the silver doped TiO_2 nanoparticles (Fig. 1b to f) coincides with the of pure TiO_2 and show no diffraction due to the silver species, thus suggesting that silver particles are well dispersed on the TiO_2 surface. Doping with silver does not disturb the crystal structure of anatase TiO_2 indicating that the silver dopants are merely placed on the surface on crystal without being covalently anchored into the crystal lattice. There are no diffraction pattern charactertics of the silver in the XRD patterns. Hence these metal sites are expected to be below the visibility limit of x-ray analysis. The diffraction patterns of pure TiO_2 and Ag/TiO_2 samples showed considerable line width, indicating small particles. The crystal size of each sample is calculated from the full width at half maximum (FWHM) of the (101) diffraction peak using Scherrer's equation [26].

$$D = K\lambda/\beta \cos\theta \qquad (1)$$

Where D is the average crystal size of the sample, λ the X-ray wavelength (1.54056 Å), β the fullwidth at half maximum (FWHM) of the diffraction peak (radian), K is a coefficient (0.89)and θ is the

diffraction angle at the peak maximum. The results are demonstrated in Table 1. All prepared samples are in nano-size range, from 9.80 to 11.00 nm, and all Ag/TiO_2 samples show equal or larger crystal size compared to pure TiO_2.The increased particle size may be explained by the fact that the ionic radius of Ag^+ (1.15Å) is greater than that of Ti^{4+} (0.60 Å).

SEM/EDX Analysis

In order to investigate the surface morphology of the synthesized Ag doped TiO_2 nanoparticles, SEM studies were performed. The SEM images of TiO_2 and Ag/TiO_2 samples are shown in Fig. 2. The SEM images of Ag/TiO_2 samples show that the distribution of silver on the surface of titanium dioxide is not uniform and the Ag/TiO_2 catalyst contains irregular shaped particles which may be due to the aggregation of tiny crystals. However, it can not be ruled out that some silver particles are too small to be observed at the resolution of the used microscope. The SEM images also reveal that the doping of silver metal does not leave any change in the topology of the catalyst surface.

Table 1. Average crystal size of the synthesized nanoparticles calculated by Scherrer equation

Sample	TiO_2	Ag/TiO$_2$(0.17)	Ag/TiO$_2$(0.73)	Ag/TiO$_2$(1.68)	Ag/TiO$_2$(1.62)	Ag/TiO$_2$(1.98)
Crystal size (nm)	9.80	9.80	9.90	10.80	10.70	11.00

Table 2. Elemental analysis of synthesized nanoparticles obtained by EDX analysis

Sample	Nominal Ag (wt%)	Ag (wt%)	Ti (wt%)	O (wt%)
TiO_2	0.00	0.00	40.44	47.27
Ag/TiO$_2$(0.17)	0.50	0.17	30.33	56.03
Ag/TiO$_2$(0.73)	1.00	0.73	40.19	46.03
Ag/TiO$_2$(1.68)	2.00	1.68	36.19	50.52
Ag/TiO$_2$(1.62)	2.50	1.62	43.19	42.79
Ag/TiO$_2$(1.98)	3.00	1.98	42.85	44.54

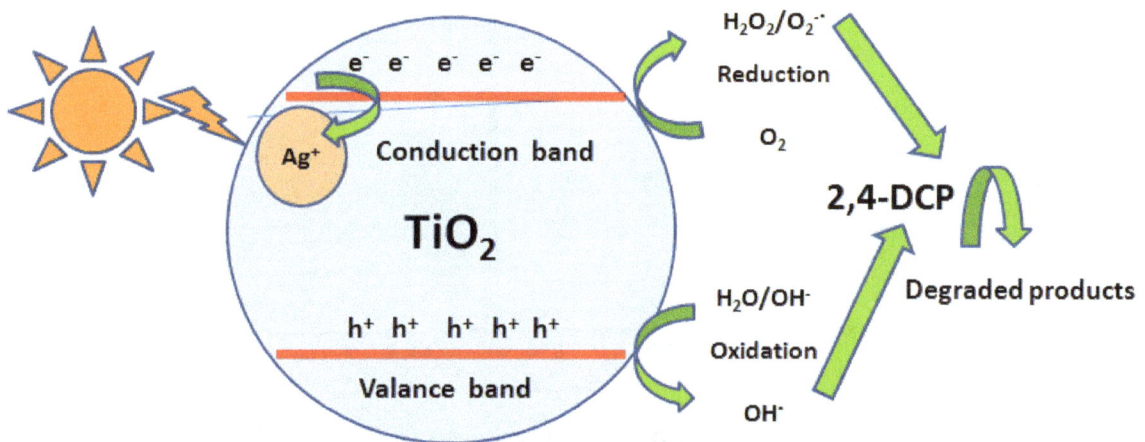

Fig. 5. Our proposed mechanism for photocatalytic degradation of 2,4-DCP

Also Fig. 2 show the hardness and density of nanoparticles increase as silver content increase in TiO2 lattice [27]. Spongy and porous structure causes more surface area at high hardness; that surely would be more efficient for absorption and photocatalytic applications [28].Table 2 is the result of elemental determination for the synthesized samples that obtained by EDX analysis. According to Table 2, doped silver amounts are lower than nominal amounts; it can be explained by the loss of silver nitrate during the hydrothermal synthesis or possibly TiO2 network does not have the more capacity to dope higher amounts of silver.

FTIR Analysis

The FTIR spectra of the synthesized samples are shown in Fig. 3. The peaks observed at ~3400, 2930 and 2850 cm^{-1} are attributed to the Ti – OH bond [29] . The spectra shows relatively strong band at ~ 1630 cm^{-1} observed for all the photocatalysts which is due to the OH bending vibration of chemisorbed

and/or physisorbed water molecule on the surface of the catalysts. The strong band in the range of 700 –500 cm^{-1} is attributed to stretching vibrations of Ti –O–Ti bond [30]. The FTIR spectra of Ag/TiO$_2$ samples revealed a peak at about 1385 cm^{-1}, which was not observed for the undoped TiO$_2$. The intensity of this peak increased with the increase of silver precursor (AgNO$_3$) in the composite samples. The peak at 1385 cm^{-1} was assigned tentatively to the interaction between Ag and TiO$_2$ particles [31].

Photocatalytic Degradation of 2,4-DCP

To evaluate the photocatalytic activity of the synthesized samples and find out the optimum content of Ag doping, a set of experiments for 2,4-DCP degradation with an initial concentration of 40 mg L^{-1} under visible light was carried out in aqueous suspension using TiO$_2$ or Ag/TiO$_2$ catalysts with a Ag content between 0.17 wt% and 1.98 wt%, and the experimental results are shown in Fig. 4. The experimental results demonstrated

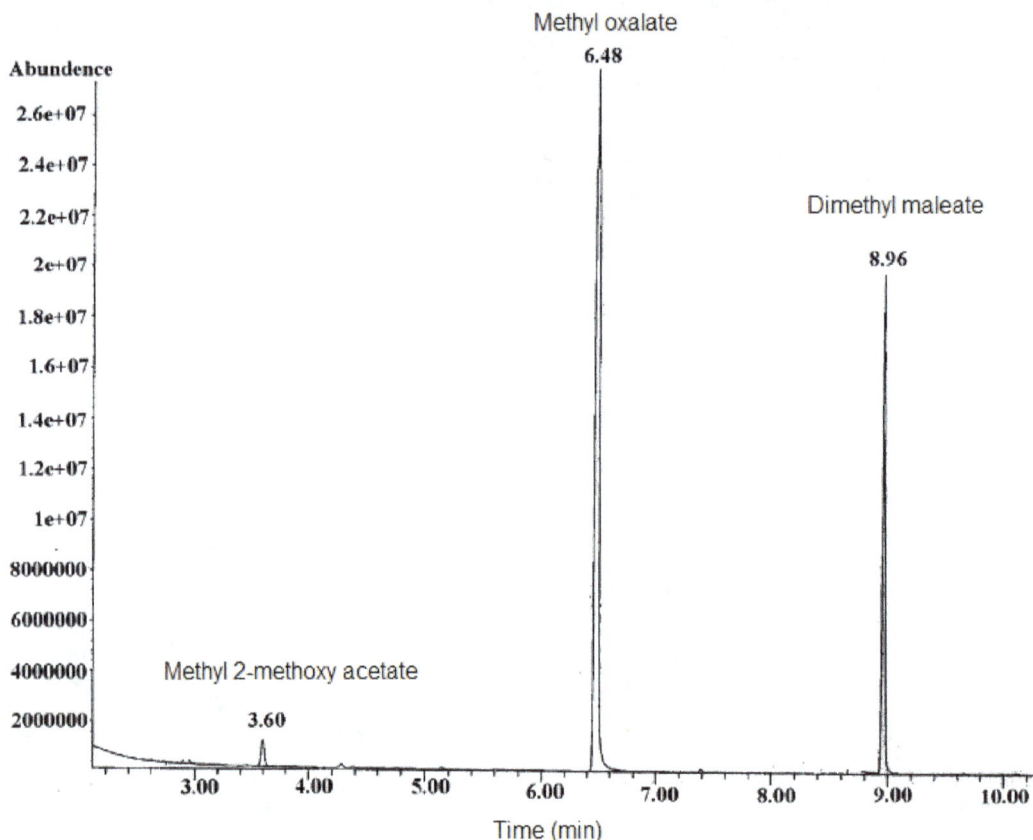

Fig. 6. GC-Chromatogram of the final products of the photocatalytic degradation of 2,4-DCP by Ag/TiO$_2$(1.68)

that among the prepared samples, the Ag/TiO$_2$ (1.68) catalyst achieved the highest efficiency of the 2,4-DCP degradation (95% degradation obtained after 180 min irradiation). A further more Ag content on TiO$_2$ seems to be detrimental to the 2,4-DCP photodegradation efficiency. From the Fig. 4, the results show that the photocatalytic activity of pure TiO$_2$ is lower than that of Ag/TiO$_2$ samples. It implies that the Ag dopant promotes the charge pair separation efficiency for TiO$_2$ catalysts. It may be explained that at the Ag content below its optimum level (1.68 wt%), the Ag particles deposited on the TiO$_2$ surface can act as electron-hole separation centers [32]. The electron transfer from the TiO$_2$ conduction band to Ag particles at the interface is thermodynamically possible because the Fermi level of TiO$_2$ is higher than that of Ag metal [33]. This results in the formation of a Schottky barrier at metal semiconductor contact region and improves the photocatalytic activity of TiO$_2$. On the contrary, at the Ag content beyond its optimum value, the Ag particles can also act as recombination centers,

thereby decreasing the photocatalytic activity of TiO$_2$. It has been reported that the probability for the hole capture is increased by the large number of negatively charged Ag particles on TiO$_2$ at high Ag content, which reduces the efficiency of charge separation [34,35]. In this study, the Ag/TiO$_2$ samples under visible light demonstrated a considerable degradation of 2,4-DCP in aqueous solution. We will discuss about the mechanism of photocatalytic degradation in next section.

Mechanism of Photocatalytic Degradation

Fig. 5 shows our proposed mechanism for photocatalytic degradation of 2,4-DCP. It is acknowledged that the radius of the Ag$^+$ and Ag atom is much larger than the Ti^{4+}, so very unlikely for Ag$^+$ and Ag atom to enter the TiO$_2$ lattice [36]. In fact, the Ag atoms were in direct contact with the TiO$_2$ nanoparticles. Because the work function of the metal Ag is higher than that of TiO$_2$, electrons are removed from the TiO$_2$ particles to the vicinity of the Ag particles. This results in the formation of Schottky barriers at the Ag–TiO$_2$ contact region and results in charge separation [37] i.e. the electronic interaction occurring at the contact region between the metal deposits and the semiconductor surface [38]. The Ag deposits act as electron traps immobilizing the photogenerated electrons in the traps and shortly transferring them to oxygen to form highly oxidative species such as O$_2^{-\cdot}$. This type of electron scavenging by Ag metal is reported to be a faster process compared to the electron transfer to oxygen (or) recombination with holes [38] since, the trapping of electron by Ag metal from TiO$_2$ occurs at a faster rate when compared to the electron transfer from TiO$_2$ to O$_2$ [39,40].

More specifically, Schottky barriers facilitate the electron transfer from TiO$_2$ nanoparticles (with high Fermi level) to Ag (with low Fermi level), resulting in higher transferring efficiency of electrons [41]. The effect of Schottky barriers in our particular case was evidenced by the fact that all the Ag-loaded TiO$_2$ had higher photocatalytic efficiencies than Ag free TiO$_2$. Firstly, the Schottky barriers facilitate the photoelectron movement to a certain direction, from the semiconductor TiO$_2$ nanoparticles to adjacent Ag atom or cluster. Sample Ag/TiO$_2$(1.68) (Fig. 4) has the highest photocatalytic activity and can be attributed to the fact that Ag was distributed at surface of the TiO$_2$. For this sample, the Schottky barriers attract the electrons and the holes will be left on the surface layer of the TiO$_2$. This attraction greatly inhibits the recombination of electron–hole pairs. The enriched holes in the surface layer can directly degrade 2,4-DCP, at the same time, it can oxidize H$_2$O adsorbed on the surface and produce hydroxyl radicals (OH$^\cdot$) that can degrade organic matters effectively.

Identification of Intermediate Products

The intermediate species formed during degradation of 2,4- DCP, were identified by GC–MS. The major reaction intermediates identified in an aliquot withdrawn after 200 min following a degradation condition specified as in Fig. 4. Presence of these intermediates (Fig. 6) supports our proposed mechanism which is based on OH radicals (Fig. 5). The hydroxyl radicals attack 2,4-DCP converting it to chlorocatechol and then to chlorobenzoquinone. Subsequently, hydroxyl groups would break the aromatic rings

of chlorobenzo-quinone transferring them into simple acids like oxalic acid, acetic acid, etc . as the final products [42,43]. In addition to identifying the organic intermediates, chloride ions were also detected and identified as one of the final products of the photocatalytic degradation. The amount of Cl$^-$ in the reaction media at the end of the photocatalytic experiment almost equals the amount of chlorine present in 2,4-DCP indicating essentially complete degradation.

CONCLUSION

TiO$_2$-based photocatalysts containing different amounts of silver were synthesized and characterized by several techniques successfully. XRD analysis confirmed all the prepared samples consist of pure anatase phase. The XRD and SEM/EDX data showed that the incorporation of silver in TiO$_2$ network increases the crystal size. Among the Ag/TiO$_2$ samples, the Ag/TiO$_2$ (1.68) photocatalyst exhibited the highest photocatalytic activity under visible light. GC-MS analysis showed, the hydroxyl radicals attack 2,4-DCP converting it to chlorocatechol and then to chlorobenzoquinone. Subsequently, hydroxyl groups would break the aromatic rings of chlorobenzo-quinone transferring them into simple acids like oxalic acid, acetic acid, etc. as the final products

ACKNOWLEDGEMENTS

The authors wish to acknowledge the financial support of the University of Tehran.

CONFLICT OF INTEREST

The authors declare that there are no conflicts of interest regarding the publication of this manuscript.

REFERENCES

1. Serra, F., M. Trillas, J. Garcia and X. Domenech, 1994. Titanium dioxide-photocatalized Oxidation of 2,4-dichlorophenol, Journal of Environmental Science and Health . A. 29 (7) : 1409-1421

2. Theurich, J., M. Lindner and D.W. Bahnemann,1996. Photocatalytic Degradation of 4-Chlorophenol in Aerated Aqueous Titanium Dioxide Suspensions: A Kinetic and Mechanistic Study, Langmuir. 12: 6368-6376.

3. Manahan, S., 1994. Environmental Chemistry., CRC Press., Boca Raton, Florida.

4. Bandara, J., J.A. Mielczarski, A. Lopez and J. Kiwi, 2001. Sensitized Degradation of Chlorophenols on Iron Oxides Induced by Visible Light: Comparison with Titanium Oxide, Applied Catalysis B: Environmental. 34 (4) 321-333.

5. Howard, ph., 1991. Handbook of Environmental Degradation Rates, Lewis Publishers, MI.

6. Sabbaghi S. and F. Doragh, 2016. Photo-Catalytic Degradation of Methylene blue by ZnO/SnO2 Nanocomposite, Journal of Water and Environmental Nanotechnology. 1(1): 27-34.

7. Hoffman, M.R., S.T. Martin, W. Choi and D.W. Bahnemann, 1995. Environmental Applications of Semiconductor Photocatalysis, Chemical Reviews 95: 69-96.

8. Byun, D., Y. Jin, J. Kim, K. Lee and P. Hofmann, 2000. Photocatalytic TiO2 Deposition by Chemical Vapor Deposition, Journal of Hazardous Materials, 73: 199-206.

9. Fox, M.A. and M.T. Dulay, 1993. Heterogeneous Photocatalysis. Chemical Reviews, 93: 341–347.

10. Mohamed, M.M., W.A. Bayoumy and M. Khairy, 2008. Structural Features and Photocatalytic Behavior of Titania and Titania Supported Vanadia Synthesized by Polyol Functionalized Materials, Microporous and Mesoporous Materials, 109 (1–3): 445–457.

11. Mohamed, M.M., I. Othman and R.M. Mohamed, 2007. Synthesis and Characterization of MnOx/TiO2 Nanoparticles for Photocatalytic Oxidation of Indigo Carmine Dye, Journal of Photochemistry and Photobiology A: Chemistry, A 191 (2–3): 153–161.

12. Mohamed, M.M., T.M. Salama and I. Othman, 2007. Synthesis and Characterization of Mordenites Encapsulated Titania Nanoparticles: Photocatalytic Degradation of meta-Chlorophenol, Journal of Molecular Catalysis A: Chemical, 273:198-210

13. M.M. Mohamed, M.M., W.A. Bayoumy, M. Khairy and M.A. Mousa, 2006. Synthesis and Structural Characterization of TiO2 and V2O5/TiO2 Nanoparticles Assembled by the Anionic Surfactant Sodium Dodecyl Sulfate, Microporous and Mesoporous Materials, 97:66–77.

14. Zakeeruddin, S.M., Md.K. Nazeeruddin, R. Humphry-Baker, P. Pechy, P. Quagliotto, C. Barolo, G. Viscardi and M. Gratzel, 2002. Design, Synthesis, and Application of Amphiphilic Ruthenium Polypyridyl Photosensitizers in Solar Cells Based on Nanocrystalline TiO2 Films,Langmuir, 18: 952–954.

15. Burnside, S., J.E. Moser, K. Brooks, M. Gratzel and D. Cahen, 1999. Nanocrystalline Mesoporous Strontium Titanate as Photoelectrode Material for Photosensitized Solar Devices: Increasing Photovoltage through Flatband Potential Engineering, Journal of Physical Chemistry, B 103: 9328–9332.

16. Elder, S.H., Y. Gao, X. Li, J. Liu, D.E. McCready and C.F. Windisch Jr., 1998. Zirconia-Stabilized 25-Å TiO2 Anatase Crystallites in a Mesoporous Structure, Chemistry of Materials, 10: 3140–3145.

17. Mu, W., J.M. Herrmann and P. Pichat, 1989. Room Temperature Photocatalytic Oxidation of Liquid Cyclohexane into Cyclohexanone over Neat and Modified TiO2, Catalysis Letter 3:73–84

18. Kararkitsou, K.E. and X.E. Verykios, 1993. Effects of Altervalent Cation Doping of Titania on its Performance as a Photocatalyst for Water Cleavage, Journal of Physical chemistry, 97:1184–1189.

19. Zang, Y. and R. Farnood, 2008. Photocatalytic Activity of AgBr/TiO2 in Water under Simulated Sunlight Irradiation, Applied Catalysis B: Environmental, 79: 334–343.

20. Kakuka, N., N. Goto, H. Ohkita and T. Mizushima, 1999. Silver Bromide as a Photocatalyst for Hydrogen Generation from CH3OH/H2O Solution, Journal of Physical Chemistry B 103:5917–5919

21. Rodrigues, S., S. Uma, I.N. Martyanov and K.J. Klabunde, 2005. AgBr/Al-MCM-41 Visible-light Photocatalyst for Gas-phase Decomposition of CH3CHO, Journal of Catalysis, 233: 405-410.

22. Ebrahimian A, M.A. Zanjanchi, H. Noei, M. Arvand and Y. Wang, 2014. TiO2 Nanoparticles Containing Sulphonated Cobalt Phthalocyanine: Preparation, Characterization and Photocatalytic Performance. Journal of Environmental Chemical Engineering, 2(1):484-94.

23. Ogawa, H. and A. Abe, 1981. Preparation of Tin Oxide Films from Ultrafine Particles, Journal of Electrochemical Society, 128: 685-689.

24. Baiju, K.V., P. Shajesh, W. Wunderlich, P. Mukundan, S.R. Kumar and K.G.K. Warrier, 2007. Effect of Tantalum Addition on Anatase Phase Stability and Photoactivity of Aqueous Sol–Gel Derived Mesoporous Titania, Journal of molecular Catalysis A. 276: 41-46.

25. Srivatsa, K.M.K., M. Bera and A. Basu, 2008. Pure Brookite Titania Crystals with Large Surface Area Deposited by Plasma Enhanced Chemical Vapour Deposition Technique, Thin Solid Films, 516: 7443-7446.

26. Khan, M. and W. Cao, 2013. Cationic (V, Y)-codoped TiO2 with Enhanced Visible Light Induced Photocatalytic Activity: A Combined Experimental, Theoretical Study, Journal of Applied Physics, 114: 183514.

27. Perumal, S., C. Gnana Sambandam and A. Peer Mohamed, 2014. Synthesis and Characterization Studies of Solvothermally Synthesized Undoped and Ag-Doped TiO2 Nanoparticles Using Toluene as a Solvent, Journal of Engineering Research and Applications, 4(7):184-187

28. Younas H., I.A. Qazi, I. Hashmi, M.A. Awan, A. Mahmood and H.A. Qayyum, 2014. Visible Light Photocatalytic Water Disinfection and Its Kinetics Using Ag-Doped Titania Nanoparticles, Environmental Science and Pollution Research, 21(1):740-52.

29. Wang Y., Y. Huang, W. Ho, L. Zhang, Z. Zou and S. Lee, 2009, Biomolecule-Controlled Hydrothermal Synthesis of C–N–S-tridoped TiO2 Nanocrystalline Photocatalysts for NO Removal under Simulated Solar Light Irradiation, Journal of Hazardous Materials, 169: 77-87.

30. Bae E. and W. Choi, 2003, Highly Enhanced Photoreductive Degradation of Perchlorinated Compounds on Dye-Sensitized Metal/TiO2 under Visible Light, Environmental Science and Technology, 37: 147–152.

31. García-Serrano J., E. Gómez-Hernández, M. Ocampo-Fernández and U. Pal, 2009, Effect of Ag Doping on the Crystallization and Phase Transition of TiO2 Nanoparticles, Current Applied Physics, 9:1097–1105.

32. Hermann, J.M., H. Tahiri, Y. Ait-Ichou, G. Lassaletta, A. R. Gonzalez-Elipe and A. Fernandez, 1997. Characterization and Photocatalytic Activity in Aqueous Medium of TiO2 and Ag-TiO2 Coatings on Quartz, Applied Catalysis B: Environmental, 13: 219-228.

33. Scalfani, A. and J. M. Hermann, 1998. Influence of Metallic Silver and of Platinum-Silver Bimetallic Deposits on the Photocatalytic Activity of Titania (Anatase and Rutile) in Organic and Aqueous Media,Journal of Photochemistry and Photobiology A: Chemistry, 113: 181-188.

34. Vamathevan, V., R. Amal, D. Beydoun, G. Low and S. McEvoy, 2002. Photocatalytic Oxidation of Organics in Water Using Pure and Silver-Modified Titanium Dioxide Particles, Journal of Photochemistry and Photobiology A: Chemistry, 148: 233-245.

35. Li, X.Z. and F. B. Li, 2001. Study of Au/Au3+-TiO2

Photocatalysts toward Visible Photooxidation for Water and Wastewater Treatment, Environmental Science and Technology, 35: 2381-2387.

36. Rao, K.V.S., B. Lavedrine and P. Boule, 2003. Influence of Metallic Species on TiO2 for the Photocatalytic Degradation of Dyes and Dye Intermediates, Journal of Photochemistry and Photobiology A: Chemistry, 154: 189-195.

37. Awazu, K., M. Fusimaki, C. Rockstuhp, J. Tominaga, H. Murakami, Y. Onki and N. Yoshida, 2008. A Plasmonic Photocatalyst Consisting of Silver Nanoparticles Embedded in Titanium Dioxide, Journal of American Chemical Society, 130 (5): 1676–1683.

38. Rengifo-Herrera, J.A., K. Pierzchała, A. Sienkiewicz, L. Forro, J. Kiwi and C. Pulgarin, 2009. Abatement of Organics and Escherichia coli by N, S co-Doped TiO2 under UV and Visible Light. Implications of the Formation of Singlet Oxygen (1O2) under Visible Light, Applied Catalysis B: Environmental, 88 : 398–406.

39. Cheng, Z. and Y. Li, 2007. What Is Responsible for the Initiating Chemistry of Iron-Mediated Lipid Peroxidation: An Update, Chemical Reviwes, 107: 748-765.

40. Gole, J.M., J.D. Stout, C. Burda, Y. Lou and X. Chen, 2003. Highly Efficient Formation of Visible Light Tunable TiO2-xNx Photocatalysts and Their Transformation at the Nanoscale, Journal of Physical Chemistry B, 108: 1230–1237.

41. Sen, S., S. Mahanty, S. Roy, O. Heintz, S. Bourgeois and D. Chaumont, 2005. Investigation on Sol–Gel Synthesized Ag-Doped TiO2 Cermet Thin Films, Thin Solid Films, 474: 245–253

42. Minabe T., D.A. Tryk, P. Sawunyama, Y. Kikuchi, K. Hashimoto and A. Fujishima, 2000, TiO2- mediated Photodegradation of Liquid and Solid Organic Compounds, Journal of Photochemical and Photobiology A Chemistry, 137 :53–62.

43. Chaliha S. and K.G. Bhattacharyya, 2009, Fe(III)-, Co(II)- and Ni(II)-Impregnated MCM41 for Wet Oxidative Destruction of 2,4-dichlorophenol in Water, Catalysis Today 141: 225–233.

3

Catalytic oxidation of naphtol blue black in water: Effect of Operating Parameters and the Type of Catalyst

Ouahiba Bechiri*, Mostefa Abbessi

Laboratory of Environmental Engineering, Department of Process Engineering, Faculty of Engineering, University of Annaba, Annaba, Algeria

ABSTRACT

The main objective of this work is to study the oxidation of naphthol blue black (NBB) in aqueous solution by hydrogen peroxide using a recyclable Dawson type heteropolyanion [$H_{1.5}Fe_{1.5}P_2W_{12}Mo_6O_{61}.23H_2O$] as catalyst. The effects of various experimental parameters of the oxidation reaction of the dye were investigated. The mineralization of the dye was investigated by the total organic carbon (TOC) measurement in optimum conditions.

The influence of the catalyst nature (Dawson- type iron-substituted heteropolyanion) and (Dawson- type copper-substituted heteropolyanion) on the oxidation process was investigated. The catalytic oxidation using a recyclable heteropolyanions as catalysts is an economically and environmentally friendly process to remove the toxicity of the recalcitrant compounds in water.

Keywords: Dye removal, Naphthol blue black (NBB), Water treatment, Catalytic oxidation.

INTRODUCTION

Among the water consuming industries in large quantity, the textile industry is found at the top of the list, constituting a major source of water pollution. The waters released by the textile mills usually are highly concentrated colorants, usually with little to no degradable potential, making biological treatments hardly applicable. Therefore, it is necessary to find alternative techniques of biodegradable efficiency and that are also cost-effective. The last twenty years there has been much work published that is devoted to the emergence of new treatment processes; among which is advanced oxidation processes.

The advanced oxidation process (AOP) is an alternative method for the degradation of many organic pollutants [1, 2, 3]. (AOP)s are oxidation processes which generate hydroxyl radicals

(OH·) that are very effective at degrading organic pollutants because of their strong oxidizing capabilities. One of them is the homogeneous Fenton process, which is widely studied as an alternative for the treatment of industrial waste water containing non-biodegradable organic pollutants [4, 5, 6].

But often this method needs ultra filtration for the separation of catalyst; it is an especially big problem when applying it for treating large waste streams. The use of a recyclable heteropolyanions as catalysts in oxidation of organic dyes by hydrogen peroxide may provide the best alternative approach to solve this problem.

The heteropolyanions, molecular oxides to the properties, are many and varied, in both the homogeneous phase and the heterogeneous phase. These compounds, which are fully minerals, are generally easy to synthesize from simple and little polluting reagents [7, 8].

Heteropolyanions with Dawson structure [9] may be promising catalysts in homogeneous and heterogeneous systems because their redox and acidic properties can be controlled at both the atomic and molecular levels. Dawson-type heteropolyanions are formidable catalysts, which have proved their effectiveness in many reactions of oxidation [10, 11, 12]. Recently, there has been considerable interest in the use Dawson-type heteropolyanions as environmentally catalysts due to their unique properties such as high solubility in polar solvents and fairly high thermal stability in solid state, low cost, ease of preparation and ease of recyclability [13].

In this study, we report a detailed discussion on the oxidative degradation of Naphthol blue black (NBB) in aqueous solution containing hydrogen peroxide and a Dawson-type heteropolyanion as catalyst.

This reaction is part of the depollution of water, in particular to the treatment of discharged water by the textile industry, soiled by the organic dyes. The choice of the naphthol blue black is dictated by the fact that it is an azo dye which presents a high toxicity to the environment because of the presence of phenolic, anilino, naphthalene and sulfonated groups, (see Fig.1)

NBB is an industrially important acidic diazo dye, which has a high photo- and thermal- stability [14].

Due to its high degree of reaction to light, the commercial grades of naphthol blue black (NBB) are widely used in the textile industry for dyeing wool, nylon, silk and textile printing. Other industrial use includes coloring of soaps, anodized aluminum and casein, wood stains and writing ink preparation [14].

The degradation and removal of NBB dye was investgated by several authors in the literature

Ferkous *et al.* [15] used ultrasound for the degradation on NBB dye, in this study (5 mg L^{-1}) NBB

was completely destroyed after 45 min of sonication. The photoelectrochemical degradation of NBB dye using different semiconductor electrodes was studied. A higher photoelectrocatalytic activity has been observed for WO_3 film electrodes, prepared by electrodeposition, than for TiO_2 nanoparticulate film electrodes [16]. The heterogeneous photocatalysis degradation of NBB in the presence of zirconia-supported Ti-substituted Keggine –type polyoxometalates [17] And synthesized nanocomposite polyaniline-coated oxide (PTO) [18] was also performed. Moreover, we have investigated the degradation of NBB dye by H_2O_2 using Dawson-type Fe(III)-substituted heteropolyanion $(\alpha_2P_2W_{12}Mo_5O_{61}Fe)^{7-}$ as catalyst. This compound was synthesized by the addition of iron on the lacunary heteropolyanion $(\alpha_2P_2W_{12}Mo_5O_{62})^{10-}$[19]. The optimal values of operating parameters during the oxidation of the NBB dye by the Fe(III)$P_2W_{12}MO_5/H_2O_2$ system are pH: 3, [NBB]$_0 =$ 10 mg/L, Catalyst $(\alpha_2P_2W_{12}Mo_5Fe)^{7-}$ mass : 0.3g, $[H_2O_2]_0 = : 2m$ M.

In this work, we have investigated the removal of NBB dye from water by H_2O_2 using $HFe_{2.5}P_2W_{12}Mo_6O_{62}, 22H_2O$ (HPA Fe^{3+}). This catalyst was synthetized by the addition of Fe^{3+} ions to the Dawson acid form $H_6P_2W_{12}.Mo_6O_{62}24H_2O$[20]. The influence of different parameters such as the initial pH, the initial H_2O_2 concentration, the catalyst mass, and the initial dye concentration have also been studied. The mineralization of the dye was investigated by the total organic carbon (TOC) measurement in optimum conditions.

Even the effect of using the heteropolyanion (HPA Fe^{3+}) as a catalyst on the oxidation of NBB has been compared with a copper substituted heteropolyanion $[H_{1.2}Cu_{2.4}P_2W_{12}Mo_6O_{61.}21H_2O]$ (HPACu).

EXPERIMENTAL SECTION
Reagents
The catalyst $HFe_{1.5}P_2W_{12}Mo_6O_{61}$ $22H_2O$ was prepared starting from $H_6P_2W_{12}Mo_6O_{62}$ $24H_2O$ according to the following protocol [20].

5g (1.2 mmol) of $H_6P_2W_{12}Mo_6O_{62}$ were dissolved in 20 ml of water at room temperature and 0,541g (3.56 mmol) of solid $FeCl_2$ $6H_2O$ was then added. The mixture was stirred for 10 min. Dark yellow powder of (HPA Fe^{3+}) was obtained after five days by slow evaporation.

The heteropolyanion precursor $H_6P_2W_{12}Mo_6$ $O_{62}24H_2O$ was synthesized according to published procedure [21].

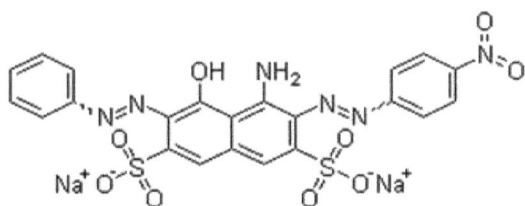

Fig. 1: Developed formula of Naphtol blue black (NBB)

Naphthol blue black (abbreviation: NBB; class: azo,C.I. number:13025, molecular formula: $C_{14}H_{14}N_3SO_3Na$) was used as a compound model. It is also known as [Noir amido 10 B, Acid Black 1, Buffalo Black NBR]. Naphtol blue black was supplied by Fluka. Its molecular structure is shown in (Fig. 2). (H_2O_2 35%, W/W) was obtained from Merck. All other reagents (NaOH , HCl, H_2SO_4, HNO_3 and H_3PO_4) that are used in this study were analytical grade.

Procedure - analysis

The initial concentration of NBB solution was 30 mg L^{-1} for all experiments, except for those carried out to examine the effect of initial dye concentration. In all experiments 100 mL of NBB solution containing the appropriate quantity of catalyst and H_2O_2 was magnetically stirred at room temperature. The pH of the reaction was adjusted by using 0.1N of acids (H_2SO_4, HNO_3, HCl and H_3PO_4) or NaOH aqueous solutions. The NBB concentration is measured by means of a 6705 UV visible spectrophotometer JENWAY. The wave length corresponding to the maximum absorbance is λ_{max}=620 nm [14]. The resolution of the wavelength and bandwidth, were 1nm and 0.5 nm. The cells used during the experiments were made of 1 cm thick quartz.

The effects of operational parameters on NBB oxidation

The oxidation of NBB by H_2O_2 using (HPA Fe^{3+}) as catalyst has been studied according to the following factors: initial pH of the solution, mass of the catalyst, H_2O_2 concentration and the dye concentration.

The oxidation efficiency (discolouration) was determined as it is shown below:
DE= (C$_i$-C$_f$)/C$_i$ .100 [19, 20].

DE: Discolouration efficiency ;
C$_i$: Initial dye concentration ;
C$_f$: Final dye concentration.

The effect of solution pH

In order to find the optimum pH for the oxidation of NBB, a series of experiments at initial pH values in the range 3-8 was conducted. For more acidic pH (<3) there is a risk of dimerization of the catalyst [7], while for pH above 10, the catalyst is likely to deteriorate [7]. Fig. 2 shows the variation of the discolouration efficiency in function of time, under the following experimental conditions: (NBB concentration is 30 mg L^{-1}, [H_2O_2]=0.005m M, catalyst mass=0.05g).

The results presented in (Fig. 2) show that the optimum pH value for NBB oxidation by H_2O_2 using Wells-Dawson-type heteropolyanion iron substituted as catalyst is achieved at pH 3 (DE=82.37% after 70 min of treatment). A similar behavior was observed by several studies reported in the literature [19, 20]. This result can be explained by the stability of the catalyst at this pH. It has also been shown that the catalytic efficiency of the Fe^{3+}/ H_2O_2 system towards the oxidation of organic dyes is better at pH= 3 than the other pH [22, 23].

At neutral pH, the discolouration efficiency increases (51.36% is reached after 89 min of treatment). Previous studies [24, 25] showed that the addition of iron (Fe^{3+}) to the heteropolyanionic matrix extends the working pH range of the Fe^{3+}/ H_2O_2 system up to neutral pH.

H_2O_2 molecules are unstable in alkaline solution [26-27] and therefore, the degradation of dye decreases in alkaline solution (DE=33% after 89 min of treatment).

The optimal value is chosen pH = 3.

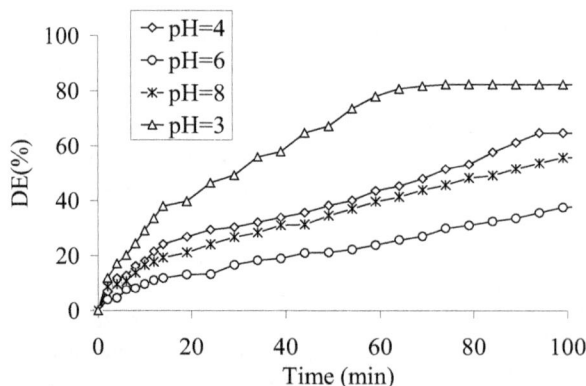

Fig. 2: Effet of solution pH on NBB oxidation (C$_i$=30mg/L, [H_2O_2]=0.005m M, catalyst mass=0.05g)

The effect of the nature of the acid used to adjust the pH

To evaluate the influence of these anions such as SO_4^{2-}, NO_3^-, Cl^-, PO_4^{3-} on the oxidation of NBB dye by a catalytic system (HPA Fe^{3+})/H_2O_2, we adjusted the pH of an aqueous solution of NBB by different acids H_2SO_4, HNO_3, HCl and H_3PO_4 at previously established optimum pH=3.

Fig. 3 shows the effect of these acid ions (chloride, sulphate, nitrate, and phosphate) on the dye oxidation. Depending on the nature of the acids, the discoloration efficiency is about 82.37%, 70%, 57.8% and 11.6 % in the presence of HCl, H_2SO_4, HNO_3 and H_3PO_4 acids respectively after 70 min of treatment.

It appears that the presence of phosphate ions inhibits the oxidation. These results agree with those found at the degradation of other organic pollutants [28]. The inhibitory effect of phosphate ions may be due to the catching of ·OH radicals according to the following equation:

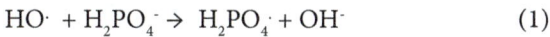

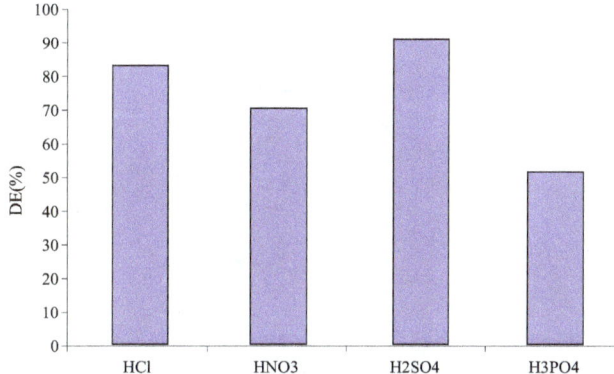

$$HO· + H_2PO_4^- \rightarrow H_2PO_4· + OH^- \quad (1)$$
$$HO·+PO_4^{3-} \rightarrow OH^-+ PO_4^{2-} \quad (2)$$

The effect of catalyst mass

It was shown [19] that (HPA Fe^{3+}) can catalyze the decomposition of H_2O_2. The reaction of H_2O_2 with a complex containing Fe^{3+} result in the reduction of Fe^{3+} to Fe^{2+} with apparition of $HO_2·$.

The action of H_2O_2 on the complex of Fe^{2+} leads to the generation of hydroxyl radicals OH·. These hydroxyl radicals cause the degradation of the dye.

In agreement with the mechanism proposed below, we can propose the following mechanism:

$$(HPA\ Fe^{3+}) + H_2O_2 \rightarrow (HPA\ Fe^{2+}) + HO_2· \quad (3)$$

$$(HPA\ Fe^{2+}) + H_2O_2 \rightarrow (HPA\ Fe^{3+}) + 2OH· \quad (4)$$

The catalyst mass is one of the critical parameters in catalytic oxidation process. In the present study, the influence of different catalyst mass [m((HPA Fe^{3+})) = [0g – 0.08g]on the decolorization efficiency of NBB is illustrated in (Fig.4). The concentration of hydrogen peroxide is fixed as 0.05mM, and NBB concentration is 30 mg L^{-1}.

It can be seen from the results that the decolorization efficiency of NBB oxidation increase when increasing the catalyst mass. This is due to the

Fig. 3: Effect of the nature of the acid used to adjust the pH (pH = 3, C_i=30mg/L, $[H_2O_2]$=0.005m M, catalyst mass=0.05g)

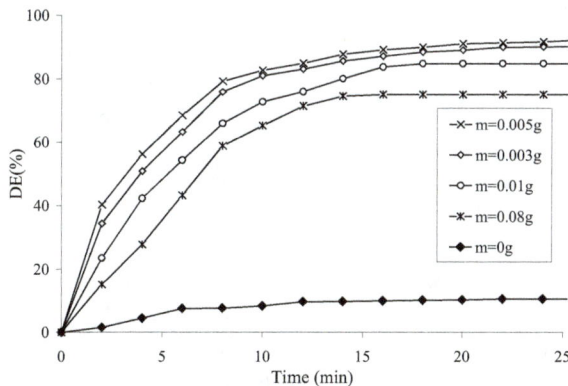

Fig. 4: Effect of catalyst mass on NBB oxidation (pH=3, $[C_i]$= 30mg/L, $[H_2O_2]$=0.005mM).

fact that (HPA Fe^{3+}) plays a very important role in the decomposition of H_2O_2 to generate the OH$\cdot$. The lower degradation capacity of the catalyst at small mass (0g-0.005g) is probably due to the lowest of OH$\cdot$ radicals producing a variable for oxidation, for higher mass of catalyst(0.005g-0.08g), there is a decrease of the decolorization efficiency. The decrease of the decolorization efficiecy of (NBB) oxidation by the increase the catalyst mass can be explained by the presence of the reaction (5) which enters in competition, at higher (HPA Fe^{3+}) mass, with (NBB) oxidation reaction :

$$HPA\ Fe^{2+} + OH\cdot \rightarrow HPA\ Fe^{3+} + OH^- \qquad (5)$$

Consequently, a mass of (HPA Fe^{3+}) of 0.005g was chosen throughout this work.

The effect of initial H_2O_2 concentration

The effect of H_2O_2 concentration is an important parameter for NBB degradation and for the decolorization efficiency. This effect was studied by varying the H_2O_2 concentration from 0.003 mM to 0.2 mM in the following optimal conditions: pH=3, NBB concentration is 30 mg L^{-1}, catalyst mass=0.005g).

According to the results shown above (Fig.5), the critical H_2O_2 concentration for the degradation of 30 mg L^{-1} NBB is about 0.08mM.

The activation of hydrogen peroxide by homogeneous catalysts was attributed to the formation of highly active hydroxyl radicals [29]. High concentrated H_2O_2 solution undergoes self quenching of $\cdot$OH radicals, with formation of hydro peroxyl radicals HO$_2\cdot$. Although HO$_2\cdot$ Is an effective oxidant itself, its potential oxidation is much lower than that of $\cdot$OH radicals [30].

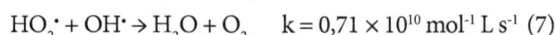

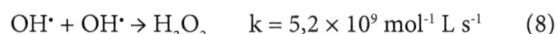

$$H_2O_2 + OH\cdot \rightarrow H_2O + HO_2\cdot \quad \cdots k = 2{,}7 \times 10^7 \text{ mol}^{-1} \text{L s}^{-1} \quad (6)$$

$$HO_2\cdot + OH\cdot \rightarrow H_2O + O_2 \quad_{\cdots} k = 0{,}71 \times 10^{10} \text{ mol}^{-1} \text{L s}^{-1} \quad (7)$$

$$OH\cdot + OH\cdot \rightarrow H_2O_2 \quad_{\cdots} k = 5{,}2 \times 10^9 \text{ mol}^{-1} \text{L s}^{-1} \quad (8)$$

The effect of the NBB concentration:

The study of the initial concentration effect of NBB dye on the oxidation kinetics was carried out from a concentration of 10 mg/L to 50 mg/L.

From Fig.6, we can note that the oxidation kinetics decreases with increasing the initial concentration of the dye. This result is in agreement with existing literature [30].

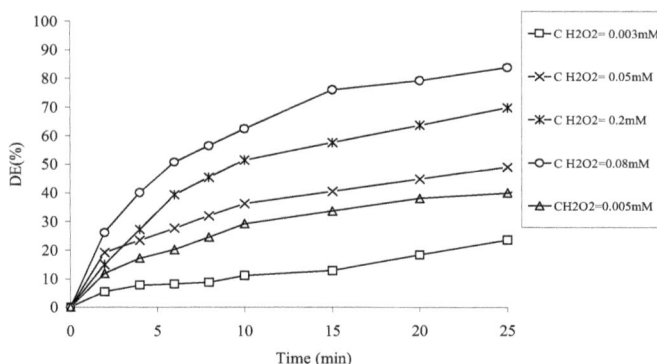

Fig. 5: Effet of initial H_2O_2 concentration on NBB oxidation (pH=3, [C$_i$]= 30mg/L, catalyst mass=0.005g)

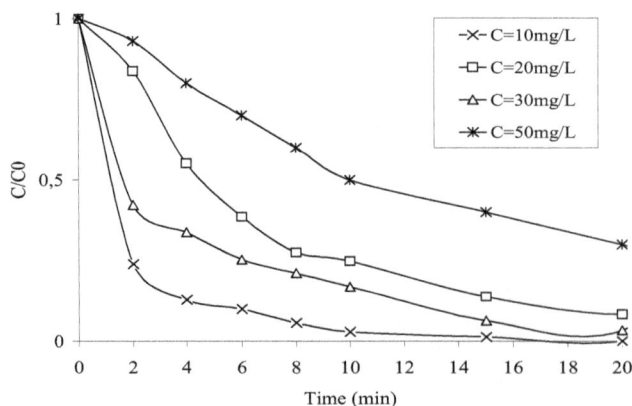

Fig. 6: Effect of initial dye concentration on NBB oxidation (pH=3, [catalyst mass=0.005g, [H_2O_2]=0.01mM

This phenomenon can be explained by the fact that increasing the initial concentration of dye leads to an increase in the number of molecules of (NBB) , while the number of the radicals hydroxyls remain constant (H_2O_2 concentration and catalyst kept constant), thereby causing a decrease in the discoloration efficiency [31].

UV- Vis absorbance spectra of dye before and after oxidation

The UV-vis absorbance spectra of the NBB before and after oxidation are shown in Fig. 7. In general, the absorbance at 400-700 nm corresponds to the n/p*transition of the azo and hydrazone forms, which is the origin of the color of azo dyes and is used to monitor the decoloration. The absorbance at 200-400 nm was attributed to then/p* transitions in benzene and naphthalene rings of azo dyes [32]. These four characteristic bands were markedly weakened during the degradation reaction, tending to disappears completely after 80 minutes, without the appearance of new absorption bands in the visible or ultraviolet regions due to destruction of the chromophoric and auxochromic structures by oxidation reaction.

The mineralization study of NBB

A complete mineralization of the dye molecules is always a major concern in catalysis because if this is not sufficiently accomplished, it may result in the formation of even more toxic intermediates.

Therefore, it is always desirable to degrade the dye molecules into smaller and less toxic species such as carbon dioxide, water and ionic species. Total organic carbon (TOC) analysis which measures the amount of carbon chemically bound.

The mineralization of aqueous NBB solution can be monitored by measuring the TOC evolution during oxidation process.

The TOC removal ratio (TOC) is defined as follows:

$$TOC (\%) = (1-TOC\ t) / TOC$$

The TOC values as a function of the time which is shown in Table 1.

As Table 1 shows, TOC decreased with the increasing reaction time. TOC removal was obtained at 250 min. This signifies a fairly high degree of complete mineralization of NBB which is essential for efficient dye pollution treatment.

A comparative study of oxidation of NBB in the presence of [$H_{1,2}Cu_{2,4}P_2W_{12}Mo_6O_{61,}\ 21H_2O$]

In this study we compared the catalytic activity of an iron substituted Dawson-type

Table1. TOC removal ratio on the miniralization of NBB solution at different reaction time. (pH=3, [catalyst mass] =0.005g, [H_2O_2]=0.01mM, [NBB]=30mg/L).

Time (min)	TOC removal ratio (%)
30	23.84
60	67.2
90	75.26
260	91.35

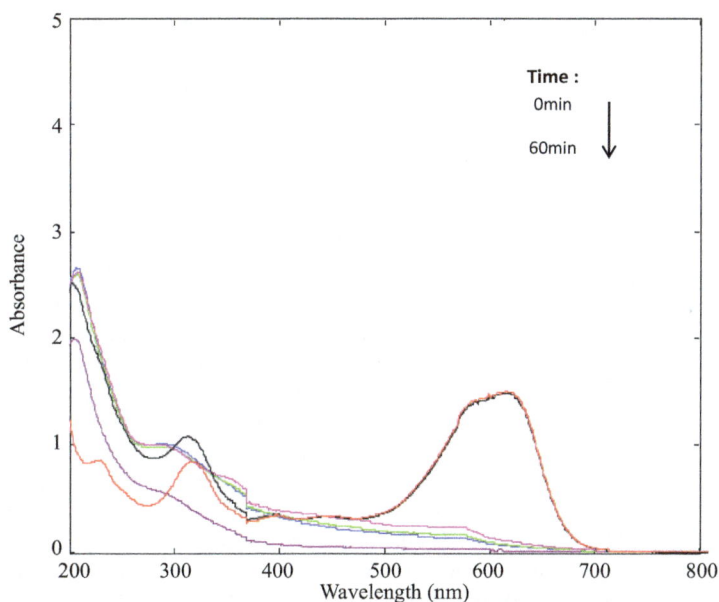

Fig. 7 : UV-Vis spectra of NBB water solutions during the treatment process with Fe(III) $P_2W_{12}MO_6/H_2O_2$ system. (pH=3, [catalyst mass=0.005g, [H_2O_2]=0.01mM.

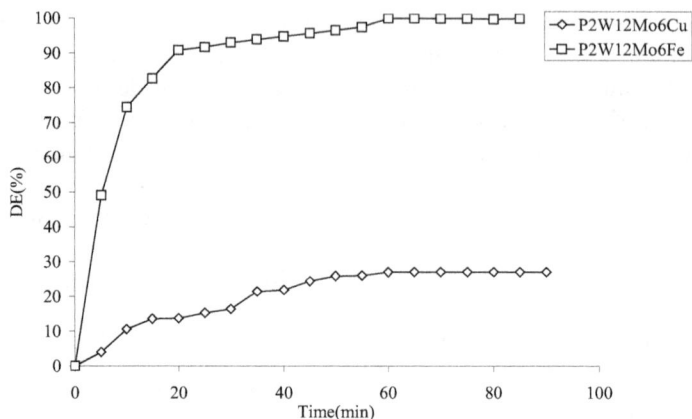

Fig. 8: Comparison of oxidation of NBB using Dawson- type iron -substituted heteropolyanion and Dawson-type copper -substituted heteropolyanion as catalysts Experimental condition: [pH=3, [catalyst mass=0.005g, $[H_2O_2]$=0.01mM.

heteropolyanion (HPA Fe^{3+})] with a copper substituted heteropolyanion [$H_{1.2}Cu_{2.4}P_2W_{12}Mo_6O_{61}$ 21H_2O]. The compound [$H_{1.2}Cu_{2.4}P_2W_{12}Mo_6O_{61}$ 21H_2O] was prepared, by the addition of Cu^{2+} ions to the Dawson acid form $H_6P_2W_{12}Mo_6O_{62}$ 24H_2O according to the methods described in the literature [33].The experiments were conducted under the same conditions as previously mentioned, by taking into consideration the optimised parameters. The results are illustrated in (Fig. 8).

Several Cu-containing systems for homogeneous catalytic decomposition of H_2O_2 have recently been demonstrated [32, 35].

Through these results, it is clear that the iron substituted heteropolyanion (HPA Fe^{3+}) is more effective compared to that of copper substituted heteropolyanion (HPA Cu^{2+}).

All compounds used catalyzed the decomposition of hydrogen peroxide and the formation of hydroxyl radicals ˙OH. The extent of peroxide to hydroxyl radical conversion was, however different from the particular substituted heteropolyanions.

The findings of this study are consistent with other results reported in the literature [23]

CONCLUSIONS

The oxidation of an azo dye (NBB), in an aqueous solution, by H_2O_2 in presence Dawson-type iron -substituted heteropolyanion (HPA Fe^{3+}) as catalyst was studied.

The optimum conditions had been determined, and it was found that the efficiency of the degradation obtained after 15 minutes of reaction, was about 100%.

The optimal parameters were: initial pH=3; $[H_2O_2]_0$=0.08 mM; catalyst mass=0.005g; for a concentration of dye $[NBB]_0$=30 mg L^{-1}.

Total organic carbon (TOC) analysis revealed degree of complete mineralization of naphtol blue black (91.35 % TOC removal after 260 min) which minimizes the possible formation of toxic degradation by-products such as the aromatic amines.

(HPA Fe^{3+}) is more effective compared to that of copper substituted heteropolyanion (HPA Cu^{2+}).

ACKNOWLEDGMENTS

This work was supported by the Engineering Environmental Laboratory of Badji Mokhtar University (Annaba-Algeria.)

CONFLICT OF INTEREST

The authors declare that there is no conflict of interests regarding the publication of this manuscript.

REFERENCES

1. Zhang, Y.Y., He, C., Deng, J., Tu, Y.-T., Liu, J.K., Xiong, Y, 2009. Photo-Fenton-like catalytic activity of nano-lamellar $Fe_2V_4O_{13}$ in the degradation of organic pollutants. Res Chem Intermed, 35: 727–737.

2. Ji, F., Li, C., Zhang, J., Deng, L, 2011.Efficient decolorization of dye pollutants with $LiFe(WO_4)_2$ as a reusable heterogeneous Fenton-like catalyst. Desalination, 269: 284- 290.

3. Tamimi, M., Qourzal, S., Barka, N., Assabbane , A., Ait-Ichou, Y, 2008. Methomyl degradation in aqueous solutions by Fenton's reagent and the photo-Fenton system. Sep. Puri. Tech, 61: 103–108.

4. Duesterberg, C.K., Mylon, S.E., Waite, T.D, 2008. pH effects on iron-catalyzed oxidation using Fenton's reagent. Environ. Sci. Technol, 42: 8522- 8527.

5. Sun, J.H., Shi, S.H., Lee, Y.F., Sun, S.P, 2009. Fenton oxidative decolorization of the azo dye Direct Blue 15 in aqueous solution. Chem. Eng. J, 155: 680-683.

6. Anipsitakis, G.P., Dionysiou, D.D, 2004. Radical generation by the interaction of transition metals with common oxidants. Environ. Sci. Technol, 38: 3705 - 3712.

7. Pope, M. T.(1983). Heteropoly and Isopoly Oxometalates , Springer : New York.

8. Ciabrini, P., Contant, R., Fruchart. J.M,1983. Heteropolyblues: Relationship between metal-oxygen-metal bridges and reduction behaviour of octadeca(molybdotungsto) diphosphate anions. Polyhedron, 111: 229-1233.

9. Dawson, B, 1953. The structure of the 9 (18) –heteropoly anion in potassium 9(18)-tungstophosphate, $K_6 (P_2W_{18}O_{62})$ .14 H_2O. Acta. Cristallogr, 6:113-126.

10. Hu, J., Burns, R. C, 2002. Homogeneous-phase catalytic H_2O_2 oxidation of isobutyraldehyde using Keggin, Dawson and transition metal-substituted lacunary heteropolyanions. J. Mol. Cat A: Chem, 184: 451–464.

11. Kuznetsova, N., Kirillova, N.V., Kuznetsova, L.I., Smirnova, M. Y.,   Likholobov, V.A, 2007.Hydrogen peroxide and oxygen oxidation of aromatic compounds in catalytic systems containing heteropolycompounds. J. Hazard. Mate, 146 569-576.

12. Keita, B., Fssaadi, K., Belhouari, A., Nadjo, L., Contant, R., Justum, Y, 1998. Catalysis of the oxidation of NADH by heteropolyanions: a kinetic study. C. R. Acad. Sci, 13: 43-350.

13. Alimohammadi, K., Sarrafi , Y., Tajbakhsh, M , 2008. $H_6P_2W_{18}O_{62}$: A green and reusable catalyst for the synthesis of 3,3-diaryloxindole derivatives in water. Monatsh Chem, 139:1037–1039.

14. Şaşmaz, S., Gedikli, S., Aytar, P., Güngörmedi, G., Evrim, A. Ç., Ünal, H. A., Kolankaya, N, 2011. Decolorization Potential of Some Reactive Dyes with Crude Laccase and Laccase-Mediated System. Appl Biochem Biotechnol, 163: 346–361.

15. Ferkous, H., Merouani, S., Hamdaoui, O, 2016. Sonolytic degradation of naphthol blue black at 1700 kHz: effects of salts, complex matrices and persulfate. J. Water Process Eng, 9: 67–77.

16. Luo, J., Hepel, M, 2001 . Photoelectrochemical degradation of naphthol blue black diazo dye on WO_3 film electrode. Electrochimica Acta, 46: 2913–2922.

17. Jiang, C., Guo, Y., Hu, C., Wang, C, 2004. Photocatalytic degradation of dye naphthol blue black in the presence of zirconia-supported Ti-substituted Keggin-type polyoxometalates. Materials Research Bulletin, 39: 251-261.

18. Mamba, G., Yangkou Mbianda, X., Kumar Mishra, A, 2015. Photocatalytic degradation of the diazo dye naphthol blue black in water using MWCNT/Gd,N,S-TiO$_2$ nanocomposites under simulated solar light.

19. Bechiri, O., Abbessi, M., Samar, ME, 2013. Decolorization of organic dye (NBB) using $Fe(III)P_2W_{12}Mo_5/H_2O_2$ system. Desalination and water treatment, 51: 31-33.

20 Bechiri, O., Abbessi, M.,Belghiche , R., Ouahab, L, 2014. Wells-Dawson Polyoxometelates $[HP_2W_{18-n}Mo_nO_{62}]Fe_{2.5}$, xH_2O; n = 0, 6: Synthesis, spectroscopic characterization and catalytic application for dyes oxidation. Comptes rendus Chimie, 17: 135–140.

21. Ciabrini , J.P ., Contant, R., Fruchart, M, 1983. Heteropolyblues: relationship between metal-oxygen-metal bridges and reduction behaviour of octadeca(molybdotungsto)diphosphate anions. Polyhedron, 21: 229-1233.

22. Lee, C., D. L, Sedlak, 2009. A novel homogeneous Fenton-like system with FeIII- phosphotungstate for oxidation of organic compounds at neutral pH values. J. Mol. Catal. A: Chem, 311: 1-6.

23. Bechiri, O., Abbessi, M., Ouahab, L, 2012. The oxidation study of methyl orange dye by hydrogenperoxide using Dawson-type heteropolyanionsas catalysts. Res. Chem. Intermed, 39:2945-2954.

24. Fan, X., Hao, H., Wang, Y., Chen, F., Zhang, J, 2013. Fenton-like degradation of nalidixic acid with Fe^{3+}/H_2O_2. Environ Sci Pollut Res In, 20: 3649-3656.

25. bou-Gamra, Z. M, 2014. Kinetic and thermodynamic study for Fenton –Like oxidation of Amaranth Red Dye. Advances in Chemical Engineering and Science, 4: 285-291.

26. Tangestaninejad, S., Moghadam, M., Mirkhani, V, 2008. Sonochemical and visible light induced photochemical and sonophotochemical degradation of dyes catalyzed by recoverable vanadium-containing polyphosphomolybdate immobilized on TiO$_2$nanoparticles, Ultrason. Sonochem. 15: 815.

27. Tabatabaee , M., Roozbeh, M., Roozbeh, M, 2011. Catalytic effect of lucunary heteropolyanion containing molybdenum and tungsten atoms on decolorization of direct blue 71. Chinese Chemical Letters, 22: 1501–1504.

28. Dalhatou, S., Pétrier, C., Laminsi, S., Baup, S, 2015. Sonochemical removel of naphtol blue black azo dye: influence of parameters and effect of mineral ions. Int. J. Environ. Sci. Technol, 12:35–44.

29. Strukul, G, 1992. Catalytic oxidations with Hydrogen Peroxide as Oxidant, Kluwer Academic, Dordrecht, The Netherlands.

30. Jung Fan, H., Tsuen Huang, S., Hsin Chung, W., Lyan Jan, J., Yu Linc, W., Chang Chen, C, 2009. Degradation pathways of crystal violet by Fenton and Fenton-like systems: Condition optimization and intermediate separation and identification. J. Hazard. Mater, 171: 1032–1044.

31. Ntampegliotis, K., Riga, A., Karayannis, V., Bontozoglou, V., Papapolymerou, G, 2006. Decolorization kinetics of Procion H-exl dyes from textile dyeing using Fenton-like reactions. J. Hazard. Mate , 136: 75–84.

32. Baldrian , P., Merhautova, V., Gabriel, J., Nerud, F., Stopka, P., Hruby, M., Beneš˘, J, 2006. Decolorization of synthetic dyes by hydrogen peroxide with heterogeneous catalysis by mixed iron oxides. Appl. Catal B: Environ. 66 : 258–264.

33. Belghiche, R., Bechiri, O., Abbessi, M., Golhen, S., Le Gal , Y., Ouahab, L, 2009 . 2D and 3D Polymeric Wells-Dawson Polyoxometallates : Synthesis, Crystal Structures   and Cyclic Voltammetry of $[M(H_2O)_4]_x[H_{6-2x}P_2W_{18-n}Mo_nO_{62}]$ (M = Cu^{II}, Co^{II}, Ni^{II}). Inorg.Chem. 48: 6026-6033.

34. Pecci, L., Montefoschi, G., Cavallini, D, 1997. Some new details of the copper-hydrogen peroxide interactionBiochem. Biophy. Res. Commun. 235: 264-267.

4

A comparative study for the electrochemical regeneration of adsorbents loaded with methylene blue

Ines Bouaziz[1,2], Morched Hamza[2], Ridha Abdelhedi[2], André Savall[1], Karine Groenen Serrano[1,]*

[1] *Laboratory of Chemical Engineering, University of Toulouse, CNRS, INP, 118 route de Narbonne 31062 Toulouse cedex, France*
[2] *Laboratory of Electrochemistry and Environment, National Engineering School, Sfax University, BP 1173, 3038 Sfax, Tunisia*

ABSTRACT

The electrochemical regeneration of methylene blue-saturated activated carbon, Nyex®1000 and sawdust has been studied and the performances in terms of capacity of adsorbent regeneration have been compared in this work. The adsorption isotherms were investigated. The results showed that the adsorption of methylene blue onto the investigated adsorbents obeyed Langmuir's model. The electrochemical oxidation of methylene blue beforehand adsorbed was studied using a boron doped diamond anode. The electrochemical regeneration efficiencies, under the same experimental conditions, of the activated carbon and Nyex®1000 were significantly less than 100% which were much lower to that of sawdust. Indeed the electrolysis tends to activate the sawdust because all the regeneration efficiencies obtained, whatever the applied current intensity, are higher than 100 %. Increasing treatment time would also result in a better regeneration of sawdust. This study confirmed that the coupling adsorption onto sawdust and electrochemical degradation is a potential technique for the efficient elimination of low concentration organic dyes from wastewater.

Keywords: Electrochemical regeneration, adsorption, methylene blue, Sawdust, Nyex®1000, Activated carbon, wastewater treatment.

INTRODUCTION

Pollution caused by industrial wastewaters has become a common problem for many countries. The search for alternative wastewater treatment technologies has been intensified due to a growing public concern about health and related environmental issues of trace levels of pollutants [1-2]. Dyes are one of the common organic pollutants in wastewaters. Over 7×10^5 tones and approximately 10,000 different types of dyes and pigments are produced world wide annually [3]. Most of these compounds are visually detected even in low concentrations, carcinogenic,

mutagenic, and can cause a severe health hazard to human beings.

Due to the low biodegradability of colored compounds, a conventional biological treatment process is not very effective. Dyes laden wastewaters are usually treated by physical or chemical processes such as adsorption [4], membrane filtration [5], ion exchange [6], ozonation [6], Fenton oxidation [6] and so on. However, these methods have different color removal capabilities, produce secondary waste and involve capital costs [6].

In recent years, the electrochemical method for wastewater treatment has attracted a great deal of attention, especially regarding the electrochemical oxidation [7-11]. This process was applied successfully to the total destruction of different persistent organic pollutants and in particular of dyes. It has been determined that the nature of the anode material is the main factor that affects the process [7]. In fact, with the use of nanostructured boron doped diamond (BDD) anode, many refractory organic compounds can be completely mineralized with high efficiency due to electrogenerated hydroxyl radicals from water discharge without addition of any reagent [9,11]. Although electrochemical oxidation is an attractive means for wastewater treatment, the low current efficiency is still a critical problem for the treatment of dilute solutions because of mass-transfer limitations. Therefore, organic pollutants in industrial wastewater should be concentrated to obtain a large quantity of organic pollutants before carrying out electrochemical degradation. Consequently, a single electrochemical process alone may not be adequate for the treatment of organic compounds. Hence, the researchers are attempting a combination of two or more treatment methods to overcome these limitations.

In this study, the adsorption has been chosen as preconcentration step and the electrochemical process allows the desorption and destruction of the adsorbed organic matter, restoring the adsorptive capacity of adsorbent.

Electrochemical regeneration of activated carbon (AC), the most popular adsorbent, has been widely investigated [12-16]. The first report on the electrochemical regeneration of activated carbon was by Owen and Barry [12] who achieved regeneration efficiencies of up to 61%. Whilst other researchers have demonstrated that efficiencies of electrochemical regeneration of activated carbon were often significantly less than 100% [13-20]. Moreover, the rate of adsorption and desorption of organic compounds from loaded adsorbent is often governed by intra-particle diffusion which requires long adsorption and regeneration periods [13]. Research in the development of alternative adsorbents has been performed. Over the last few years, Brown et al. [17-19] have worked on an alternative approach to adsorption and electrochemical regeneration based on a novel, non-porous, highly-conducting carbon-based adsorbent material (called Nyex®). It was shown that this adsorbent can be rapidly and fully electrochemically regenerated with low energy consumption.

In most cases, the regeneration of adsorbent by electrochemistry was studied in the presence of sodium chloride. However, using NaCl the main drawback is the formation of hazardous organochloride by-products during the electrolysis. That is why sodium sulfate has been chosen in this study.

We have recently studied the coupling of adsorption onto sawdust, chosen as a by-product of furniture industry without commercial value, and electrochemical oxidation in the presence of Na_2SO_4 [20]. The electrochemical oxidation of methylene blue adsorbed onto sawdust led simultaneously to its degradation and sawdust regeneration for the next adsorption. It was also observed that multiple adsorption and electrochemical regeneration cycles led to an enhancement of adsorption capacity of the sawdust.

The efficiency and performance of the different electrochemical regeneration methods cannot be directly compared to each other if the lay out of the electrochemical reactor, the adsorbate, and/or electrodes are different. To the best of our knowledge, no comparison including the electrochemical regeneration of AC, Nyex® and sawdust has been reported to date. This work is intended to fill this gap and make a comparative study between electrochemical regeneration of the three organic saturated adsorbents under the same experimental conditions. For this purpose, the treatment of aqueous solution of methylene blue (MB) by coupling adsorption onto commercial activated carbon, Nyex®1000 and sawdust, and electrochemical oxidation on BDD anode has been studied.

MATERIALS AND METHODS
Materials
Chemicals
MB, a cationic dye with a chemical formula of $C_{16}H_{18}N_3$ and molecular weight of 319.85 g mol^{-1}, was chosen as a model organic contaminant. MB was purchased from Merck, and it was used as received without further purification. MB solutions were prepared with ultra pure water.

All other chemicals used were of analytical grade (Acros Organics). Na_2SO_4 (0.1M) was used as the supporting electrolyte.

Adsorbents
Three adsorbents were used in this study: sawdust, commercial A.C. and Nyex®1000.

The red wood sawdust, a low cost material with BET surface equals to 0.4m^2/g was obtained from a furniture factory in Sfax, Tunisia.

The commercial A.C. was supplied by Merck (reference 2514). The specific surface area of this adsorbent is 980m^2/g [21].

Both adsorbents were washed with distilled water several times to remove the suspended particles, then dried in a hot air oven at 373 K. The dried adsorbents were sieved and the final sizes of the particles retained were in the range of 0.5 to 1.12 mm and 0.4 mm for sawdust and A.C., respectively.

The third adsorbent used in this work was a graphite intercalation compound, Nyex˚1000 supplied by Arvia Technology Ltd. The specification for Nyex˚1000 provided by Arvia Technology Ltd and mentioned by Mohamed et al. [19] indicates a carbon content of 95 wt%, with particle diameters in the range of 100 to 700 μm. The BET surface area of Nyex˚1000 was found to be 1m^2/g, which is not far to that of sawdust and very small compared to the commercial activated carbon. Its conductivity is about 0.16 Ω^{-1}cm^{-1}.

Experimental methods
Adsorption experiments

The adsorption isotherms studies were performed by batch adsorption technique.

In the case of Nyex˚, various known quantities of adsorbent were added to 100 mL of 100 mg/L of MB solution in a 500 mL flask at 303 K.

The adsorption isotherms of sawdust were investigated, at three different temperatures 303, 313 and 323 K, by adding 0.5 g of sawdust into 50 ml of different initial concentrations of MB. For all experiments, the agitation time was one hour to reach equilibrium before analysis.

The adsorptive capacity, q, defined as the mass of adsorbate per gram of adsorbent (mg/g), can be calculated from the initial (C_0) and final (C_f) concentrations of MB according to Eq.1:

$$q = \frac{(C_0 - C_f)V}{m} \qquad (1)$$

Where V is the volume of solution used and m is the weight of adsorbent.

To reduce the number of adsorption steps and make sure that the adsorbent has reached saturation before electroxidation experiments, the saturation of adsorbent has been performed in a (30 cm x 1.5 cm) glass column, packed with a known mass of adsorbent sandwiched between two layers of glass wool was used. MB solution with the initial concentration of 300 mg/l (in the case of sawdust and activated carbon) and 100 mg/l (in the case of Nyex) was pumped into the column at a constant flow rate of 0.3 mL/min.

Desorption experiments

Desorption experiments were conducted to quantify the long-term non-electrochemical desorption of MB. For simple desorption studies, the loaded adsorbent after drying at room temperature was contacted with 200 mL of 0.1mol/L Na$_2$SO$_4$ solution. The solution was stirred at a temperature of 303 K until the release process reached equilibrium.

Electrolysis
Batch reactor coupling with adsorption

The loaded adsorbent was removed from the column of adsorption and stirred with Na$_2$SO$_4$ solution. Immediately, a constant current was applied in the same reactor at 303 K using a Meteix d.c. power supply. The mixture was stirred constantly during the electrochemical degradation of adsorbed and released MB. A nanostructured BDD electrode with a geometric area of 7 cm^2 was used as anode. The counter electrode was a cylindrical mesh of platinum (67.5 cm^2).

The regeneration efficiency of the adsorbent, R_e, by electrolysis on BDD anode was defined as the ratio between the capacity of adsorption of the sawdust after the electro-regeneration, q_r and the initial capacity of adsorption q_i:

$$R_e = \frac{q_r}{q_i} \times 100 \qquad (2)$$

Where q_r is the capacity of adsorption of the regenerated adsorbent (mg of MB/g of adsorbent) and q_i is the initial capacity of adsorption of fresh adsorbent under the same adsorption conditions.

It should be noted that for the batch desorption and electrochemical experiments, 1g of loaded adsorbent was contacted with 200 ml of Na$_2$SO$_4$ solution in the case of sawdust and activated carbon. Whereas due to the low adsorption capacity of Nyex˚ and in order to obtain the same range of desorbed MB concentration, 16 g of MB saturated Nyex˚ were used in 200 ml of Na$_2$SO$_4$. Moreover, a high concentration of MB solution was used in this study to reach the saturation of adsorbent in short time. All the experiments have been performed at natural pH.

Analytical techniques

The concentration of MB in the solution was determined using a Hewlett-Packard UV-visible spectrophotometer at a wavelength of 660 nm. A previously established linear Beer–Lambert relationship was used to determine the concentration. In a range of 0-8 mg/L of MB, the molar extinction coefficient obtained was: $\varepsilon= 6.3 \times 10^4$ L mol^{-1} cm^{-1} at 660 nm at 298 K. The relative standard deviation has been determined as 2%.

RESULTS AND DISCUSSION

Adsorption

Batch adsorption studies at 303 K have been performed previously to investigate the kinetic and equilibrium isotherm of MB adsorption onto sawdust [14]. Experimental data obtained at 303 K were compared with models of Langmuir (Fig.1) and Freundlich (Supplementary materials). These models suggest interactions between the adsorbed molecules, assuming a heterogeneous surface with a non-uniform distribution of heat of adsorption on the surface in the case of Freundlich model. The Langmuir model supposes a monolayer with a homogeneous distribution of sorption sites and sorption energies without interactions between the adsorbed molecules. In order to optimize the design of a sorption system to remove MB from wastewater, it is important to establish the most appropriate correlation from the equilibrium curve. Both models are commonly used for investigation of the sorption of a variety of dyes on sawdust and activated carbon.

The linear form of Langmuir equation is given by Eq. 3:

$$\frac{C_e}{q_e} = \frac{C_e}{q_m} + \frac{1}{K_L q_m} \tag{3}$$

C_e is the concentration of the adsorbate in the solution at equilibrium (mg/L), q_e and q_m are the capacity of adsorption (mg g^{-1}) at equilibrium and maximum, respectively and K_L is the Langmuir isotherm constant (L mg^{-1}). The parameters K_L and q_m can be obtained by plotting C_e/q_e versus C_e.

As shown in Fig. 1, the adsorption isotherms of MB onto Nyex° and sawdust fit very well with the theoretical Langmuir isotherms ($R^2 > 0.996$ for sawdust and Nyex) whereas low correlation coefficients ($R^2 < 0.949$) were obtained using the model of Freundlich. This result is similar to that found by Mohammed et al. [22] for the adsorption of Acid Violet onto Nyex°1000.

As the adsorption isotherms of MB onto activated carbon were extensively studied in the literature [23], the maximum adsorption capacity of MB onto activated carbon was determined, in this study, with the column adsorption only. Consequently, the highest value of the quantity of MB adsorbed was obtained for the commercial activated carbon (285 mg/g) followed by the sawdust (23 mg/g) and then by Nyex° (1.4 mg/g) which is close from that observed for the sorption of Acid violet onto the same adsorbent [23]. The high adsorption capacity of the activated carbon is due to its very high surface area (980m²/g). However

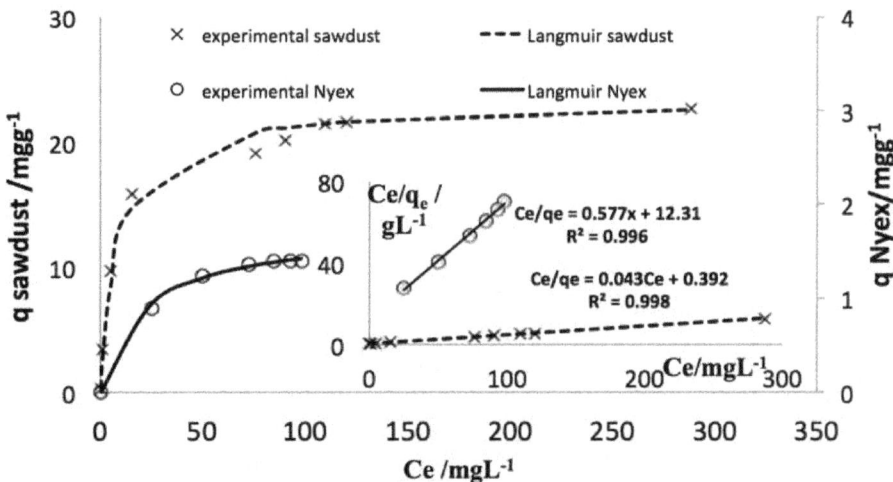

Fig. 1: Adsorption isotherms of MB onto sawdust and Nyex® at 303 K. Comparison of experimental data (symbols) with Langmuir model (lines). Inset panel: Linear form of Langmuir isotherms. Operating conditions: T=303 K. For Nyex®: C°=100 mg/L; V=100 mL; m$_{nyex}$=0.5-8 g. For sawdust:m$_{sawdust}$ = 0.5g, V=50mL, C°=35-512 mg/L

the sorption capacities of sawdust and Nyex˙ are inversely proportional to their specific surface area (0.4 and 1m²/g, respectively). The comparison of Langmuir isotherm constants for adsorption of MB calculated in this work with those determined by Hameed et al. [23] for activated carbon are shown in Table 1.

Table 1 shows that the adsorption capacity (q_m) of the different adsorbents for MB is comparable to the maximum adsorption obtained from adsorption isotherms (Fig. 1) and/or column adsorption. However, if the adsorptive capacity (q_m) is normalized with the specific area, it is found that the sawdust is able to adsorb the highest mass of dye per surface unit (q_m = 58 mg/m²), against Nyex (q_m = 1.7 mg/m²) and activated carbon (q_m = 0.3 mg/m²).

These results again show that sawdust, having the lowest surface area, can be an attractive option for the removal of dyes from dilute industrial effluents. It indicates that the adsorption capacities depend not only on the specific surface area or porosity but also on the surface chemistry of the adsorbents. To evaluate the influence of the temperature of the adsorption capacity of the sawdust, adsorption isotherms were performed at 313 and 323 K and shown in Fig. 2. It can be observed that adsorption decreases with an increasing temperature. Same results were obtained by many authors for the adsorption of dyes on various adsorbents [24-25]. This can be explained by the exothermicity and spontaneity of the adsorption process and by the weakening of bonds between dye molecules and active sites of adsorbents at high temperatures.

All adsorption isotherms are fitted well with the Langmuir model. The Langmuir constants and correlation coefficients calculated from the linear form of Langmuir equation are given in Table 2.

Table 1: Comparison of Langmuir constants and desorption parameters for adsorption of methylene blue onto sawdust, Nyex® and activated carbon at 303 K.

Adsorbent	Langmuir constants			Desorption parameters	
	q_m (mg/g)	K_L (L/mg)	R^2	Desorbed amount of MB (%)	[MB] desorbed (mgL^{-1})
Sawdust	23.2	0.11	0.998	33	38
Nyex®1000	1.7	0.05	0.996	10	10.5
Activated carbon	294.1[23]	0.13[23]	0.999 [23]	0.17	2.3

Table 2: Langmuir parameters for the adsorption of methylene blue onto sawdust at 303, 313 and 323 K. Operating conditions: see Fig. 2

Temperature	q_m (mg/g)	K_L (L/mg)	r^2
303 K	23.2	0.11	0.998
313 K	16.1	0.10	0.997
323 K	13.1	0.09	0.999

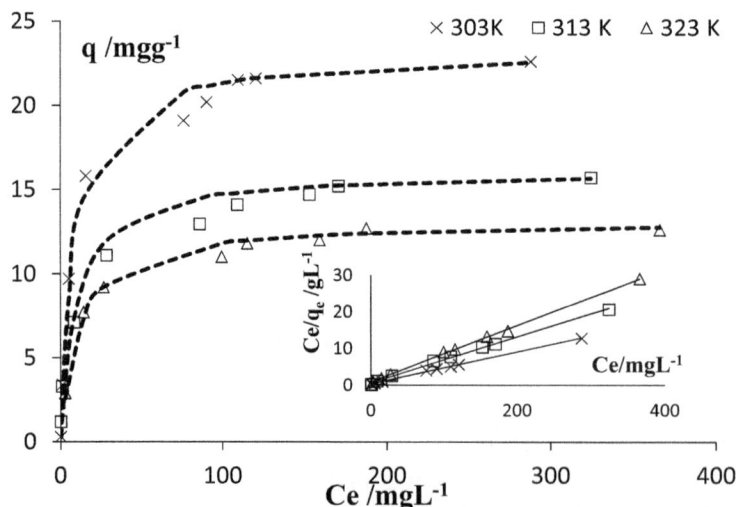

Fig. 2: Adsorption isotherms of MB onto sawdust at three different temperatures 303, 313 and 323 K. Comparison of experimental data (symbols) with Langmuir models (lines). Inset panel: Linear form of Langmuir isotherms. Operating conditions: msawdust=0.5g, V=50 mL, C°= 12-600 mg/L.

Thermodynamic parameters of MB adsorption onto the sawdust samples, including free energy change ($\Delta G°$), enthalpy change ($\Delta H°$) and entropy change ($\Delta S°$), were calculated using the following equations:

$$\Delta G°=-RT \ln K \qquad (4)$$

$$\Delta G° = \Delta H° - T\Delta S° \qquad (5)$$

Where T is the temperature (K), R is universal gas constant (8.314 J mol^{-1} K^{-1}), and K (l g^{-1}) is an equilibrium constant defined as the product of the Langmuir constants q_m and K_L.

The values of $\Delta H°$ and $\Delta S°$ were calculated from the intercept and slope of $\Delta G°$ versus T according to Eq. (5) by linear regression analysis. The results are listed in Table 3.

Due to the values of the free energy obtained in Table 3, one can deduce that physisorption of MB takes place on sawdust. Indeed, it's known that for physisorption, the change in free energy for adsorption of MB onto sawdust ranges between -0.4 kJ mol^{-1}and -2.36 kJ mol^{-1}. It is known that, the absolute magnitude of the change in free energy for physisorption is between -20 kJ mol^{-1}and 0 kJ mol^{-1} and for chemisorption the corresponding range is -80 kJ mol^{-1}-400 kJ mol^{-1} [26]. The negative value of $\Delta H°$ and $\Delta S°$ confirm that the adsorption process is exothermic and renders the molecule distribution more orderly in the adsorbent than in the solution.

Desorption

Batch desorption experiments of MB from loaded sawdust in Na$_2$SO$_4$ aqueous solution have been previously performed [20]. Similar experiments were carried out with the commercial A.C. and Nyex˙ in order to compare the results and quantify the long term desorption of MB from these three adsorbents in the Na$_2$SO$_4$ solution without electrolysis. The comparison is shown in Fig. 3 at 303 K.

The calculated desorption rates are shown in Table 1. In the case of activated carbon, the desorption equilibrium was reached in 20 min and the desorbent amount of MB was 0.44 mg/g corresponding to 0.17% of the adsorbed dye. The low desorption capacity of activated carbon suggests that chimiosorption is involved. The desorbed amount of MB from the Nyex˙ at equilibrium, after 10 min, was relatively higher, 0.13 mg/g corresponding to 10% of the MB adsorbed. The highest value of the desorbed amounts of the dye, at the equilibrium in 15 min, was obtained from the sawdust and was close to 7.6 mg/g, accounting for about 33% of the adsorbed amount of MB.

Table 3: Thermodynamics parameters for the adsorption of methylene blue onto sawdust at different temperatures. The negative values of $\Delta G°$ indicate the spontaneous adsorption of MB onto sawdust samples.

Temperature	$\Delta G°$ (kJ mol^{-1})	$\Delta H°$ (kJ mol^{-1})	$\Delta S°$ (kJ mol^{-1} K^{-1})
303 K	-2.36	-32	-0.1
313 K	-1.23		
323 K	- 0.4		

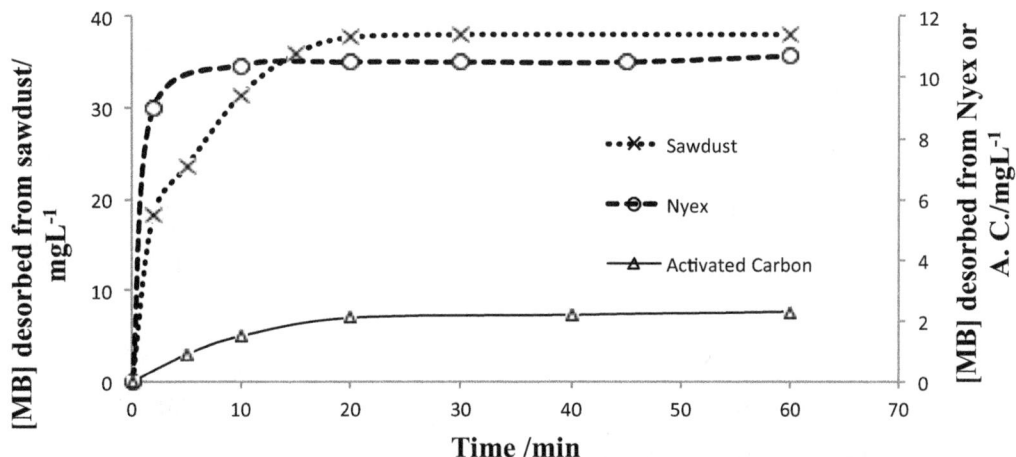

Fig. 3: Desorption kinetics of MB from sawdust, Nyex® and Activated Carbon at 303 K. Operating conditions : 1 g of saturated sawdust or A.C. and 16 g of Nyex®, V=200 mL, [Na$_2$SO$_4$]= 0.1 mol/L, T=303 K.

Electrochemical regeneration

The adsorbent after saturation with MB was contacted with Na_2SO_4 solution. Immediately, a constant current was applied in the same reactor at 303 K and the temporal variation of the concentration of MB in solution has been measured. Fig. 4 shows the comparison of the concentration profile for a single desorption (full symbols) and for the electrolysis (empty symbols). It appears that the speed of appearance of MB and the MB concentration in the solution are less important with electrolysis. This difference may be explained by the fact that a part of desorbed MB is oxidized on BDD surface. Consequently the instantaneous variation of MB concentration during the electrolysis results on the speed difference of MB desorption from the sawdust and oxidation of MB desorbed.

The efficiencies of the electrochemical regeneration on the three adsorbents have been compared in Fig. 5 at different applied current intensities and different regeneration time. One can observe that for a regeneration time of 600 min for an applied current density of 0.215 A/cm², the regeneration efficiencies of activated carbon and Nyex° were low, only 35 and 44%, respectively, in comparison with the one of sawdust (240%).

The limitation in the electrochemical performance of activated carbon can be attributed to the nature of the activated carbon and, more specifically, by its characteristics of micropores structures. Indeed, the narrow microporosity that favors MB uptake may also

hinder subsequent MB desorption (as demonstrated in paragraphs 3-2), and consequently the regeneration efficiency achieved by electrochemistry. These trends have been observed by other researchers. For example Narbaitz et al. [27] have found that the electrochemical regeneration reached only 8-15% in the case of adsorbed natural organic material (NOM) on granular activated carbon. The authors have been attributed the low efficiency to the high irreversibility of NOM adsorption onto A.C.

Fig. 5 shows also that the regeneration efficiency of Nyex cannot exceed 50% whatever the treatment time in the range of 120 to 616 min and the applied current densities used. The low electrochemical regeneration efficiencies were surprising in comparison with the ones obtained by Brown and co-workers [17-19,22]: they have demonstrated that the Nyex°, which considered as the first adsorbent specifically designed for electrochemical regeneration, can be rapidly and fully regenerated by electrochemical oxidation. For example, they observed a regeneration efficiency of 100% in 10 min by passing a charge of 25 C g⁻¹ and suggested that the organic compounds were oxidized during regeneration without being released. However, it should be noted that it is difficult to compare with our results because the design of the electrochemical cell is different (they used a bed of loaded Nyex) whereas in the present work, the Nyex particles were in suspension in the electrochemical batch cell. Moreover amount of sodium chloride were added

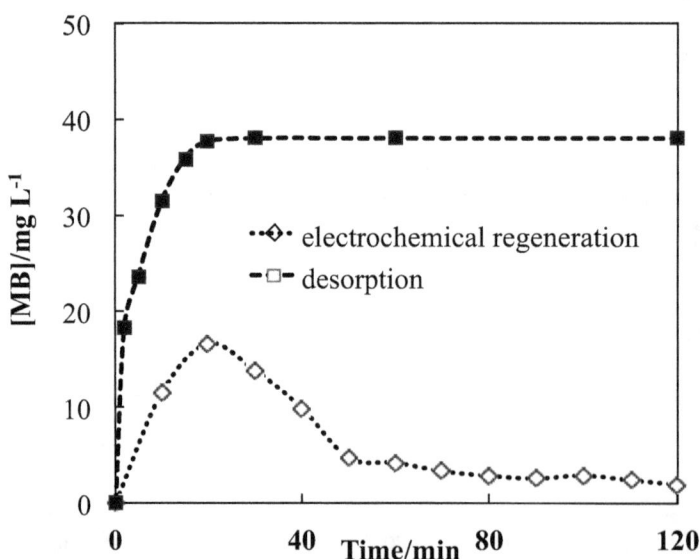

Fig. 4: Comparison of variation of MB concentration during desorption (full symbol) and electrochemical regeneration (empty symbol). [Na₂SO₄] =0.1 M, m_sawdust. = 1 g, q° = 21 mg/g. anode: BDD, i = 0.215 A/cm², V = 100 mL, T = 303 K, anode DDB , cathode Platinum mesh.

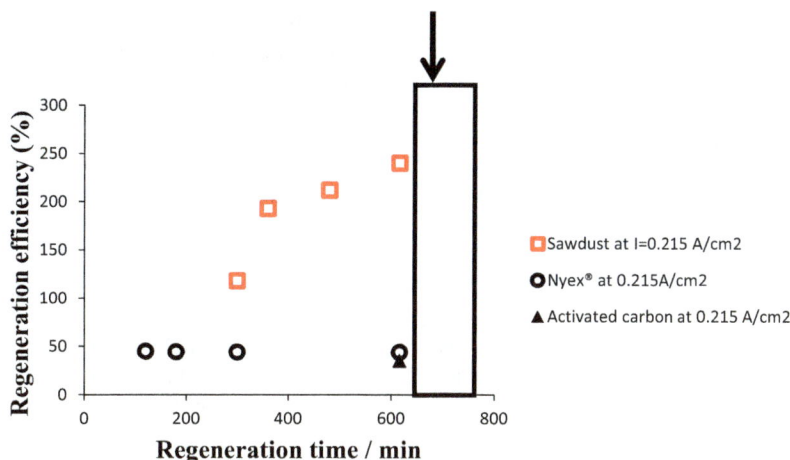

Fig. 5: Comparison of regeneration efficiency of different adsorbent with electrolysis time and applied current intensity. Operating conditions: m_{ads} = 1g for AC and sawdust, 16 g for Nyex®, V= 200 mL, [Na$_2$SO$_4$] =0.1 M, T=303 K.

in their cell and electrogenerated hypochlorite ions can react chemically with MB.

The electrochemical treatment of saturated sawdust shows efficiencies higher than 100% with an increase of the regeneration efficiency with regeneration time. This result is similar to that found in our previous work for the lowest current density (0.05 A/cm^2) [20]. Thus, this present work confirms our earlier studies showing that the modification of adsorption of sawdust depends on the electrical charge passed. However in this study, it was observed that the regeneration efficiency increases slightly for regeneration time more than 480 min. It seems that it tends to reach a maximum value. These data also confirm the hypothesis suggested in the previous work that the modification of the surface chemistry of sawdust causes the increase in adsorptive capacity after regeneration [20]. Moreno-Castilla has evidenced that the adsorption capacity depends on the surface chemistry of the adsorbent [27]. Hence, the characterization of surface functional groups of sawdust before and after electrochemical treatment is the subject of on-going research.

As final remark, it appears clearly that, under our experimental conditions, the electrochemical regeneration is largely attributed to the desorption of the organic substances from the adsorbents, the desorption is enhanced under polarization.

CONCLUSION

In order to treat dilute wastewater, the coupling of adsorption and electrochemical oxidation can be considered as a promising approach. It is the first time that a study is devoted on the suitability of the electrochemical regeneration of three adsorbents: Activated carbon, Nyex® and sawdust.

The thermodynamic study shows that the isotherms of the three adsorbents follow the Langmuir model. The maximum adsorption capacity obtained for MB onto the virgin sawdust reached 23 mg/g which is 16.4 times higher than the one obtained onto Nyex® and 12.4 times lower than the one obtained onto AC. The desorption experiments confirm that on A.C. the low desorption occurs due to an intra-particles diffusion controlled process. However on sawdust, 33 % of the MB adsorbed amount was desorbed after 15 minutes of contact with the electrolyte. The desorbed amount from Nyex® corresponds to 10 %. The study of the electrochemical regeneration of adsorbents has evidenced that only sawdust can be completely regenerated and even more: the capacity of adsorption is enhanced after electrolysis. By contrast, whatever the electrical charge used, it appears that electrolysis cannot prevent the deterioration of the adsorption performances of Nyex and A.C.

This present work has confirmed that the coupling adsorption onto sawdust and electrochemical treatment is a potential technique for an efficient elimination of low concentration organic dyes. However, the development of an effective electrochemical reactor with a very small volume is needed to improve the performance of electrochemical oxidation (to increase the average current efficiency). Further work is also needed to characterize surface chemistry of the sawdust after regeneration, optimize the various parameters and reduce the regeneration charge passed per gram of sawdust.

CONFLICT OF INTEREST

The authors declare that there is no conflict of interests regarding the publication of this manuscript.

REFERENCES

[1] Banat I.M., Nigam P., Singh D., Marchant R. , 1996. Microbial decolourization of textile dyes containing effluents: a review. Bioresour. Technol. 58: 217-227.

[2] Deegan A. M., Shaik B., Nolan K., Urell K., Oelgemöller M., Tobin J., Morrissey A., 2011. Treatment options for wastewater effluents from pharmaceutical companies, Int. J. Environ. Sci. Tech., , 8(3): 649-666.

[3] Iqbal M.J., Ashiq M.N., 2007. Adsorption of dyes from aqueous solutions on activated charcoal, J. Hazard. Mater., B139: 57–66.

[4] Walker G. M., Weatherley L. R. 1997. Adsorption of acid dyes on to granular activated carbon in fixed beds, Water Res., 31: 2093–2101.

[5] Bes-Piá A., Mendoza-Roca J. A, Roig-Alcover L., Iborra-Clar A., Iborra-Clar M. I., Alcaina-Miranda M. I., 2003. Comparison between nanofiltration and ozonation of biologically treated textile wastewater for its reuse in the industry, Desalination, 157: 81-86.

[7] Robinson T, McMullan G, Marchant R, Nigam P., 2001. Remediation of dyes in textile effluent: a critical review on current treatment technologies with a proposed alternative. Bioresour. Technol., 77: 247-255.

[7] Belhadj Tahar N., Savall A., Mechanistic aspects of phenol electrochemical degradation by oxidation on a Ta/PbO$_2$ anode, 1998. J. Electrochem. Soc., 145: 3427-3434.

[8] Panizza M., Barbucci A., Ricotti R., Cerisola G., 2007. Electrochemical degradation of methylene blue, Sep. Purif. Technol. 54: 382-387.

[9] Weiss E., Groenen-Serrano K., Savall A., 2008. A comparison study of electrochemical degradation of phenol on boron doped diamond and lead dioxide, J. Applied Electrochem., 38: 329-337.

[10] Martínez-Huitle, C.A., Brillas E. Decontamination of wastewaters containing synthetic organic dyes by electrochemical methods. A general review, 2009, Appl. Catal. B: Environ., 87: 105–145.

[11] Sirés, I., Brillas E., Hamza M., Abdelhedi R., 2009. Decolorization and degradation of the triphenylmethane dye Methyl Violet by electrochemical oxidation using boron-doped diamond and Pt anodes, J. Electroanal. Chem., 627: 41-50.

[12] P.H. Owen, J.P. Barry, Environics Inc., Huntington Beach, California, 1972. Report number PB 239156.

[13] Narbaitz R. M., Cen J., 1994. Electrochemical regeneration of granular activated carbon, Water Res., 28:1771-1778.

[14] García-Otón M., Montilla F., Lillo-Ródenas M.A., Morallón E., Vázquez J. L., 2005. Electrochemical regeneration of activated carbon saturated with toluene, J. Appl. Electrochem. 35: 319-325.

[15] Narbaitz R. M., Karimi-Jashni A., 2009. Electrochemical regeneration of granular activated carbons loaded with phenol and natural organic matter, Environ. Technol., 30: 27–36.

[16] Weng C. H., Hsu M. C., 2008. Regeneration of granular activated carbon by an electrochemical process, Separation and Purification Technology, 64: 227–236.

[17] Brown N. W., Roberts E. P. L., Garforth A. A., Dryfe R. A. W, 2004. Electrochemical regeneration of a carbon-based adsorbent loaded with crystal violet dye, Electrochim. Acta 49: 3269–3281.

[18] Brown N. W., Roberts E. P. L., Chasiotis A., Cherdron T., Sanghrajka N., 2004. Atrazine removal using adsorption and electrochemical regeneration, Water Res., 38: 3067–3074.

[19] Mohammed F. M., Roberts E. P. L., Hill A., Campen A. K., Brown N. W., 2011. Continuous water treatment by adsorption and electrochemical regeneration, Water Res., 45: 3065-3074.

[20] Bouaziz I., Chiron C., Abdelhedi R., Savall A., Groenen Serrano K., 2014.Treatment of dilute methylene blue-containing wastewater by coupling sawdust adsorption and electrochemical regeneration, Environ. Sci. Pollut. Res. 21 : 8565-8572.

[21] Ayral C., Elimination de polluants aromatiques par oxydation catalytique sur charbon actif , PhD Thesis, University of Toulouse, Toulouse, France, 2009.

[22] Mohammed F. M., Roberts E. P. L., Campen A. K., Brown N. W., 2012. Wastewater treatment by multi-stage batch adsorption and electrochemical regeneration, J. Electrochem. Sci. Eng., 2: 223-236.

[23] Hameed B. H., Ahmad A. L., Latiff K. N. A., 2007. Adsorption of basic dye (methylene blue) onto activated carbon prepared from rattan sawdust, Dyes and Pigments, 75: 143-149.

[24] Hamdaoui O., 2006. Batch study of liquid-phase adsorption of methylene blue using cedar sawdust and crushed brick, J. Hazard. Mater., B135: 264–273.

[25] Chern J. M., Wu C. Y., 2001. Desorption of dye from activated carbon beds: effects of temperature, pH, and alcohol. Water Res., 35: 4159–4165.

[26] Yu Y., Zhuang Y. Y., Wang Z. H., 2001. Adsorption of water-soluble dye onto functionalized resin, Journal of Colloid and Interface Science, 242: 288-293.

[27] Narbaitz R. M., McEwen J., 2012. Electrochemical regeneration of field spent GAC from two water treatment plants. Water Res., 46: 4852–4860.

[28] C. Moreno-Castilla, 2004. Adsorption of organic molecules from aqueous solutions on carbon materials, Carbon , 42: 83-94.

5

Removal of nickel (II) from aqueous solution by graphene and boron nitride nanosheets

Jafar Azamat

Department of Chemical Engineering, Ahar Branch, Islamic Azad University, Ahar, Iran

ABSTRACT

Molecular dynamics simulations were carried out to study the removal of Ni^{2+} as a heavy metal from the water by the functionalized graphene nanosheet (GNS) and boron nitride nanosheet (BNNS). Nickel causes asthma, conjunctivitis and inflammatory reactions and nickel salts act as emetics when swallowed; therefore, removal of nickel is necessary from the aqueous solutions. The systems were comprised of a nanosheet (GNS or BNNS) with a pore in its center that it is containing an aqueous ionic solution of nickel chloride. For the removal of Ni^{2+} from an aqueous solution, the pores of nanosheet were functionalized by passivating each atom at the pores edge and then an external electric field was applied along the z-axis of the simulated system. To justify the passage of ions through the pores, the potential of the mean force (PMF) of ions was calculated. To evaluate the properties of the system, the ion retention time and the radial distribution functions of species were measured. Based on the findings of this study, these nanostructure membranes can be recommended as a model for removal of heavy metals.

Keywords: Ni^{2+}; GNS; BNNS; PMF; Heavy metal.

INTRODUCTION

Many toxic heavy metals are frequently discharged into the environment in the form of industrial wastewater, causing serious soil as well as water pollution. The wastewater discharged directly into natural water bodies is highly harmful to the aquatic ecosystems. Among these heavy metal species, Ni (II) has been frequently identified as a contaminant because of its high toxicity, mobility and enormous use in different industries, including electroplating and the manufacture of steel, pigments and storage batteries [1]. Although Ni (II) plays a pivotal role as a micronutrient in the synthesis of vitamin B12, its affluence over tolerance limits can cause cancer of the lungs, nose and bones as well as nausea, cyanosis, rapid respiration and etc. For these reasons, the necessity to removal of Ni (II) from industrial effluents has been emphasized

before they are discharged [2-4]. The main source of nickel pollution in the water was derived from processes such as galvanization, smelting, mining, dyeing operations, batteries manufacturing and metal finishing. Trace amounts of nickel are beneficial to human organism as an activator of some enzyme systems, but if it is beyond the scope of normal levels, different types of diseases have been occurred. Evidently it is most important to remove these toxic metals from wastewater prior to their discharge into natural water bodies as they are non-biodegradable and persistent. Technologies, such as chemical reduction, ion-exchange, adsorption, reverse osmosis, electro dialysis and nanofiltration have been proposed for the removal of heavy metals [5-10]. However, most of these processes are relatively complicated or do not work efficiently to reduce metal concentrations below regulatory standards. Thus, the removal

* Corresponding Author Email: jafar.azamat@yahoo.com

of heavy metal ions in an effective way is of great importance from the viewpoints of both scientific research and engineering application.

Recently, membrane techniques have been introduced to remove heavy metal ions from water. Nanosheet membranes have been also used for the removal of various heavy metals [11], because they can separate ions and molecules. Ion separation using nanosheet membranes is done via pores created in them [12]. Graphene nanosheet (GNS) and boron nitride nanosheet (BNNS) are examples of nanostructured membranes. GNS is a carbon based nanomaterial with layers of carbon atoms densely packed in a honeycomb crystal lattice composed of two equivalent carbon sub-lattices [13]. GNS have some significant advantages in the preparation of carrier matrix due to their high surface area and remarkable mechanical stiffness and it exhibits a number of intriguing unique properties, such as high surface area of over 2600 m^2/g, large surface-to-volume ratio, high room temperature carrier mobility and conductance quantization. GNS has a wide range of possible membrane applications because of its ultimate thinness, flexibility, chemical stability and mechanical strength [14]. Recently, ion and gas separation through GNS by molecular dynamics (MD) simulation method have been studied [15-18]. The results of these works show that GNS with pores in its center can be efficiently utilized for separation technology.

For further investigation, BNNS was used also for the removal of nickel ions. A BNNS with a very high specific surface area exhibits excellent sorption performance for a wide range of oils, solvents, and dyes from water [19]. This membrane has unique properties, including a wide energy band gap, electrical insulation, ultraviolet photoluminescence, high thermal conductivity and stability, and high resistance to oxidation and chemical inertness [20-24]. This easy recyclability makes BNNS a good candidate for ion separation.

Perfect GNS and BNNS are impermeable to ions because there are no pores, and the electron density of their aromatic rings is enough to repel ions trying to pass through them. To pass ions through them, the drilling of pores is required. Functionalized pore in a nanosheet is obtained by passivating each atom at the pore edge using chemical functional groups. Herein, for nickel removal from an aqueous solution using a functionalized pore in center of GNS or BNNS, external electric field applied to the simulated system. We expect that our findings can be used to aid the design of nanostructured membranes for the removal of heavy metals for water treatment by nanostructure membranes such as GNS and BNNS.

SIMULATION METHOD

Fig. 1 displays an image of the simulation system. The size of the simulation box was $3 \times 3 \times 6$ nm^3. The box constituted of 1700 water molecules with 0.3 M $NiCl_2$ and a GNS or BNNS as a membrane was inside the box.

The full geometric optimization of a functionalized GNS and BNNS was calculated by the density functional theory method to obtain atomic charges and their optimized structures.

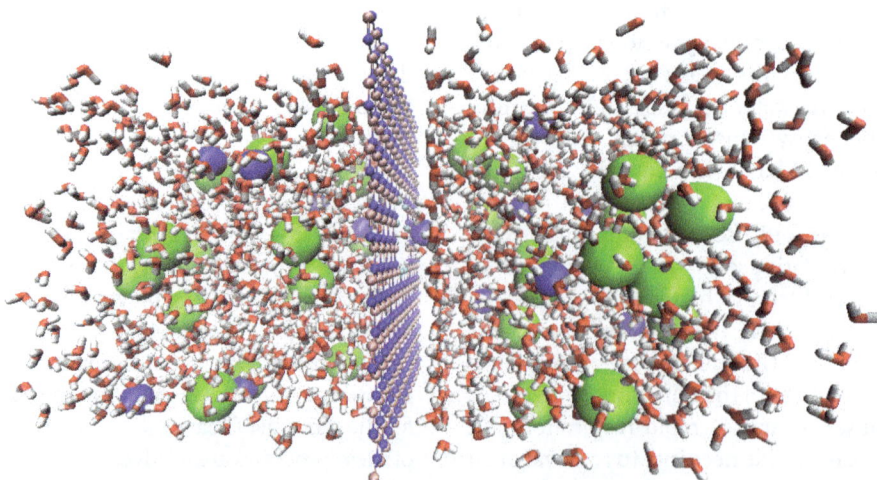

Fig. 1: A snapshot of the simulated system. The BNNS membrane with a functionalized pore in its center is located in the middle of simulation box (red: O, white: H, blue: Ni, and green: Cl).

These calculations were done using the GAMESS [25] at the B3LYP level of theory with 6-311G basis sets. The obtained results from DFT calculations for the functionalized membranes are given in Table 1.

As shown in Fig. 2, there were 377 carbon atoms and 9 fluorine atoms in the GNS and 183 nitrogen, 183 boron and 12 fluorine atoms in the BNNS. During the simulations, these membranes were held fixed. The diameter of the pores in the GNS and BNNS were about 0.6 nm and 0.8 nm, respectively.

Periodic boundary conditions were applied in the three directions. Water was modelled by using the simple point charge model [26]. Intermolecular interactions were described by the 12-6 Lennard-Jones (LJ) potential together with a Coulomb potential. The system was initially equilibrated at 298 K with a coupling time of 0.1 ps^{-1} for 1000 ps. In the modelling of sieving properties, our typical simulation runs were 5 ns long and obtained in the isobaric ensemble at the atmospheric pressure under applied electric field. The range of applied electric field was from 0 to 30 V, which is similar to other experimental studies in which the voltage is used to remove heavy metals [27]. The cutoff distance for nonbonding interactions was set up at 12 Å, and the particle mesh Ewald summations method was used to model the system's electrostatics [28]. During simulations, all the GNS and BNNS atoms were held in fixed positions. A time step of 1 fs was employed. At the beginning of each simulation run, water molecules rapidly filled the pore of membranes. Then after a certain period of time, ions started enter the pore. All MD simulations were carried out at constant

Table 1: Partial charges of GNS and BNNS atoms obtained from DFT calculations.

Atom	Charge (q)
Nitrogen	-0.4
Boron	0.4
Nitrogen bonded to Fluorine	0.25
Boron bonded to Fluorine	0.25
Fluorine of BNNS	-0.25
Carbon	0
Carbon bonded to Fluorine	0.29
Fluorine of GNS	-0.29

volume using the NAMD 2.10 [29] package with the CHARMM27 force field [30]. A Langevin thermostat and a hybrid Nose-Hoover Langevin piston were used to maintain the temperature and pressure of the system at 298 K and 1 bar, respectively. In this research, as in previous works [31-38], all analyses were performed using VMD 1.9.2 software [39]. The force field parameters for GNS and BNNS were obtained from [36,40] and nickel ions were modelled by using the parameters from reference [41].

The ion permeation from functionalized pores can be explained by calculating the potential of the mean force (PMF) [42]. The PMF was calculated by sampling the force experienced by ions that were placed at several positions along the z-axis of the box. In this work, each sampling window was run for 1 ns. The PMF of the ions was calculated using umbrella sampling [43] with either ion harmonically restrained in 0.1 Å steps in the axial z direction. Collective analysis was made using WHAM [44].

RESULTS AND DISCUSSION

To removal of Ni^{2+}, we used GNS and BNNS as a membrane with a functionalized pore in their

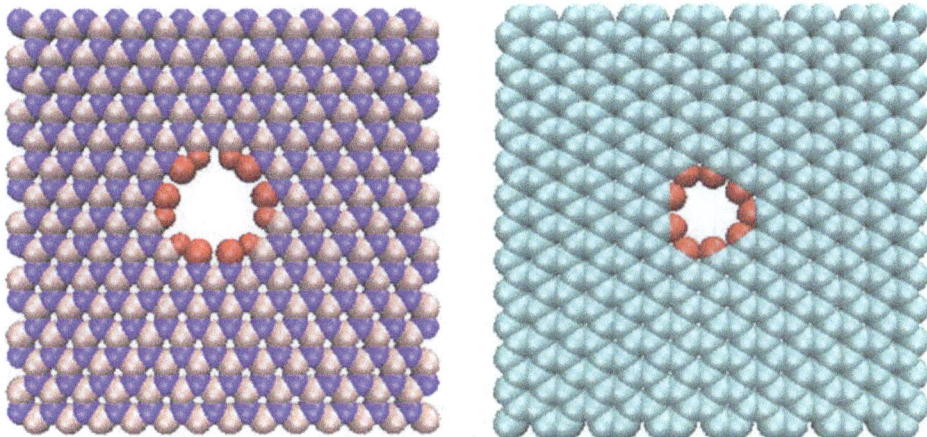

Fig. 2: Functionalized pore of membranes. Left: BNNS and right is GNS (blue represents nitrogen, silver is boron, cyan is carbon and red is fluoride).

centres under the influence of external voltage which was applied to the system, perpendicular to the pore of the membranes. With applying electric field, only nickel ions infuse through the functionalized pore of the membranes. Although GNS and BNNS have a large enough pore to accept nickel and chlorine ions, the results indicated that only Ni^{2+} permeates through these pores.

The MD results can be clarified by calculating the PMF of ions. We have done 50 ns MD simulation without applied electric field, to examine passage of ions in these conditions. It was observed that, any ions did not pass through the pores of GNS or BNNS. Therefore, we applied electric field for passage of ions from these pores. PMF can predict passing or rejecting of an ion across a path. Fig. 3 demonstrated the PMF for Ni^{2+} and Cl^- in the GNS and BNNS. As can be seen from Fig. 3, in both membranes, the energy barrier for chloride is higher than nickel. Thus, the chloride ion will not be able to cross the membrane. This is due to termination of functionalized pores of membranes by the negatively-charged atoms (fluoride atom), which favours the passage of cations.

Also, the PMF for Ni^{2+} in the BNNS is lower than the GNS. This trend leads to more Ni^{2+} permeating from the pore of the BNNS, in effect of identical applied electric field to GNS and BNNS. The Ni^{2+} permeated from functionalized pores when 5 and 7.5 V electric fields were applied to the BNNS and GNS, respectively. Fig. 4 shows the number of Ni^{2+} passing through the functionalized pore of the GNS and BNNS. With the increment of the applied electric field, the number of Ni^{2+} passing was increased. As seen in Fig. 4, Ni^{2+} passes through the pore of the BNNS more than the GNS. This is

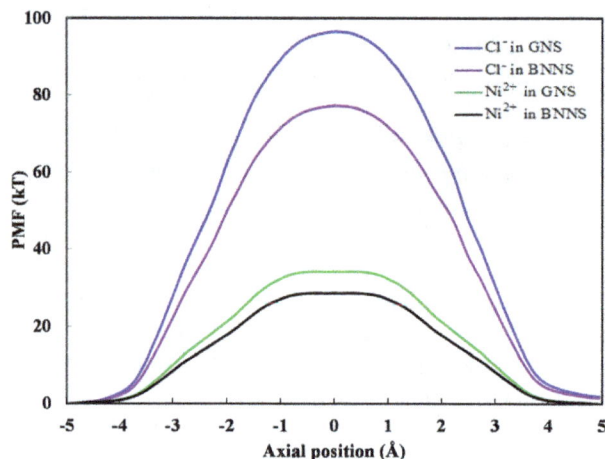

Fig. 3: The PMF for Ni^{2+} and Cl^- ions in the GNS and BNNS systems.

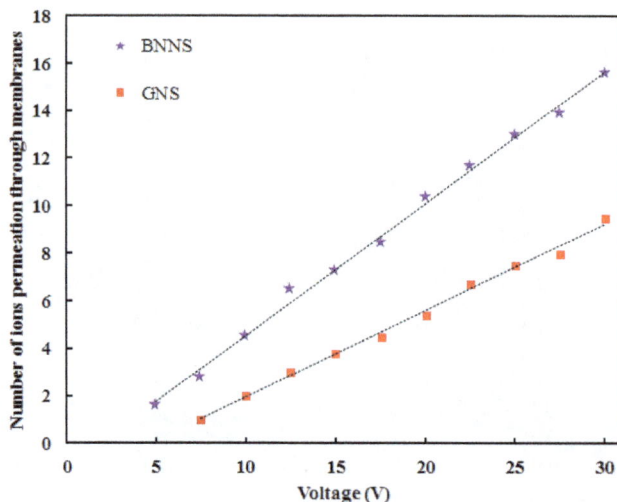

Figure 4. The number of Ni^{2+} passing through the pore of GNS and BNNS.

due to the low energy barrier of Ni^{2+} in the BNNS compared to that in the GNS.

The time required for a nickel ion to pass through the pore (retention time) is an important parameter. Fig. 5 shows the retention time of a Ni^{2+} ion as a function of applied electric field. Increasing the electric field has decreased the retention time.

Therefore, removal of heavy metal from the aqueous solution at high voltages happens quickly. The energy barrier in the PMF for Ni^{2+} in the pore of GNS is high compared to that of the BNNS pore. Thus, the retention time of the Ni^{2+} in the GNS system is greater than that of the BNNS. In the other words, nickel ions passes quickly through the BNNS pore.

To characterize the structure of water molecules around the ions, the radial distribution function (RDF) between the ion and water molecule was considered that it was obtained from the trajectory files of MD simulation. There is a clear maximum and minimum for each first RDF. These have been shown that a coordination shell of water exists around of ions. Fig. 6 shows the RDF Ni^{2+}-water in the GNS and BNNS systems. The first peak corresponds to the distribution of the neighbour water molecules around the nickel ions. For the RDF peaks of Ni^{2+} in the GNS and BNNS systems, the position and magnitude of the first maximum and minimum are almost identical. This indicates that the hydration number of this ion in both systems is identical.

On the other hand, the structure of water molecules around the membranes in both systems is similar. The water molecules have been shown a tendency to accumulate in the region at a location about ±4.5 Å from the GNS and BNNS. This structure is shown by two visible peaks in the density profile on each side (see Fig. 7).

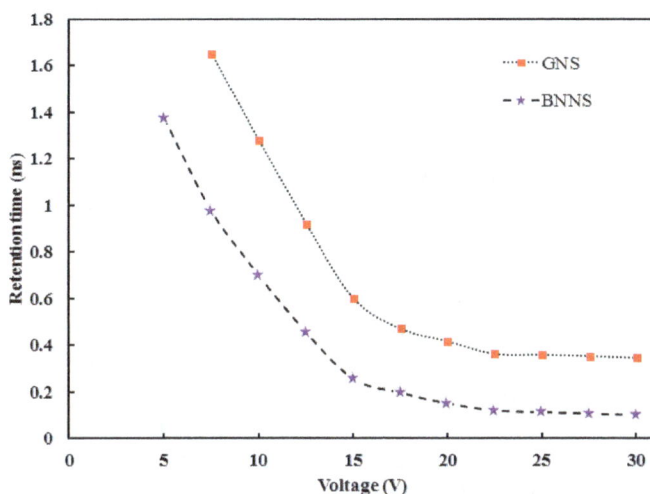

Fig. 5: Retention time for Ni^{2+} in various voltages in the GNS and BNNS.

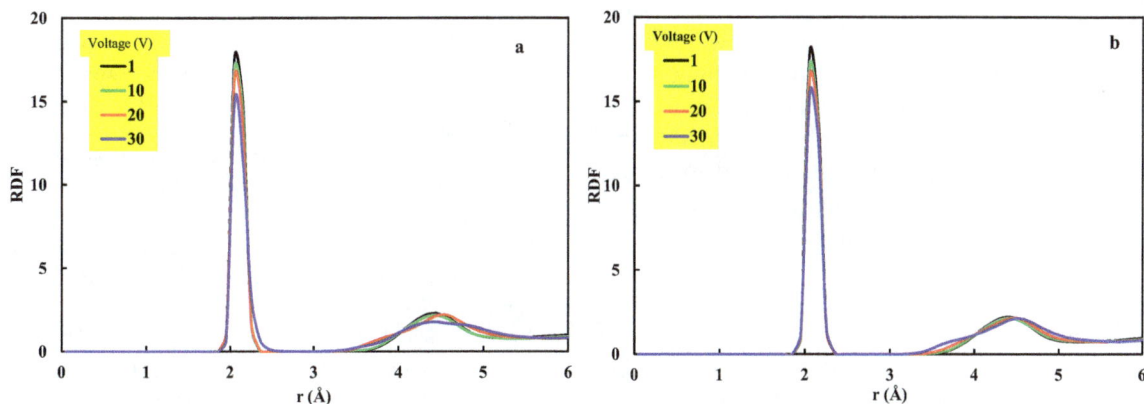

Fig. 6: RDFs under various voltages: (a) RDF Ni^{2+}- water in the GNS system, (b) RDF Ni^{2+}- water in the BNNS system.

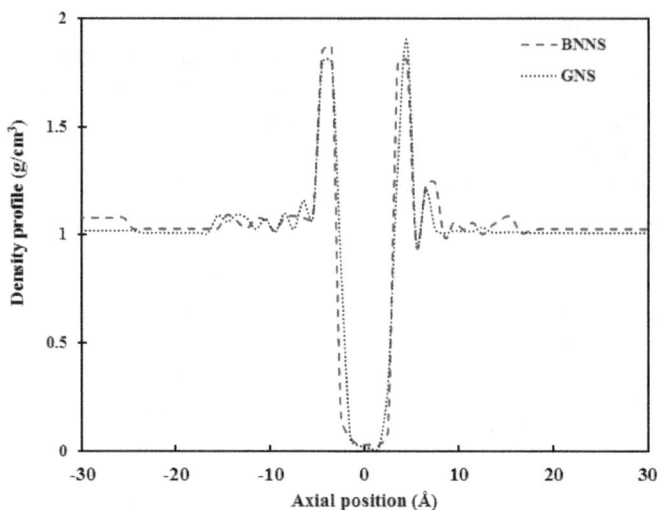

Fig. 7: Density profile of water molecules in the GNS and BNNS systems.

However, the number of Ni^{2+} passing through the GNS and BNNS is different. This indicates that structure of the water molecules were ineffective in the number of ions passing through the pores. Therefore, difference in the number of ions passing through the pores is related to the pore structure. This case is illustrated by the PMF, in which the barrier energy of Ni^{2+} ions in the BNNS is less than that of GNS. Thus, the retention time of the Ni^{2+} in the BNNS system is less than of the GNS system. Therefore, Ni^{2+} passes quickly through the pore of the BNNS.

When water molecules pass across the functionalized pores of GNS and BNNS, they form a single file structure. As can be seen in Fig. 7, the structure of water molecules on both sides of the membranes was different from water molecules in the bulk region. The density profile was calculated from histograms of the number of water molecules. The water molecules displayed a tendency to accumulate in the region at a location about ±4.5 Å from the GNS and BNNS. In fact, the density of water on both sides of the membranes was higher than that in the bulk water. This phenomenon occurred for both types of membranes. In the region far away from the membranes, the density of water was about 1 g/cm³. The layered structure of water molecules in this region is due to the non-bonded interaction of membranes atoms, and water molecules.

CONCLUSION

MD simulations technique was used to investigate the removal of nickel ions as a heavy

metal across functionalized pores of GNS and BNNS. Based on the results of PMF calculations, the GNS and BNNS were able to remove Ni^{2+} from aqueous solutions under applied electric field. Compared to the GNS, a BNNS can remove the Ni^{2+} from water efficiently. With increasing applied electric field to the considered systems, more ions could pass through the pore of membranes. In addition, with increasing applied voltage, the retention time of the ions decreased.

ACKNOWLEDGMENTS

Author thanks the Iranian Nanotechnology Initiative Council for the support provided.

CONFLICT OF INTEREST

The author declares that there are no conflicts of interest regarding the publication of this manuscript.

REFERENCES

1. Panneerselvam P., Morad N., Tan K. A., 2011. Magnetic nanoparticle (Fe_3O_4) impregnated onto tea waste for the removal of nickel (II) from aqueous solution. *J. Hazard. Mater.* 186: 160-168.
2. Babel S., Kurniawan T. A., 2003. Low-cost adsorbents for heavy metals uptake from contaminated water: a review. *J. Hazard. Mater.* 97: 219-243.
3. Wong S., Li X., Zhang G., Qi S., Min Y., 2002. Heavy metals in agricultural soils of the Pearl River Delta, South China. *Environ. Pollut.* 119: 33-44.
4. Nandi D., Saha I., Ray S. S., Maity A., 2015. Development of a reduced-graphene-oxide based superparamagnetic nanocomposite for the removal of nickel (II) from an

aqueous medium via a fluorescence sensor platform. *J. Colloid Interface Sci.* 454: 69-79.

5. Hegazi H. A., 2013. Removal of heavy metals from wastewater using agricultural and industrial wastes as adsorbents. *HBRC Journal* 9: 276-282.

6. Štandeker S., Veronovski A., Novak Z., Knez Ž., 2011. Silica aerogels modified with mercapto functional groups used for Cu(II) and Hg(II) removal from aqueous solutions. *Desalination* 269: 223-230.

7. Meena A. K., Mishra G. K., Rai P. K., Rajagopal C., Nagar P. N., 2005. Removal of heavy metal ions from aqueous solutions using carbon aerogel as an adsorbent. *J. Hazard. Mater.* 122: 161-170.

8. Chen G., 2004. Electrochemical technologies in wastewater treatment. *Sep. Purif. Technol.* 38: 11-41.

9. Borhade A. V., Kshirsagar T. A., Dholi A. G., Agashe J. A., 2015. Removal of heavy metals Cd^{2+}, Pb^{2+}, and Ni^{2+} from aqueous solutions using synthesized azide cancrinite, Na$_8$[AlSiO$_4$]$_6$(N$_3$)$_{2.4}$(H$_2$O)$_{4.6}$. *J. Chem. Eng. Data* 60: 586-593.

10. Zhang S., Cheng F., Tao Z., Gao F., Chen J., 2006. Removal of nickel ions from wastewater by Mg(OH)$_2$/MgO nanostructures embedded in Al$_2$O$_3$ membranes. *J. Alloys Compd.* 426: 281-285.

11. Goh P. S., Ismail A. F., 2015. Graphene-based nanomaterial: The state-of-the-art material for cutting edge desalination technology. *Desalination* 356: 115-128.

12. Suk M. E., Aluru N. R., 2010. Water transport through ultrathin graphene. *J. Phys. Chem. Lett.* 1: 1590-1594.

13. He Z., Zhou J., Lu X., Corry B., 2013. Bioinspired graphene nanopores with voltage-tunable ion selectivity for Na$^+$ and K$^+$. *ACS Nano* 7: 10148-10157.

14. Taherian F., Marcon V., van der Vegt N. F. A., Leroy F., 2013. What is the contact angle of water on graphene? *Langmuir* 29: 1457-1465.

15. Konatham D., Yu J., Ho T. A., Striolo A., 2013. Simulation insights for graphene-based water desalination membranes. *Langmuir* 29: 11884-11897.

16. Cohen-Tanugi D., Grossman J. C., 2012. Water desalination across nanoporous graphene. *Nano Lett.* 12: 3602-3608.

17. Drahushuk L. W., Strano M. S., 2012. Mechanisms of gas permeation through single layer graphene membranes. *Langmuir* 28: 16671-16678.

18. Sun C., Boutilier M. S. H., Au H., Poesio P., Bai B., Karnik R., Hadjiconstantinou N. G., 2013. Mechanisms of molecular permeation through nanoporous graphene membranes. *Langmuir* 30: 675-682.

19. Lei W., Portehault D., Liu D., Qin S., Chen Y., 2013. Porous boron nitride nanosheets for effective water cleaning. *Nat. Commun.* 4: 1777.

20. Golberg D., Bando Y., Huang Y., Terao T., Mitome M., Tang C., Zhi C., 2010. Boron nitride nanotubes and nanosheets. *ACS Nano* 4: 2979-2993.

21. Mortazavi B., Rémond Y., 2012. Investigation of tensile response and thermal conductivity of boron-nitride nanosheets using molecular dynamics simulations. *Physica E* 44: 1846-1852.

22. Lei W., Zhang H., Wu Y., Zhang B., Liu D., Qin S., Liu Z.,

Liu L., Ma Y., Chen Y., 2014. Oxygen-doped boron nitride nanosheets with excellent performance in hydrogen storage. *Nano Energy* 6: 219-224.

23. Sun Q., Li Z., Searles D. J., Chen Y., Lu G., Du A., 2013. Charge-controlled switchable CO$_2$ capture on boron nitride nanomaterials. *J. Am. Chem. Soc.* 135: 8246-8253.

24. Pakdel A., Zhi C., Bando Y., Golberg D., 2012. Low-dimensional boron nitride nanomaterials. *Mater. Today* 15: 256-265.

25. Schmidt M. W., Baldridge K. K., Boatz J. A., Elbert S. T., Gordon M. S., Jensen J. H., Koseki S., Matsunaga N., Nguyen K. A., Su S., Windus T. L., Dupuis M., Montgomery J. A., 1993. General atomic and molecular electronic structure system. *J. Comput. Chem.* 14: 1347-1363.

26. Berendsen H. J. C., Grigera J. R., Straatsma T. P., 1987. The missing term in effective pair potentials. *J. Phys. Chem.* 91: 6269-6271.

27. Bazrafshan E., Mahvi A. H., Zazouli M. A., 2011. Removal of zinc and copper from aqueous Solutions by electrocoagulation technology using iron electrodes. *Asian J. Chem.* 23: 5506-5510.

28. Essmann U., Perera L., Berkowitz M. L., Darden T., Lee H., Pedersen L. G., 1995. A smooth particle mesh Ewald method. *J. Chem. Phys.* 103: 8577-8593.

29. Phillips J. C., Braun R., Wang W., Gumbart J., Tajkhorshid E., Villa E., Chipot C., Skeel R. D., Kale L., Schulten K., 2005. Scalable molecular dynamics with NAMD. *J. Comput. Chem.* 26: 1781-1802.

30. MacKerell A. D., Bashford D., Bellott, Dunbrack R. L., Evanseck J. D., Field M. J., Fischer S., Gao J., Guo H., Ha S., Joseph-McCarthy D., Kuchnir L., Kuczera K., Lau F. T. K., Mattos C., Michnick S., Ngo T., Nguyen D. T., Prodhom B., Reiher W. E., Roux B., Schlenkrich M., Smith J. C., Stote R., Straub J., Watanabe M., Wiórkiewicz-Kuczera J., Yin D., Karplus M., 1998. All-atom empirical potential for molecular modeling and dynamics studies of proteins. *J. Phys. Chem. B* 102: 3586-3616.

31. Azamat J., Khataee A., Joo S. W., 2015. Molecular dynamics simulation of trihalomethanes separation from water by functionalized nanoporous graphene under induced pressure. *Chem. Eng. Sci.* 127: 285-292.

32. Azamat J., Khataee A., Joo S. W., 2015. Removal of heavy metals from water through armchair carbon and boron nitride nanotubes: a computer simulation study. *RSC Adv.* 5: 25097-25104.

33. Azamat J., Khataee A., Joo S. W., Yin B., 2015. Removal of trihalomethanes from aqueous solution through armchair carbon nanotubes: A molecular dynamics study. *J. Mol. Graphics Modell.* 57: 70-75.

34. Khataee A., Azamat J., Bayat G., 2016. Separation of nitrate ion from water using silicon carbide nanotubes as a membrane: Insights from molecular dynamics simulation. *Comput. Mater. Sci* 119: 74-81.

35. Azamat J., Sardroodi J. J., Mansouri K., Poursoltani L., 2016. Molecular dynamics simulation of transport of water/DMSO and water/acetone mixtures through boron nitride nanotube. *Fluid Phase Equilib.* 425: 230-236.

36. Azamat J., Khataee A., Joo S. W., 2016. Molecular dynamics simulations of trihalomethanes removal from water using boron nitride nanosheets. *J. Mol. Model.* 22: 1-8.

37. Azamat J., Balaei A., Gerami M., 2016. A theoretical study of nanostructure membranes for separating Li+ and Mg2+ from Cl–. *Comput. Mater. Sci* 113: 66-74.

38. Azamat J., Khataee A., Joo S. W., 2016. Separation of copper and mercury as heavy metals from aqueous solution using functionalized boron nitride nanosheets: A theoretical study. *J. Mol. Struct.* 1108: 144-149.

39. Humphrey W., Dalke A., Schulten K., 1996. VMD: Visual molecular dynamics. *J. Mol. Graphics* 14: 33-38.

40. Sint K., Wang B., Král P., 2008. Selective ion passage through functionalized graphene nanopores. *J. Am. Chem. Soc.* 130: 16448-16449.

41. Li P., Roberts B. P., Chakravorty D. K., Merz K. M., 2013. Rational design of particle mesh ewald compatible lennard-jones parameters for +2 metal cations in explicit solvent. *J. Chem. Theory Comput.* 9: 2733-2748.

42. Kjellander R., Greberg H., 1998. Mechanisms behind concentration profiles illustrated by charge and concentration distributions around ions in double layers. *J. Electroanal. Chem.* 450: 233-251.

43. Torrie G. M., Valleau J. P., 1977. Nonphysical sampling distributions in monte carlo free-energy estimation: Umbrella sampling. *J. Comput. Phys.* 23: 187-199.

44. Kumar S., Payne P. W., Vásquez M., 1996. Method for free-energy calculations using iterative techniques. *J. Comput. Chem.* 17: 1269-1275.

6

Investigation of Conversion CO$_2$ to Fuel by TiN nanotube-Cu nanoparticle

Leila Mahdavian

Department of Chemistry, Doroud Branch, Islamic Azad University, Doroud, Iran

ABSTRACT

The CO and CO$_2$ effects are global warming, acid rain, limit visibility, decreases UV radiation; yellow/black color over cities and so on. In this study, convention of CO$_2$ and H$_2$O to CH$_4$ and O$_2$ near TiN- nanotube with Cu-nanoparticle calculated by Density Functional Theory (DFT) methods. We have studied the structural, total energy, thermodynamic properties of these systems at room temperature. All the geometry optimization structures were carried out using GAMESS program package under Linux. DFT optimized their intermediates and transient states. The results have shown a sensitivity enhancement in resistance and capacitance when CO$_2$ and H$_2$O are converted to CH$_4$ and O$_2$.

TiN-nanotube used photo-catalytic reactivity for the reduction of CO$_2$ with H$_2$O to form CH$_4$ and O$_2$ at 298K. The calculations are done in state them between of three TiN-nanotubes near Cu-nanoparticle. The calculation shown which heat reaction formation (ΔH) is endothermic for this reaction. This reaction needs to sun, photo active or other energy in the presence of visible light for doing.

Keywords: CO$_2$ and H$_2$O, CH$_4$ and O$_2$, TiN- nanotube, Waste Pollution, ZINDO/1-DFT.

INTRODUCTION

One of the biggest problems of the global environment is an excessive increase of greenhouse gases. Carbon dioxide due to higher shelf life and the most its amount in the atmosphere is the most important greenhouse gas[1, 2]. Carbon dioxide concentration in the pre-industrial era was only 250 ppm, whereas currently it is at 380 ppm[3].

Largest sources and emissions of carbon dioxide are industrial, power plants, fossil fuel (coal), the cars and etc. Preventing the release of greenhouse gases, especially carbon dioxide is the most important problem for industries and societies.

In recent years, many solutions have been proposed to reduce and converted this pollutant[4, 5]. No need to change phase, having the energy needed for regeneration, the small size of the membrane

system, the lack of loss stream, the operation is simple, environmentally friendly and low cost investment are caused, the researchers and industry men have led into the potential of membranes, such as membrane polymers, zeolites, carbon nanotubes, metal-organic frameworks (MOFs), Hydrate phase equilibrium and so on[6-8].

Among the various methods that are available for the removal of gaseous pollutants, three main methods are considered for CO$_2$ removal include chemical reactions, burning and absorption. One of the methods to eliminate carbon dioxide was implemented of gas mixture produced as fuel. Methane and methanol is major product of chemical industry and also a feedstock for many chemicals. However CO$_2$ conversion to methanol and so on is challenging, there are many methods for it [9-12].

* Corresponding Author Email: mahdavian@iau-doroud.ac.ir

TiN- nanotubes with Cu-nanoparticle have attracted great interest due to their unique electronic properties and nanometer size. Because of these unique properties, they are great potential candidates in many important applications such as nanoscale electronic devices, chemical sensors and field emitters. The effect of gas adsorption on the electrical resistance of a TiN-Cu nanoscale has received great attraction because of fast response, good sensitivity of chemical environment gases and low operating temperature [13].

As shown in Fig. 1, there are four situations in which CO_2 and H_2O can pass between TiNTs-Cu. In this work, the second situations are investigated for them. In Fig. 2, TiN- nanotubes with Cu-nanoparticle simulated by ball-and-stick models, CO_2 and H_2O converted to CH_4 and O_2.

Interactions between CO_2 and H_2O on Cu nanoparticle between TiN-nanotubes are optimized by GAMESS program package and ZINDO/1 method by semi empirical-DFT methods at room temperature was implemented for investigation of thermodynamic properties of considered systems.

THE COMPUTATIONAL METHODS

The computational approach consists of three stages: First, the optimization of geometrical structures is done for the reactions, products and the transition state, in this study theB3LYP/6-31G level of density functional theory (DFT) was used[15, 16]. Then the frequency calculations were performed for each optimized structures to obtain thermo-chemical quantities by ZINDO/1[17, 18] and IR-DFT methods by the GAMESS program package. At the end, the reaction pathway analyzed in order to confirm the obtained structures of the transition state. For removal of pollutants in the environment, we need to new catalysts, are being the most active and selectivity. Basically, the catalytic properties of these compounds are completely determined by their electronic structure. The electronic structure design is important based on

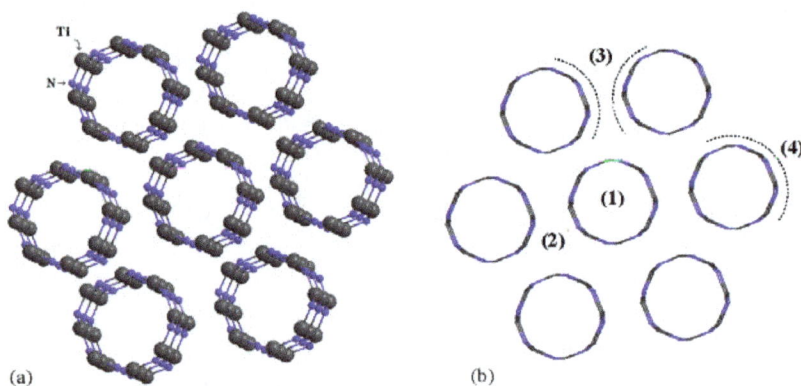

Fig. 1: There are four situations for interaction of CO_2 and H_2O in TiN-nanotube, a) Ball-and-stick, b) stick models Configuration for them.

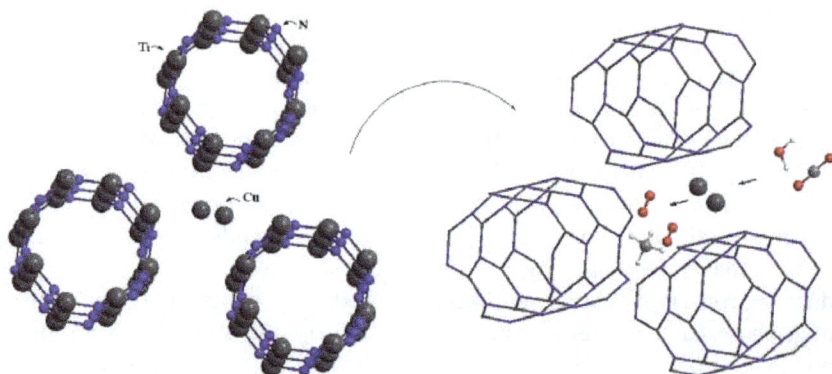

Fig. 2: Ball-and-stick models configuration: Top-view of second situation passing CO_2 and H_2O between TiN-nanotube with Cu- nanoparticle[14].

changes in the composition and physical structure. For this purpose, the ZINDO/1 calculations enable in complex systems and extended. In this way, we have provided the interaction energies of molecules with sufficient precision to describe the reactivity of transition metal and alloys.

In this study, entry of carbon dioxide and water on Cu nanoparticles between titanium nitride nanotubes have been simulated in three steps (Fig. 2), and then in steps 4 and 5 are transition state conversion of them to CH_4 and O_2. The conversion was completed in 6th step and methane and oxygen are produced, then it is excreted from TiN-NTs in 7th step. The electronic structure and the thermodynamic properties are calculated for all steps by ZINDO/1.

RESULTS AND DISCUSSION

In this research, we have developed a method for converting CO_2 into commodity chemicals, which may reduce the burden on CO_2 storage sites, in addition to providing a means to reduce anthropogenic CO_2 emissions and an inexpensive method for producing useful materials from CO_2.

This novel catalytic method for the continuous chemical converting of CO_2 is simulated and thoroughly investigated mechanistically.

The geometry optimizations of TiN nanotube with Cu nanoparticles were performed by B3LYP/6-31G level of theory (Fig. 2). The effects of CO_2 and H_2O have passed between Cu-TiN nanotubes and converted them to CH_4 and O_2 were shown in Table 1 which in this simulation, they are passing in-side to out-side between nanotubes by seven stages. The study includes conformational searches (and further refinement by DFT) and the ab initio calculation of ZINDO/1 methods in semi empirical and the dipole moments for all the steps between nanotubes. The most significant property of ZINDO/1 was finding a good correlation between the ZINDO/1 and the substitution pattern on this conversation.

In Fig. 3, the resistance (Ω) recorded for converted of CO_2 and H_2O to other chemicals on Cu-nanoparticles between TiN nanotubes have shown a sudden decrease. The E_{ele} for them is -51.17 in 5th step (Table 1), and RMS gradient (kcal/mol·Å) is different for the formation of CH_4 and H_2O in this interaction at 298K: $CO_2+2H_2O \rightarrow CH_4+2O_2$

Table 1: The thermodynamic properties of interaction CO_2 and H_2O on Cu-TiN nanotube at 298K.

Steps	E_{total} (MJ/mol)	E_{nuc} (MJ/mol)	Dipol Moment (D)	RMS kcal/mol.°A	E_{bin} (MJ/mol)	H (MJ/mol)	E_{ele} (v)
TiN-Cu	18844.55	39060.15	1.74×10^4	8335	21706.32	21775.13	-49.89
1	20843.06	40819.57	1.73×10^4	8272	23854.13	23924.84	-49.30
2	21057.41	41301.82	1.73×10^4	8317	24068.48	24139.19	-49.96
3	17675.78	41577.29	1.82×10^4	8751	20686.84	20757.56	-58.98
4	14958.85	41733.14	2.16×10^4	1.04×10^4	17969.91	18040.63	-66.07
5	21208.79	41945.84	1.74×10^4	9635	24219.85	24290.57	-51.17
6	17366.76	42840.00	1.97×10^4	1.02×10^4	20424.33	20495.74	-62.61
7	14875.68	41996.33	2.09×10^4	1.05×10^4	17933.24	18004.65	-66.93

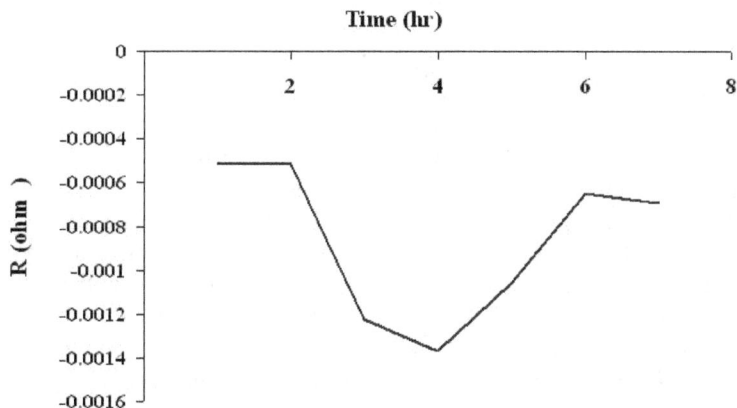

Fig. 3: The resistance (Ω), interaction CO_2 and H_2O on Cu-TiN nanotube at 298K.

Table 2: The thermodynamic properties of CO_2 and H_2O passing through Cu-TiN and convert to CH_4 and O_2.

CO_2 and H_2O passing through Cu-TiN	ΔG_{ele} (MJ/mol)	ΔH_{ele} (MJ/mol)	ΔS_{ele} (MJ/mol)	K
	-6797.78	-5884.21	3.065662	2743725

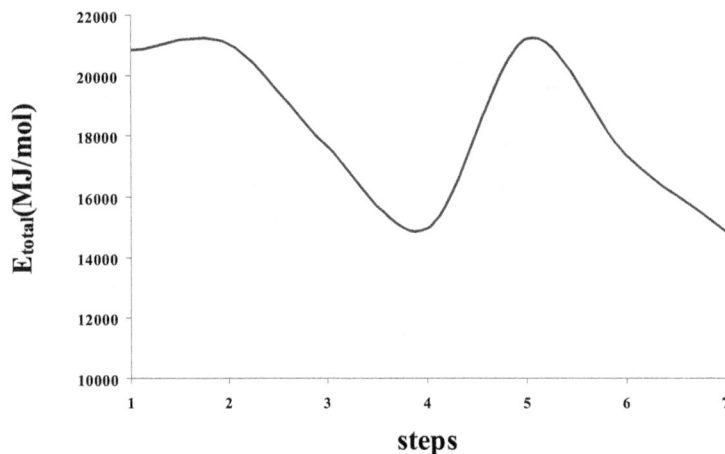

Fig. 4: The total energy (MJ/mol) of converted CO_2 and H_2O to CH_4.

The interaction between them is:

The heat of formation (enthalpy) for their interaction calculated with ZINDO/1 methods by subtracting atomic heats of formation from the binding energy. The heat of formation and binding energy is appropriate for this interaction. The H (MJ/mol) for these interactions has a minimum amount in 4th step of TiN-nanotube. The enthalpy of all steps was positive that is shown, this interaction was endothermic. Therefore, this interaction needs to ultraviolet solar radiation. The results show, the dipole moment (D) has the most amounts in the 4th step of this interaction; indicate that location between nanotubes is the perfect place to trapping of CO_2, which can convert it to other products. Table 1 shows the nuclear energy for them, that has an increasing trend.

The least amount of total energy is shown in the middle of nanotube because the field of TiN-nanotube is the largest amount in this place. The total energy for them is shown in Fig. 4. The E_{total} decreases in the middle length between nanotubes, which is a potential for converting CO_2 to CH_4. To determine sensitivity of TiN-nanotube to carbon dioxide, the electrical resistance (Ω) of this interaction is calculated and evaluated, it is shown in Fig. 3. The TiN-NT is an up hole-doped semiconductor, as can be gleaned from the current versus gate voltage curve shown in Fig. 3 (middle curve), where the resistance of them is observed to

be decreasing. Band bending induced by charging molecules causes the increase or decrease in surface conductivity responsible for the gas response signal.

Their enthalpy difference was negative; it is an exothermic and spontaneous interaction during which CO_2 is separated from the air in the environment. In Table 2, thermodynamic parameters (ΔG_{ele}, ΔH_{ele}, and ΔS_{ele}) for them between the lengths of the tubes were calculated and the results suggest that the nature of adsorption is exothermic, spontaneous and favorable. This method is best for converting CO_2 in the environment.

CONCLUSION

One of the causes of global warming is due to carbon dioxide emissions from fossil fuel combustion. The new methods have developed for reducing carbon dioxide emissions. For example, removal of carbon dioxide from the flue gases are stating by carbon dioxide-hydrogen-water (Syngas) method [19, 20].

In this study, using filters of titanium nitride nanotubes with copper nanoparticles to convert the pollutants into methane fuel was simulated and calculations were done. We choose armchair TiN-nanotube (4, 4) and it is investigated to know whether CO_2 passes between them. The interaction between them was calculated by ZINDO/1 methods. A change in the potential of all atoms of inner surface TiN-nanotube was observed in passing and convert of CO_2. According to the results, this nanofilter can be installed in automobile exhaust. Exhaust gas pressure for carbon dioxide gas entry, and convert them between nanotubes on copper nanoparticles is very suitable. Temperatures of automobile exhaust gas are higher than 700 °C, which can provide the energy required to break the double bonds C=O and bonds O-H in carbon dioxide and water on copper nanoparticles. In these calculations, the energy separation of carbon dioxide and water were lower from than the energy of methane and oxygen (Fig. 4), which can improve interaction to remove CO_2.

CONFLICT OF INTEREST

The author declares that there are no conflicts of interest regarding the publication of this manuscript.

REFERENCES

1. H. Arakawa, M. Aresta, J. N. Armor, M. A. Barteau, E. J. Beckman, A. T. Bell, *et al.*, "Catalysis research of relevance to carbon management: progress, challenges, and opportunities," *Chemical Reviews*, vol. 101, pp. 953-996, 2001.

2. M. Pervaiz and M. M. Sain, "Carbon storage potential in natural fiber composites," *Resources, conservation and Recycling*, vol. 39, pp. 325-340, 2003.

3. F. M. Orr Jr, "CO 2 capture and storage: are we ready?," *Energy and Environmental Science*, vol. 2, pp. 449-458, 2009.

4. P. Styring and K. Armstrong, "Catalytic carbon dioxide," *chimica oggi/Chemistry Today*, vol. 29, 2011.

5. K. M. K. Yu, I. Curcic, J. Gabriel, and S. C. E. Tsang, "Recent advances in CO2 capture and utilization," *ChemSusChem*, vol. 1, pp. 893-899, 2008.

6. O. K. Farha, A. Ö. Yazaydın, I. Eryazici, C. D. Malliakas, B. G. Hauser, M. G. Kanatzidis, *et al.*, "De novo synthesis of a metal–organic framework material featuring ultrahigh surface area and gas storage capacities," *Nature chemistry*, vol. 2, pp. 944-948, 2010.

7. D.-A. Yang, H.-Y. Cho, J. Kim, S.-T. Yang, and W.-S. Ahn, "CO 2 capture and conversion using Mg-MOF-74 prepared by a sonochemical method," *Energy & Environmental Science*, vol. 5, pp. 6465-6473, 2012.

8. J.-W. Du, D.-Q. Liang, X.-X. Dai, D.-L. Li, and X.-J. Li, "Hydrate phase equilibrium for the (hydrogen+ tert-butylamine+ water) system," *The Journal of Chemical Thermodynamics*, vol. 43, pp. 617-621, 2011.

9. Y. Xie, T.-T. Wang, X.-H. Liu, K. Zou, and W.-Q. Deng, "Capture and conversion of CO2 at ambient conditions by a conjugated microporous polymer," *Nature communications*, vol. 4, 2013.

10. K. Ikeue, S. Nozaki, M. Ogawa, and M. Anpo, "Characterization of self-standing Ti-containing porous silica thin films and their reactivity for the photocatalytic reduction of CO 2 with H 2 O," *Catalysis Today*, vol. 74, pp. 241-248, 2002.

11. O. K. Varghese, M. Paulose, T. J. LaTempa, and C. A. Grimes, "High-rate solar photocatalytic conversion of CO2 and water vapor to hydrocarbon fuels," *Nano letters*, vol. 9, pp. 731-737, 2009.

12. C. Duke and H. Gibson, "Encyclopedia of chemical technology," ed: Wiley, New York, 1982.

13. N. Yan, G. Li, G. Pan, and X. Gao, "TiN nanotube arrays as electrocatalytic electrode for solar storable rechargeable battery," *Journal of The Electrochemical Society*, vol. 159, pp. A1770-A1774, 2012.

14. L. Mahdavian, "Using of TiN-nanotubes and Cu-nanoparticles for conversion of CO2 to hydrocarbon fuels," *Journal of molecular modeling*, vol. 21, pp. 1-6, 2015.

15. P. Hohenberg and W. Kohn, "Inhomogeneous electron gas," *Physical review*, vol. 136, p. B864, 1964.

16. J. Huang, C. Liu, Q. Jin, H. Tong, W. Li, and D. Wu, "Density functional theory study on bond dissociation enthalpies for lignin dimer model compounds," *Journal of Renewable and Sustainable Energy*, vol. 6, p. 033116, 2014.

17. M. J. Dewar and M. L. McKee, "Ground states of molecules. 41. MNDO results for molecules containing boron," *Journal of the American Chemical Society*, vol. 99, pp. 5231-5241, 1977.

18. M. J. Dewar and W. Thiel, "Ground states of molecules. 38. The MNDO method. Approximations and parameters," *Journal of the American Chemical Society*, vol. 99, pp. 4899-4907, 1977.

19. S.-P. Kang and H. Lee, "Recovery of CO2 from flue gas using gas hydrate: thermodynamic verification through phase equilibrium measurements," *Environmental science & technology*, vol. 34, pp. 4397-4400, 2000.

20. J. Zhang, P. Yedlapalli, and J. W. Lee, "Thermodynamic analysis of hydrate-based pre-combustion capture of CO2," *Chemical Engineering Science*, vol. 64, pp. 4732-4736, 2009.

7

Trace Cd(II), Pb(II) and Ni(II) ions extraction and preconcentration from different water samples by using *Ghezeljeh* montmorillonite nanoclay as a natural new adsorbent

*Zahra Hassanzadeh Siahpoosh *, Majid Soleimani*

Department of Chemistry, Imam Khomeini International University (IKIU), Qazvin, Iran

ABSTRACT

This investigate presents the extraction-preconcentration of Lead, Cadmium, and Nickel ions from water samples using *Ghezeljeh* montmorillonite nanoclay or *"Geleh-Sar-Shoor"* (means head-washing clay) as a natural and native new adsorbent in batch single element systems. The *Ghezeljeh* clay is categorized by using Fourier Transform Infrared Spectroscopy (FT-IR), Scanning Electron Microscopy-Energy Dispersive Spectrometer Operating (SEM-EDS), X-ray Diffractometry (XRD), X-ray Fluorescence (XRF), Cation Exchange Capacity (CEC) measurements, Surface property valuation (S_{BET}) by the BET method from nitrogen adsorption isotherms and Zeta potential. According to BET theory, the specific surface area of *Ghezeljeh* nanoclay was computed as 19.8 m^2 g^{-1} whereas the cation exchange capacity was determined as 150 meq (100 g^{-1}). The results of XRD, FT-IR, XRF, zeta potential, BET surface area and CEC of the *Ghezeljeh* clay confirm that montmorillonite is the dominant mineral phase. Based on SEM images of clay, it can be seen that the distance between the plates is nm level. For all three ions, the limit of detection, the limit of quantification, dynamic linear range, preconcentration factor, and the adsorption capacity were obtained. The result of several interfering ions was considered. The *Ghezeljeh* nanoclay as a new adsorbent and experimental method were effectively used for the extraction of heavy metals (Lead, Cadmium, and Nickel) in a variety of real water samples.

Keywords: Water, Nanoclay, Solid phase extraction, Lead, Cadmium, Nickel.

INTRODUCTION

The elimination of toxic heavy metals from aqueous environmental samples has received significant attentions in latest years due to gathering in living tissues, and consequent bio exaggeration in the food chain improving their poisonousness [1]. Nickel (Ni), lead (Pb), and cadmium (Cd) are among the toxic heavy metals [2,3]. Because of the damaging properties of extreme intakes of heavy metals ions, it is required to define their trace in water and food samples [4,5]. Solid-phase extraction (SPE) is an attractive enrichment-separation

manner for heavy metal ions. It is trouble-free, high preconcentration factor, time- and price-saving, and can be straight used in microliter volumes without any sample loss [6]. For the subtraction of numerous metal ions in natural waters and a variety of food samples, different conventional and nonconventional adsorbents have been stated, such as red mud [7], activated carbon [8], tree fern [9], sewage sludge [10], sawdust [11], silica [12], bone char [13], rice husk [14], bagasse fly ash [15], resin [16], polymetallic sea nodules [17], modified zeolite [18], spirogyra bioadsorbent [19], and etc.

* Corresponding Author Email: z.hassanzadeh.s@gmail.com

However, these extractors are frequently non-selective or exceedingly expensive. High specific surface area, chemical and mechanical stability, layered structure, high cation exchange capacity (CEC), affinity to hold water in the interlayer sites, and the existence of Bronsted and Lewis acidity have made clays exceptional adsorbent materials [3]. Dias et al. [20] used 2-mercaptobenzothiazole loaded on clays for SPE of Hg(II), Pb(II), Zn(II), Cd(II), Cu(II), and Mn(II) from an aqueous solution. Akcay and Kurtulmus [21] examined the adsorption position for uranium on Turgutlu and Kula clays. Krikorian and Martin [22] used adjusted clays for the SPE of copper(II), cadmium(II), silver(I), nickel(II), and lead(II) ions. Mohamed et al. [23] used Aswan clay from Egypt for speciation and preconcentration of Cr(III) and Cr(VI) from synthetic solution and tannery wastewater. Tuzen et al. [4] used Celtek clay as adsorbent for the separation-preconcentration of metal ions from environmental samples. Turan [24] studied the uptake of trivalent chromium ions from aqueous solutions using kaolinite. Bhattacharyya and Gupta [6] explored kinetic and thermodynamic exclusion of Cu(II) by natural and acid-activated clays.

In this study, a solid phase extraction (SPE) in batch equilibrium procedure was used to extraction Ni^{2+}, Pb^{2+}, and Cd^{2+} ions using *Ghezeljeh* montmorillonite nanoclay as a native new adsorbent in batch single component systems. *Ghezeljeh* montmorillonite nanoclay is exceptionally cheap; its price is $ 0.1/kg. There is no requirement to recover the clay due to its low price. The industrial water organization by using clay materials as an adsorbent is reasonable according to its low price. Only Soleimani and Hassanzadeh in the Imam Khomeini International University (IKIU) have used *Geleh-Sar-Shoor* for the extraction of metal ions [25-28]. It is interesting to mention that the *Ghezeljeh* nanoclay (*Geleh-Sar-Shoor*) was used in olden Persia to clean the body, hair, and also to bathe dead bodies prior to the funeral ("*Geleh-Sar-Shoor*" means head-washing clay). The *Ghezeljeh* clay was characterized using FT-IR, SEM-EDS, XRF, XRD, BET surface area, cation exchange capacity (CEC) and Zeta potential. The adsorbent was readied using the *Galehouse* method for the SPE of Cd^{2+}, Ni^{2+}, and Pb^{2+} ions. The influence of the quantity of adsorbent, eluent characteristics, pH and type of buffer solutions, shaking time, desorption time, centrifugation time, sample volume, and concentration of the sample solution

were examined to optimize the procedure. Finally, the presented technique was effectively used for the extraction of Cd^{2+}, Ni^{2+}, and Pb^{2+} ions in real different water samples.

Clay

Clays are hydrous aluminum silicates which are categorized as either 1:1 or 2:1 clay minerals. The sheets in these clays are held together by weak Van Der Waals forces creating it easy for other chemicals to enter the interlayer region. Many 2:1 clay minerals have permanent negative charge due to isomorphous substitution of aluminum(III) for silicon(IV) in the silica layer or magnesium(II) for aluminum(III) in the alumina layer [29]. Montmorillonite is dioctahedral clay of the smectite group and is composed of alumino-silicate layers. The Silica Tetrahedral (T) (Si^{4+} in tetrahedral coordination with O^{2-}) and alumina octahedral (O) (Al^{3+} in octahedral coordination with O^{2-}) are interconnected (via the sharing of O^{2-} at polyhedral corners and edges) in such a way that a sheet of alumina octahedral is sandwiched between two sheets of silica tetrahedral. Consequently, the composition is T-O-T (2:1) [30]. Most of the surface charges on montmorillonite are produced by isomorphous substitution or non-ideal octahedral occupancy. These permanent negative charges are distributed along the mineral basal surfaces and are well-adjusted by absorbing aqueous cations, such as Na^+, K^+, Ca^{2+} and Mg^{2+}. These cations can be exchanged with other cations in solution and the exchange reactions are non-specific, stoichiometric and encompass the creation of surface outer-sphere complexes. In montmorillonite, the edge sites account for a much lesser fraction of the exchange capacity. The adsorption of metals to these sites includes the creation of surface inner-sphere complexes analogous to the interaction of these metals with the surfaces of oxide minerals [31,32]. Subtraction of metal cations by clay minerals is organized by restrictions such as charge characteristics of the clay [33,34]. The exchange manners exhibited depend on several factors, like the physicochemical features of solid and cation (such as ionic radius, charge size, hard–soft acid–base properties, hydration volume and hydration enthalpy of cation), existence of challenging ions, temperature, ionic strength and investigational situations containing time of reaction, concentration of ions, and pH of the medium [33-44].

MATERIALS AND METHODS
Reagents and solutions

All the reagents were bought from the German company of Merck: acids, bases, hydrogen peroxide, sodium acetate, sodium citrate, nitrate salts of copper, silver, lead, chromium, nickel, cobalt, cadmium, sulfate salts of aluminum, manganese, zinc, magnesium, chloride salts of sodium, potassium, iron, calcium, and ammonium. Since the reagents were of the highest purity, they were applied without any additional purification.

The element standard solutions were prepared by diluting a stock solution of 1000 mg L^{-1} of the specified element using doubly distillated water. A citrate-citric acid buffer solution was readied using 0.1 M citric acid solution at pH 2-3. Acetate buffer solution was used by combining appropriate volumes of 0.1 M acetic acid and 0.1 M sodium acetate at pH 4-6. Phosphate buffer solution was prepared using 0.1 M phosphoric acid at pH 7. Ammonium buffer solution was organized by mixing suitable amounts of 0.1 M ammonia and 0.1 M ammonium chloride at pH 8-10. The pH of the buffer solutions was adjusted by adding 1 M NaOH or HCl, as needed.

The *Ghezeljeh* montmorillonite nanoclay was collected from *Ghezeljeh*, a village 18 km west of the city of Tafresh in Iran. The different real water samples used in the experiments were collected from Caspian Sea (Iran), Karun River (inside and outside the city of Ahvaz, Iran), Persian Gulf (Iran), well water (Herat, Afghanistan), Haryrood River (Afghanistan), and tap water (Herat, Afghanistan).

Instrumentation

A model 420A digital Orion pH meter (Gemini, the Netherlands) equipped with a combined glass electrode was used for pH adjustments. An ultrasonic water bath (Bandelin, Berlin, Germany) was applied to disperse and disaggregate the *Ghezeljeh* montmorillonite nanoclay. Agitation of the system was carried out on a mechanical shaker (Flask shaker SF1 Scientific model, STUART, Britain). X-Ray Diffraction (XRD) data were attained using an Ital Structures diffractometer (GNR, Novara, Italy), with Cu Kα radiation (40 kV/30 mA, λ= 1.542 A°). Fourier Transform Infrared (FT-IR) study was carried out using Tensor Bruker MIR-T27 (Germany) having a standard mid-IR DTGS detector.

To quantitative measurements of Ni(II), Pb(II), and Cd(II) ions in the standard solutions, a GBC 902 flame atomic absorption spectrometry (FAAS), (Dandenong, Victoria, Australia 3175) with deuterium background corrector and an air-acetylene flame was applied. The working conditions in the FAAS spectrometer were adjusted according to the standard guidelines of the manufacturer. But, the analysis of real water samples were achieved with a Varian 735-ES inductively coupled plasma atomic emission spectrometry (ICP-AES), (Mulgrave, Australia). X-ray fluorescence (XRF) of the sample has been investigated using XRF Analysis Instruments (Philips Magix Pro, Netherlands). A scanning electron microscope (SEM) (LEO 1450 VP, Thornwood, N.Y., USA) with variable pressure secondary electron detector and energy dispersive spectrometer operating (EDS) at 30 kV (Oxford INCA software, High Wycombe, U.K.) were applied for SEM-EDS analysis. Zeta potential measurements were carried out on a Zetameter ZetaCAD (CAD Instruments, France). The specific surface areas were studied with the BET way using a Belsorp mini II instrument (BelJapan, Japan).

Preparation of the adsorbent

The adsorbent was organized using the *Galehouse* way [45]. Natural *Ghezeljeh* montmorillonite nanoclay was initially handled with 0.1 M of acetic acid to remove carbonates, and then with 30% H_2O_2 to disregard mineral and organic impurities. The *Ghezeljeh* montmorillonite nanoclay was carefully washed with doubly distilled water to reject traces of acetic acid and hydrogen peroxide. The treated nanoclay was spread and disaggregated in doubly distilled water through an ultrasonic water bath. The resultant suspension was moved to a measuring cylinder and allowed to stand for 3 h, 26 min, 6 sec for sedimentation. The fine part (< 2 μm) was removed and then located in an electric vacuum oven at 50°C for 72 h to be dehydrated. Then, it was placed in a desiccator for following experimentation [25-28].

Solid phase extraction (SPE) procedure
Adsorption step

Adsorption tests were carried out using batch technique at room temperature. First, a 50 mL solution containing nickel or lead or cadmium ions were moved into an Erlenmeyer flask. Then, 10 mL of a proper buffer solution was added followed by 0.5 min of agitation. Then, 0.5 g of the *Ghezeljeh* montmorillonite nanoclay was added. The mixture was shaken for 10 min by means of a mechanical

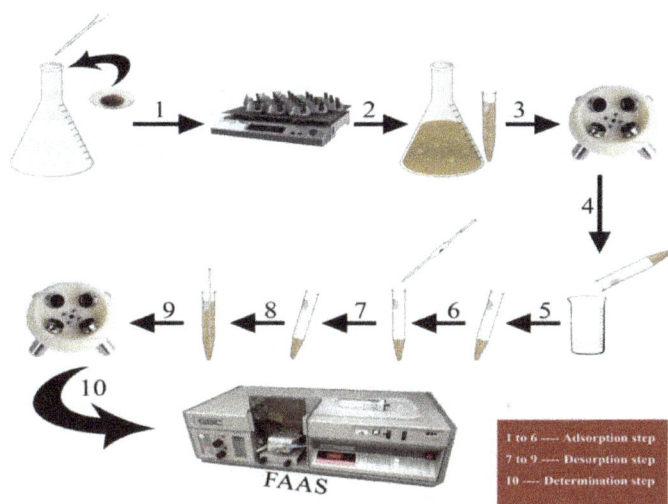

Fig. 1: Schematic diagram of the Adsorption, desorption and determination procedure.

shaker. The liquid part was disconnected from the solid part via centrifugation at 3500 rpm for 5 min. The supernatant was decanted (Fig. 1).

Desorption step

To elute the analytes adsorbed on to the *Ghezeljeh* montmorillonite nanoclay, 10 mL of 3M HCl solution was added to the solid phase under stirring for 0.5 min. Then, the suspension was permitted to stand for 10 min and was centrifuged at 3500 rpm for 30 min. The supernatant (10 mL) was collected to determine its nickel or lead or cadmium ions concentration. To optimize the experimental conditions, these steps were repeated three times. The equivalent process was used to the blank solution (Fig. 1).

Physicochemical characterization
SEM study

Scanning Electron Microscopy (SEM) is a powerful technique used in micro imaging of a diversity of surfaces. The *Ghezeljeh* montmorillonite nanoclay sample was covered with Au under vacuum in argon atmosphere (Fig. 2a). Based on SEM images of the *Ghezeljeh* montmorillonite clay, it can be seen that the distance between the plates is nm level.

XRD study

X-ray diffractograms were attained for the 2θ angles ranging from $2°$ to $40°$ 2θ at room temperature. The *Ghezeljeh* nanoclay was treated with ethylene glycol, an organic compound which steadily intercalates itself into the lattice of the clay. The

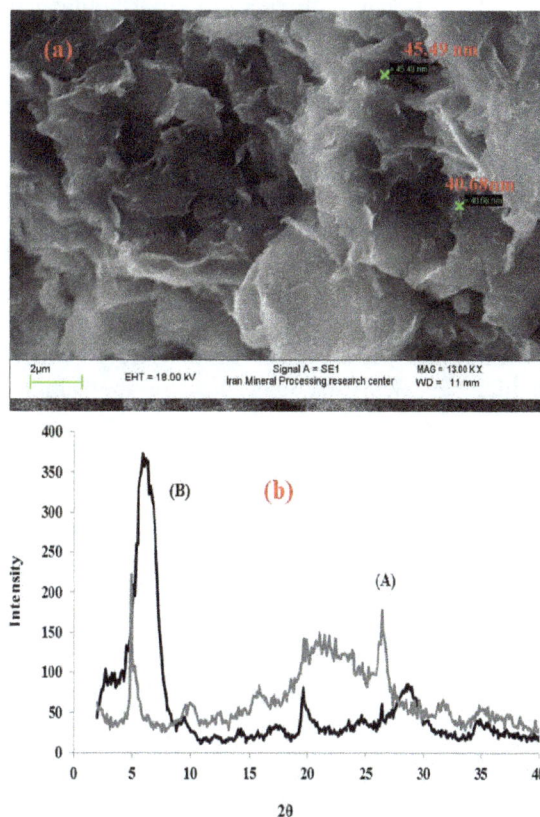

Fig. 2: (a) SEM image (b) The XRD patterns of *Ghezeljeh* montmorillonite nanoclay (A) treated with ethylene glycol (B) Untreated.

structural properties of the *Ghezeljeh* nanoclay were observed before and after treatment with ethylene glycol. The X-Ray diffraction analysis exposed that the *Ghezeljeh* nanoclay sample was mainly composed of montmorillonite minerals (Fig. 2b) [30].

FT-IR study

To prepare the *Ghezeljeh* montmorillonite nanoclay sample for FT-IR spectroscopy, an electric vacuum oven was applied to dehydrated (at 50°C for 6 h) and cool the nanoclay in a desiccator. A FT-IR spectrum was recorded in the range of 400-4000 cm^{-1} using the KBr pellet technique. FT-IR spectrum of untreated *Ghezeljeh* montmorillonite nanoclay (Fig. 3a) displays the bands at 3626 cm^{-1} in OH stretching region, which are assigned to hydroxyl groups coordinated to octahedral cations (Al^{3+} cations). The maximum intensive band at 1035 cm^{-1} is attributed to Si-O in-plane stretching and 529 cm^{-1} is due to Si-O bending vibrations. The shoulder at 1113 cm^{-1} shows Si-O out-of-plane stretching vibration. The broad bands at 3440 cm^{-1} and 1639 cm^{-1} are the stretching and bending vibrations for the hydroxyl groups of water molecules present in the clay. Montmorillonite had two characteristic FT-IR regions [45], (i) 3500–3750 cm^{-1} (due to the surface structural OH groups of layered aluminosilicates and adsorbed water) and (ii) 400–1150 cm^{-1} (due to lattice vibrations). Subsequently, the FT-IR analysis confirmed that *Ghezeljeh* nanoclay was chiefly composed of montmorillonite minerals [30].

XRF and EDS studies

The technique of XRF spectroscopy is similar to EDS in that an X-ray spectrum is achieved which signifies an elemental fingerprint of the sample. The main difference between XRF and EDS is the excitation energy. XRF applies an X-ray beam to yield characteristic X-rays, while EDS applies an electron beam. XRF gives the total composition of a sample, Instead, the EDS data are an average of some local compositions which are dependent on the locations at which the analysis is achieved. One of the advantages of XRF is the capability to identify major, minor, and trace levels of an element, however, EDS is restricted to major and minor elemental concentrations. Subsequently, the detection limit for XRF is about 10 part per million (p.p.m.) and EDS is about 1% [47-50]. The chemical composition of the *Ghezeljeh* montmorillonite nanoclay was determined with XRF and EDS. Table1 and Fig. 3b proves chemical composition of this clay [51].

Table 1: XRF-analysis of the *Ghezeljeh* montmorillonite nanoclay.

Oxides	%
SiO_2	54.47
Al_2O_3	20.92
MgO	3.65
SO_3	0.32
K_2O	1.82
CaO	1.14
TiO_2	0.37
Fe_2O_3	3.13
PbO	0.16
SrO	0.10
ZrO_2	0.05
As_2O_3	0.02
L.O.I	13.86

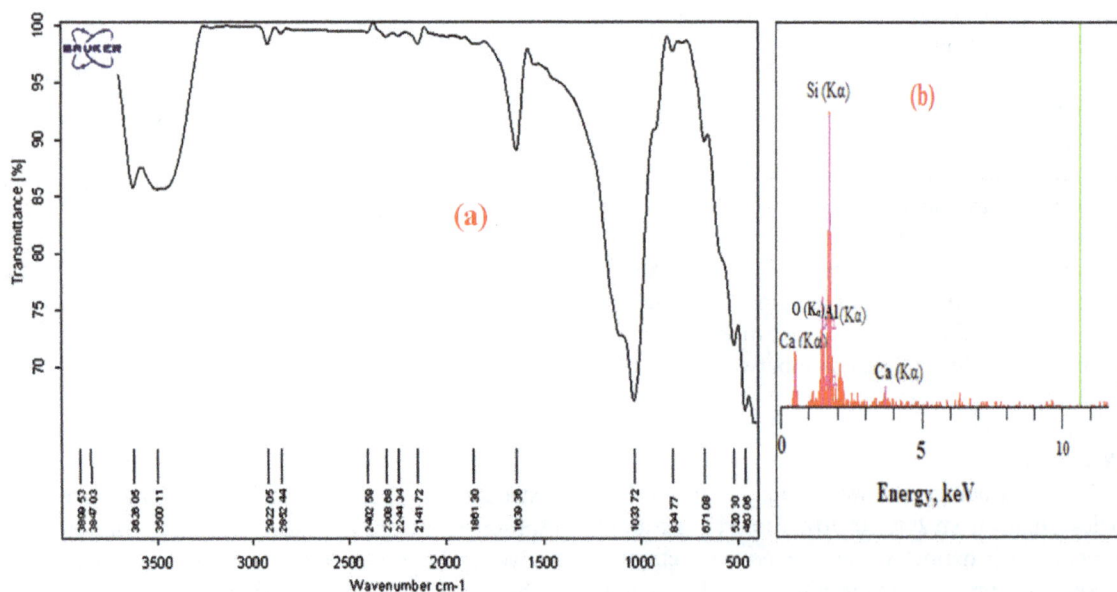

Fig. 3: (a) FT-IR spectrum (b) EDS spectrum of untreated *Ghezeljeh* montmorillonite nanoclay.

Cation exchange capacity (CEC)

The cation exchange capacity (CEC) is the number of equivalents of exchangeable charge per mass of clay, which is equivalent with the layer charge [52]. The CEC of the *Ghezeljeh* montmorillonite nanoclay was calculated with 0.01 M Cu-triethylentetramine [53,54]. The CEC value of 160.0 meq $(100 \text{ g})^{-1}$ for the *Ghezeljeh* montmorillonite nanoclay was found, and the very large CEC value approves well with the CEC values for Montmorillonite described in the literature [55].

Surface area

The specific surface area (S_{BET}), pore volume and pore radius of the *Ghezeljeh* montmorillonite nanoclay were derived from N_2 adsorption isotherms measured at liquid nitrogen temperature (at 77 K) using a Belsorp mini II instrument (BelJapan, Japan). Humidity and vapors on the solid surface or entered in the open pores were rejected by heating under vacuum at 100°C for 12 h prior to the surface area measurements. The *Ghezeljeh* montmorillonite nanoclay owns a specific surface area 90.916 $m^2 g^{-1}$, pore volume of 0.147 $cm^3 g^{-1}$ and pore radius of 4.8 nm [42,56].

Zeta potential measurement

The zeta potential of the *Ghezeljeh* nanoclay was attained from electrophoretic mobility measurements at 21°C, performed using Zetameter apparatus (ZetaCAD instruments) and the measured Zeta potential value is showed in Table 2.

RESULTS AND DISCUSSION

In order to extraction-preconcentration nickel, lead, and cadmium ions from real samples, standard solutions were subjected to SPE. To optimize the method, the result of adsorbent quantity, eluent characteristics (type, concentration, and volume), pH and type of the buffer solutions, shaking time, sample volume, and initial nickel, lead, and cadmium ions concentration were examined on the adsorption recovery. The effect of desorption time and centrifugation time were also investigated to progress the recovery of method.

Effect of type and pH of the buffer solutions

To examine the result of pH on adsorption of Ni(II), Cd(II), and Pb(II) ions onto the *Ghezeljeh* montmorillonite nanoclay, pH was changed in the range of 2 to 10 at room temperature by using buffer solutions. The buffer capacity of a buffering agent is at a local maximum when pH = pK_a, and this is where the maximum buffer action can be achieved. The pH variations relatively gradually in the buffer region, pH = $pK_a \pm 1$, for this reason the suitable range is almost $pK_a \pm 1$. Subsequently, A citrate-citric acid buffer solutions at pH 2-3, acetate buffer solutions at pH 4-6, phosphate buffer solutions at pH 7 and ammonium buffer solutions at pH 8-10 were organized. The consequences are showed in Fig. 4a. Ni(II), Cd(II), and Pb(II) ions were optimally adsorbed on the *Ghezeljeh* montmorillonite nanoclay at pH 5-6. Afterward, in all the experiments, the pH was kept as 5.5 by

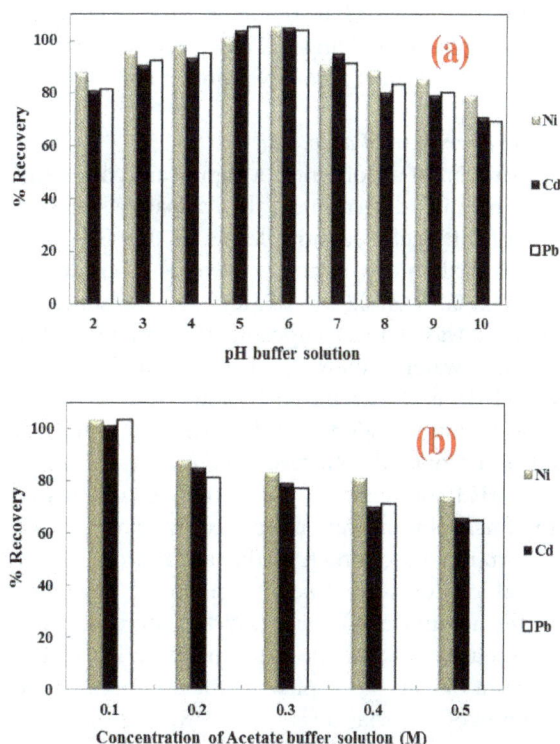

Fig. 4: Effect of (a) pH buffer solution (b) concentration acetate buffer solution on the recoveries of analytes (containing 20 µg of Ni(II), 4.5 µg of Cd(II), and 62.5 µg Pb(II) ions; 25°C; 10 mL 3M HCl (eluent); 0.5 g nanoclay; n=3).

Table 2: The Zeta potential measurement of the *Ghezeljeh* nanoclay

Sample	Dielectric Constant	Electric Field (V cm^{-1})	Mean Mobility (µm s^{-1}/V cm^{-1})	Mean Zeta Potential (mV (T=21.31°C))
Ghezeljeh nanoclay	79.810	6.940	-1.880	-25.970(pH=5.64)

using acetate buffer solution. Clays are recognized to have a negative surface charge in solution, the surface charge changes with changing the pH, and the adsorption of charged species is affected (attractive forces between the positively charged metal ion and the negatively charged clay surface). At low pH values, where there is an excess of H_3O^+ ions in solution, a competition exists between the positively charged hydrogen ions and metal ions for the available adsorption sites on the negatively charged clay surface. However at pH values higher than 6, Ni(II), Cd(II), and Pb(II) ions being precipitated from the solution in the form of hydroxides.

Effect of concentration of buffer solution

To examine the effect of concentration acetate buffer solution on adsorption of Ni(II), Cd(II), and Pb(II) ions onto the *Ghezeljeh* montmorillonite nanoclay, concentration of acetate buffer solution in the ranges of 0.1 to 0.5 M at pH 5.5 at room temperature were changed. Fig. 4b displays that the extreme fraction of recovery is gotten at 0.1 M.

Effect of amount of adsorbent

Amount of adsorbent is a significant parameter because it determines the capacity of an adsorbent. Eight quantity levels of the *Ghezeljeh* montmorillonite nanoclay were considered: 0.1, 0.2, 0.3, 0.4, 0.5, 1, 1.5, and 2g. The standard solution was 60 mL composing of 50 mL of doubly distilled water containing 20 μg of Ni(II), 4.5 μg of Cd(II), and 62.5 μg Pb(II) ions, and 10 mL of buffer solution added. In order to elute the analytes adsorbed onto the *Ghezeljeh* nanoclay, 10 mL of 3 M HCl solution was used. Analyte contents of the final solution were calculated by flame atomic absorption spectrometry. The adsorption of the metal ions onto the *Ghezeljeh* nanoclay enhanced as the amount of the nanoclay was increased. For subsequent runs of the experiment, 0.5 g of the amount of clay was applied as the optimum of the amount of nanoclay level [3,6,57,58] (Fig. 5a).

Effect of eluent characteristics

To attain suitable eluent, HCl and HNO_3 solutions were applied at various concentrations (1-5 M) with varying volumes (5-15 mL) for the elution of Ni(II), Cd(II), and Pb(II) ions adsorbed on to the *Ghezeljeh* nanoclay. The adsorbed ions were readily eluted (desorbed) from the nanoclay only when 10 mL of 3 M HCl solution was used.

Effect of shaking time

The influence of shaking time (contact time) for the adsorption of Ni(II), Cd(II), and Pb(II) ions on to the *Ghezeljeh* montmorillonite nanoclay was dignified after 5, 10, 15, 20, and 30 min of shaking by using a mechanical shaker. It was observed that after 5 min, adsorption was completed. Consequently, the metal–clay interactions reached equilibrium in less than 10 min and it was very fast. It showed that, the adsorption positions on the *Ghezeljeh* nanoclay minerals were swiftly covered by the Ni(II), Cd(II), and Pb(II) ions. The contact time of 10 min was further kept in the measurements.

Effect of volume of the standard solution

To achieve high preconcentration factor (P.F.), due to the low concentration of Ni(II), Cd(II), and Pb(II) ions in real sample, four quantities of 60, 120, 300, and 600 mL of the feed volumes were investigated. It was found that recovery was over 95% at quantitative up to 300, 300, and 120 mL of sample volumes for Ni(II), Pb(II), and Cd(II) ions,

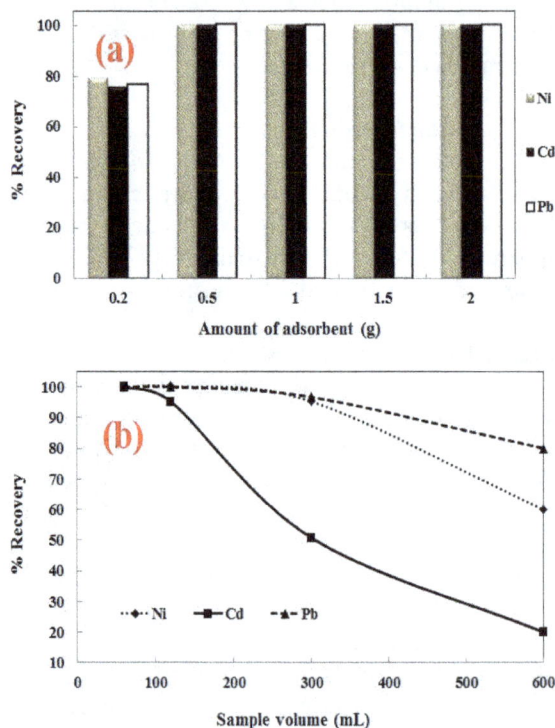

Fig. 5: Effect of (a) amount of *Ghezeljeh* montmorillonite nanoclay (containing 20 μg of Ni(II), 4.5 μg of Cd(II), and 62.5 μg Pb(II) ions; 25°C; 10 mL 3M HCl (eluent); pH 5.5; Acetate buffer solution; n=3), (b) sample volume on the recoveries of analytes (containing 20 μg of Ni(II), 4.5 μg of Cd(II), and 62.5 μg Pb(II) ions; 10 mL 3M HCl (eluent); pH 5.5; Acetate buffer solution; 0.5 g nanoclay, 25°C; n=3).

Respectively. But it declined to below 95% when the feed volumes exceeded 300, 300, and 120 mL. In this study, the final solution volume to be measured by FAAS was 10 mL, therefore the preconcentration factors are of 30, 30, and 12 for Ni(II), Pb(II), and Cd(II) ions, Respectively. The consequences are documented in Fig. 5b.

Effect of initial metal ions concentration

The adsorption capacity of an adsorbent is defined as the largest extent of metal adsorbed on to 1 g of the adsorbent [3]. In order to find out the adsorption capacity of the *Ghezeljeh* montmorillonite nanoclay, 0.5 g of the nanoclay was added to diverse experiment solutions containing 0.02018, 0.2018, 0.4036, and 0.6054 mg of Ni(II) ions (Fig. 6a); 0.0625, 0.1251, 0.6256, and 1.251 mg of Pb(II) ions (Fig. 6b); 0.0045, 0.020, 0.036, and 0.072 mg of Cd(II) ions (Fig. 6c),. The adsorption capacities of the *Ghezeljeh* montmorillonite nanoclay were designed to be 0.806, 0.250, and 0.040 mg g^{-1} for Ni(II), Pb(II), and Cd(II) ions, respectively (relative error smaller than ±5%). At lower concentrations, a large number adsorption locations on the *Ghezeljeh* montmorillonite nanoclay are available and but it is more problematic to find permitted adsorption locations at high concentrations of metal ions.

Desorption time

Desorption time is defined as the length of time an eluent is in contact with the adsorbent having metal ions. The desorption time in this

technique is studied by computing recovery of Ni(II), Pb(II), and Cd(II) ions from the *Ghezeljeh* montmorillonite nanoclay after 5, 10, 15, 20, and 30 min of contact between HCl solution and the nanoclay. Desorption time of 10 min was found to lead to the highest degree of desorption. This value was applied in the remaining tests.

Centrifugation time

The influence of centrifugation time on desorption of metal ions from *Ghezeljeh* montmorillonite nanoclay was examined in the time range of 5–30 min with the rotation speed of 3500 rpm. According to these experiments, 30 min centrifugation time is suitable for maximum desorption.

Interference from other ions

In order to evaluate the feasible analytical applications of the preconcentration way offered, the effect of numerous foreign ions which interfere with the determination of trace of Ni(II), Cd(II), and Pb(II) ions on *Ghezeljeh* montmorillonite nanoclay in diverse real environmental samples was examined in the optimized conditions. Ions were considered to be interfering when they produced an error larger than ±5% in the preconcentration and determination of the analyte. The ions frequently present in water do not interfere in the experimental situations applied. Some of the transition metals at milligram per liter levels did not interfere with the recovery of the analytes. These consequences display that the major matrix ions in natural water samples show no

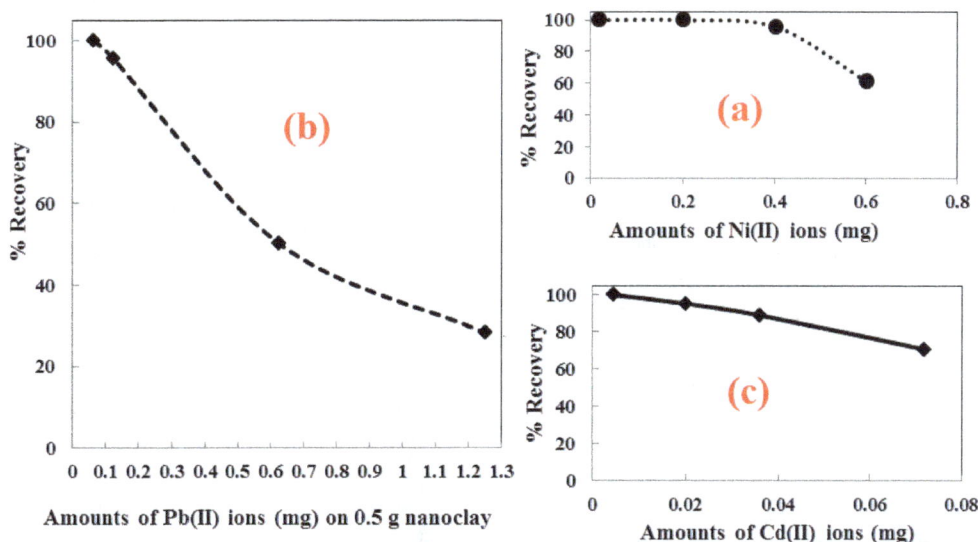

Fig. 6: Effect of initial metal ions concentration (adsorption capacity) on the recoveries of analytes (10 mL 3M HCl (eluent); pH 5.5; Acetate buffer solution; 0.5 g nanoclay; 25°C; n=3).

obvious interference with the preconcentration of Ni(II), Cd(II), and Pb(II) ions (Tables 3-5).

Figures of merit

The figures of merit for Ni(II), Pb(II), and Cd(II) ions in the current investigate were calculated under optimal experimental situations after application of the solid phase extraction way to blank solutions. Give attention to, the preconcentration factors were

30, 30, and 12 for Ni(II), Pb(II), and Cd(II) ions, respectively. The limit of detection (LOD) based on three times the standard deviations of the blank solution (k =3, n = 10) turned out to be 0.5 ng mL^{-1} for Ni(II), Pb(II), and Cd(II) ions. The value for the limit of quantification (LOQ) was 1.6 ng mL^{-1} for Ni(II), Pb(II), and Cd(II) ions. The dynamic linear range (DLR) for Ni(II) and Pb(II) were from 1.6 ng ml^{-1} to 13.3 µg ml^{-1}, and for Cd(II) from 1.6 ng

Table 3: The effect of additional metal ions on the extraction of 20 µg of Ni(II) ion (in optimum conditions; n = 3)

Ion	Added as	Ion concentration (mg L^{-1})	Salt concentration (mg L^{-1})	Recovery%	RSD%
Na$^+$	NaCl	630	1600	95	1.9
Ca^{2+}	CaCl$_2$	36	100	95	3.1
Mg^{2+}	MgSO$_4$	15	150	95	2.1
K$^+$	KCl	419	800	95	1.9
Zn^{2+}	ZnSO$_4$	250	1100	95	2.5
Fe^{3+}	FeCl$_3$	35	100	95	4.4
Mn^{2+}	MnSO$_4$	310	950	95	2.5
Al^{3+}	Al$_2$(SO$_4$)$_3$	55	700	95	3.9
Cd^{2+}	Cd(NO$_3$)$_2$	328	900	95	3.8
Co^{2+}	Co(NO$_3$)$_2$	365	1800	95	4.3
Pb^{2+}	Pb(NO$_3$)$_2$	566	905	95	4.3
Cr^{3+}	Cr(NO$_3$)$_3$	110	850	95	2.6
Cu^{2+}	Cu(NO$_3$)$_2$	215	810	95	2.8

Table 4: The effect of additional metal ions on the extraction of 4.5 µg of Cd(II) ion (in optimum conditions; n = 3)

Ion	Added as	Ion concentration (mg L^{-1})	Salt concentration (mg L^{-1})	Recovery%	RSD%
Na$^+$	NaCl	670	1700	95	1.6
Ca^{2+}	CaCl$_2$	29	80	95	4.2
Mg^{2+}	MgSO$_4$	10	100	95	2.9
K$^+$	KCl	523	1000	95	2.6
Zn^{2+}	ZnSO$_4$	228	1000	95	2.1
Fe^{3+}	FeCl$_3$	76	220	95	4.1
Mn^{2+}	MnSO$_4$	228	700	95	3.3
Al^{3+}	Al$_2$(SO$_4$)$_3$	40	500	95	3.9
Ni^{2+}	NiSO$_4$	163	800	95	2.7
Co^{2+}	Co(NO$_3$)$_2$	375	1850	95	2.2
Pb^{2+}	Pb(NO$_3$)$_2$	87	140	95	2.6
Cr^{3+}	Cr(NO$_3$)$_3$	130	1000	95	3.3
Cu^{2+}	Cu(NO$_3$)$_2$	235	890	95	3.1

Table 5: The effect of additional metal ions on the extraction of 62.5 µg Pb(II) ion (in optimum conditions; n = 3)

Ion	Added as	Ion concentration (mg L^{-1})	Salt concentration (mg L^{-1})	Recovery%	RSD%
Na$^+$	NaCl	512	1300	95	2.4
Ca^{2+}	CaCl$_2$	25	68	95	2.9
Mg^{2+}	MgSO$_4$	21	210	95	3.4
K$^+$	KCl	728	1200	95	4.2
Zn^{2+}	ZnSO$_4$	300	1300	95	2.9
Fe^{3+}	FeCl$_3$	62	180	95	3.8
Mn^{2+}	MnSO$_4$	319	980	95	4.4
Al^{3+}	Al$_2$(SO$_4$)$_3$	16	200	95	4.6
Ni^{2+}	NiSO$_4$	203	1000	95	4.1
Cd^{2+}	Cd(NO$_3$)$_2$	437	1200	95	4.7
Co^{2+}	Co(NO$_3$)$_2$	405	2000	95	3.1
Cr^{3+}	Cr(NO$_3$)$_3$	85	650	95	3.7
Cu^{2+}	Cu(NO$_3$)$_2$	264	1000	95	4.3

ml^{-1} to 8.34 µg ml^{-1}. The adsorption capacities of the *Ghezeljeh* montmorillonite nanoclay were computed to be 0.806, 0.250, and 0.040 mg g^{-1} for Ni(II), Pb(II), and Cd(II) ions, respectively.

Application to real water samples

The experimental way can be applied for the determination of Ni(II), Pb(II), and Cd(II) ions in real samples with complex media. To validate, the proposed method was used to analyze diverse natural water samples. Before the analysis, the samples were filtered through a Whatman blue band filter paper

and the pH was adjusted to optimum pH level prior than the standard addition. Spiking experiments using multiple standard additions method checked reliabilities; therefore each real water sample was spiked with three standard solutions. Ni(II), Pb(II), and Cd(II) ions level were determined by a Varian 735-ES inductively coupled plasma atomic emission spectrometry (ICP-AES). The recovery was defined as the ratio of the concentration of analytes found to the concentration of analytes spiked. The consequences are recorded in Table 6. The recoveries of the spiked standard solutions were in

Table 6: Extraction of Ni(II), Cd(II), and Pb(II) ions in different water samples (in optimum conditions; n = 3).

Sample	Nickel Added (µg mL^{-1})	Nickel Found (µg mL^{-1})	Recovery (RSD)%	Lead Added (µg mL^{-1})	Lead Found (µg mL^{-1})	Recovery (RSD)%	Cadmium Added (µg mL^{-1})	Cadmium Found (µg mL^{-1})	Recovery (RSD)%
Tap water	-	0.700	- (1.5)	-	1.110	- (1.6)	-	0.020	- (1.4)
	1.009	1.650	94 (1.8)	0.782	1.880	99 (1.8)	0.225	0.248	101 (1.8)
	2.018	2.310	80 (2.1)	1.564	2.500	89 (2.2)	0.450	0.450	96 (1.8)
	3.027	3.060	78 (1.9)	3.128	3.860	88 (2.1)	0.900	0.850	92 (1.9)
Caspian Sea	-	0.200	- (1.7)	-	1.800	- (1.7)	-	0.040	- (2.7)
	1.009	1.150	95 (1.5)	0.782	2.550	96 (2.5)	0.225	0.256	96 (2.5)
	2.018	2.070	93 (1.7)	1.564	3.130	85 (2.7)	0.450	0.445	90 (2.7)
	3.027	2.720	83 (2.6)	3.128	4.450	85 (2.9)	0.900	0.850	90 (2.6)
Karun river (inside city)	-	0.300	- (1.9)	-	1.000	- (3.1)	-	0.050	- (2.1)
	1.009	1.250	94 (2.1)	0.782	1.730	93 (3.1)	0.225	0.250	90 (3.1)
	2.018	2.110	90 (3.3)	1.564	2.370	88 (2.3)	0.450	0.450	88 (3.7)
	3.027	2.90	85 (3.2)	3.128	3.470	79 (3.2)	0.900	0.830	86 (4.2)
Karun river (outside city)	-	0.300	- (2.9)	-	2.100	- (1.9)	-	0.055	- (2.9)
	1.009	1.220	91 (2.8)	0.782	2.800	90 (2.8)	0.225	0.260	91 (3.3)
	2.018	2.120	90 (2.8)	1.564	3.470	88 (2.8)	0.450	0.460	90 (3.7)
	3.027	2.780	82 (2.6)	3.128	4.600	80 (2.6)	0.900	0.820	85 (2.2)
Persian gulf	-	0.200	- (1.8)	-	2.000	- (2.2)	-	0.070	- (1.6)
	1.009	1.080	88 (1.9)	0.782	2.700	90 (2.4)	0.225	0.270	88 (1.9)
	2.018	1.790	79 (1.9)	1.564	3.260	81 (3.8)	0.450	0.440	82 (1.8)
	3.027	2.600	79 (2.2)	3.128	4.450	78 (3.2)	0.900	0.780	78 (1.4)
well water	-	0.400	- (1.7)	-	2.300	- (1.7)	-	0.030	- (1.7)
	1.009	1.381	98 (1.6)	0.782	3.070	99 (1.6)	0.225	0.250	98 (2.6)
	2.018	2.150	87 (1.7)	1.564	3.800	96 (2.7)	0.450	0.460	95 (1.9)
	3.027	2.700	76 (2.8)	3.128	5.000	86 (2.9)	0.900	0.840	90 (2.6)
Haryrood river	-	0.500	- (1.8)	-	2.200	- (1.9)		0.05	- (1.9)
	1.009	1.420	91 (1.8)	0.782	2.900	90 (3.8)	0.225	0.260	93 (1.8)
	2.018	2.130	81 (2.1)	1.564	3.450	80 (4.1)	0.450	0.450	88 (3.1)
	3.027	2.900	79 (3.0)	3.128	4.700	80 (4.0)	0.900	0.770	80 (3.0)

Table 7: Comparison between the methods used in this research and similar studies using SPE procedures.

Analytes	Adsorbents	LOD[a]	P.F[b]	Studies
Cd, Cr, Cu, Pb, Zn	rice bran	0.56 - 1.85	100	[58]
Cu, Ni	DowexOptipore SD-2 resin	1.03 - 1.90	50	[5]
Cd, Cr, Cu, Pb,Co,Ni	Celtek clay	0.25 - 0.73	32	[4]
Cu, Pb, Zn, Cd	SNP-loaded alumina	0.21 - 0.63	83	[59]
Cu, Co, Ni, Fe, Zn, Pb	Gold nanoparticle loaded in activated carbon (Au-NP-AC)	1.5 - 2.8	30	[60]
Fe, Cr(III), Cu, Cd, Pb, Ni	Nanosilicate	0.26 – 0.55	133	[61]
Pb, Fe, Cu	Functionalized activated carbon	0.16 – 0.41	-	[62]
Pb, Cd, Ni, Cu	Gallic acid-modified silica gel	0.58 - 0.92	200	[63]
Pb, Cd, Ni, Cu, Co	Carboxylic acid (COOH) bonded to silica gel	2.1 – 17.5	80, 120	[64]
Ni, Pb, Cd	Ghezeljeh nanoclay	0.5, 0.5, 0.5	30, 30, 12	This work

[a] LOD: limit of detection (µg L^{-1})
[b] PF: preconcentration factor

the range of 76–101% with low relative standard deviations (less than 5%), which indicates that good recovery can be obtained using the *Ghezeljeh* nanoclay as adsorbent.

Comparison between this research and similar studies

The *Ghezeljeh* montmorillonite nanoclay-SPE is compared with the other SPE ways for the extraction-preconcentration of heavy metals in terms of type of analyte, adsorbent, the limit of detection (LOD), and preconcentration factor (PF). As can be realized in Table 7, the *Ghezeljeh* montmorillonite nanoclay shows a relatively low LOD. However, Up to now; Only Soleimani and Hassanzadeh in the Imam Khomeini International University (IKIU) have used *Geleh-Sar-Shoor* for the extraction of metal ions from water, and wastewater [25-28].

CONCLUSION

This examination attempted to extraction-preconcentration Ni(II), Pb(II), and Cd(II) ions from diverse real water samples using the *Ghezeljeh* montmorillonite nanoclay as a natural adsorbent. Based on SEM images of *Ghezeljeh* clay, it can be seen that the distance between the plates is nm level. The consequences of XRD, FT-IR and CEC studies of the *Ghezeljeh* nanoclay confirmed that montmorillonite was the dominant mineral phase. The specific surface area of *Ghezeljeh* nanoclay was 90.916 m^2 g^{-1} whereas the cation exchange capacity was measured as 160 meq $(100\ g)^{-1}$. For this purpose, first the adsorbent was readied using the *Galehouse* way and a number of effective parameters on extraction were optimized. The additional metal ions in the aqueous solution already containing Ni(II), Pb(II), and Cd(II) ions frequently do not have a negative effect on the recovery. The limit of detection, 0.5 ng mL^{-1}; limit of quantification, 1.6 ng mL^{-1}; preconcentration factors, 30, 30, and 12 for Ni(II), Pb(II), and Cd(II) ions, respectively; dynamic linear range (DLR) for Ni(II) and Pb(II) were from 1.6 ng ml^{-1} to 13.3 μg ml^{-1}, and for Cd(II) from 1.6 ng ml^{-1} to 8.34 μg ml^{-1}; the adsorption capacities of the *Ghezeljeh* montmorillonite nanoclay were 0.806, 0.250, and 0.040 mg g^{-1} for Ni(II), Pb(II), and Cd(II) ions, respectively. The experimental method was used to a variety of real water samples with the recovery being still significant (76-101%). The interaction Ni(II), Pb(II), and Cd(II) ions onto the *Ghezeljeh* montmorillonite nanoclay are quick and equilibrium is gotten in less than 10 min. Therefore, determination of Ni(II), Pb(II), and Cd(II) ions by the *Ghezeljeh* nanoclay is efficient, reproducible, quick and reliable in varied real samples.

CONFLICT OF INTEREST

The authors declare that there are no conflicts of interest regarding the publication of this manuscript.

ACKNOWLEDGEMENTS

The authors are grateful for the financial support of this work by the Imam Khomeini International University (IKIU), and Mines and Mining Industries Development and Renovation Organization of Iran (IMIDRO).

REFERENCES

1. Jiang, M.Q., X.Y. Jin, X.Q. Lu and Z.L. Chen, 2010. Adsorption of Pb(II), Cd(II), Ni(II) and Cu(II) onto natural kaolinite clay. Desalination, 252: 33–39.
2. Hajiaghababaei, L., A.R. Badiei, M.R. Ganjali, S. Heydari, Y. Khaniani and G.M. Ziarani, 2011. Highly efficient removal and preconcentration of lead and cadmium cations from water and wastewater samples using ethylenediamine functionalized SBA-15. Desalination, 266: 182–187.
3. Bhattacharyya, K.G. and S.S. Gupta, 2007. Adsorptive accumulation of Cd(II), Co(II), Cu(II), Pb(II), and Ni(II) from water on montmorillonite: Influence of acid activation. Journal Colloid Interface Science, 310: 411–424.
4. Tuzen, M., E. Melek and M. Soylak, 2006. Celtek clay as sorbent for separation–preconcentration of metal ions from environmental samples. Journal of Hazardous materials B, 136: 597–603.
5. Tuzen, M., M. Soylak, D. Citak, H.S. Ferreira, M.G.A. Korn and M.A. Bezerra, 2009. A preconcentration system for determination of copper and nickel in water and food samples employing flame atomic absorption spectrometry. Journal of Hazardous materials, 162: 1041–1045.
6. Bhattacharyya, K.G. and S.S. Gupta, 2011. Removal of Cu (II) by natural and acid-activated clays: An insight of adsorption isotherm, kinetic and thermodynamics. Desalination, 272: 66–75.
7. Gupta, V.K., M. Gupta and S. Sharma, 2001. Process development for the removal of lead and chromium from aqueous solutions using red mud—an aluminum industry waste. Water Research, 35: 1125-1134.
8. Leyva Ramos, R., L.A. Bernal Jacome, J. Mendoza Barron, L. Fuentes Rubio and R.M. Guerrero Coronado, 2002. Adsorption of zinc(II) from an aqueous solution onto activated carbon. Journal of Hazardous materials B, 90: 27-38.
9. Ho, Y.S., J.F. Porter and G. McKay, 2002. Equilibrium Isotherm Studies for the Sorption of Divalent Metal Ions onto Peat: Copper, Nickel and Lead Single Component Systems. Water

Air Soil Pollution, 141: 1-33.

10. Pan, S.C., C.C. Lin and D.H. Tseng, 2003. Reusing sewage sludge ash as adsorbent for copper removal from wastewater. Resources Conservation Recycling, 39: 79-90.

11. Yu, L.J., S.S. Shukla, K.L. Dorris, A. Shukla and J.L. Margrave, 2003. Adsorption of chromium from aqueous solutions by maple sawdust. Journal of Hazardous materials B, 100: 53-63.

12. Chiron, N., R. Guilet and E. Deydier, 2003. Adsorption of Cu(II) and Pb(II) onto a grafted silica: isotherms and kinetic models. Water Research, 37: 3079-3086.

13. Ko, D.C.K., C.W. Cheung, K.K.H. Choy, J.F. Porter and G. McKay, 2004. Sorption equilibria of metal ions on bone char. Chemosphere, 54: 273-281.

14. Tarley, C.R.T. and M.A.Z. Arruda, 2004. Biosorption of heavy metals using rice milling by-products: Characterisation and application for removal of metals from aqueous effluents. Chemosphere, 54: 987-995.

15. Gupta, V.K. and I. Ali, 2004. Removal of lead and chromium from wastewater using bagasse fly ash a sugar industry waste. Journal of Colloid Interface Science, 271: 321-328.

16. Gupta, V.K., P. Singh and N. Rahman, 2004. Adsorption behavior of Hg(II), Pb(II), and Cd(II) from aqueous solution on Duolite C-433: a synthetic resin. Journal of Colloid Interface Science, 275: 398-402.

17. Maity, S., S. Chakravarty, S. Bhattacharjee and B.C. Roy, 2005. A study on arsenic adsorption on polymetallic sea nodule in aqueous medium. Water Research, 39: 2579-2590.

18. Wingenfelder, U., B. Nowack, G. Furrer and R. Schulin, 2005. Adsorption of Pb and Cd by amine-modified zeolite. Water Research, 39: 3287-3297.

19. Gupta, V.K., A. Rastogi, V.K. Saini and N. Jain, 2006. Biosorption of copper(II) from aqueous solutions by Spirogyra species. Journal of Colloid Interface Science, 296: 59-63.

20. Dias, N.L., W.L. Polito and Y. Gushikem, 1995. Sorption and preconcentration of some heavy-metals by 2-mercapto-benzothiazole clay. Talanta, 42: 1031–1036.

21. Akcay, H. and F. Kurtulmus, 1995. Study of uranium sorption and desorption on some Turkish clays. Journal of Radioanalytical and Nuclear Chemistry Letters, 200: 529–544.

22. Krikorian, N. and D.F. Martin, 2005. Extraction of selected heavy metals using modified clays. Journal of Environment Science and Health, Part A Environmental Science, 40: 601–608.

23. Mohamed, O.A., S.A. Sayed, H.S. Mohamady and N.H. El-Sayed, 2005. Aswan clay as sorbent for removal of Cr(III) and Cr(VI) from synthetic solution and tannery wastewater. Journal of Society Leather Technology Chemistry, 89: 204–209.

24. Turan, P., 2007. Uptake of trivalent chromium ions from aqueous solutions using kaolinite. Journal of Hazardous materials, 148: 56–63.

25. Soleimani, M., B. Rafiei and Z. Hassanzadeh Siahpoosh, 2015. Ghezeljeh Montmorillonite Nanoclay as a Natural Adsorbent for Solid Phase Extraction of Copper Ions from Food Samples. Journal of Analytical Chemistry, 70(7): 794–803.

26. Soleimani, M. and Z. Hassanzadeh Siahpoosh, 2015b. Ghezeljeh nanoclay as a new natural adsorbent for the removal of copper and mercury ions: Equilibrium, kinetics and thermodynamics studies. Chinese Journal of Chemical Engineering, 23: 1819–1833.

27. Soleimani, M. and Z. Hassanzadeh Siahpoosh, 2016. Determination of Cu(II) in water and food samples by Na+-cloisite nanoclay as a new adsorbent: Equilibrium, kinetic and thermodynamic studies. Journal of the Taiwan Institute of Chemical Engineers, 59: 413–423.

28. Hassanzadeh Siahpoosh, Z. and M. Soleimani, 2016. Extraction of Some Divalent Metal Ions (Cadmium, Nickel and Lead) from Different Tea and Rice Samples Using Ghezeljeh Nanoclay (Geleh-Sar-Shoor) as a New Natural Sorbent. Analytical and Bioanalytical Chemistry Research, 3(2): 195-216.

29. Gustafsson, J.P., G. Jacks, M. Simonsson and I. Nilsson, 2005. Soil and Water Chemistry. Lantbruks Universitet Sveriges (SLU), KTH Arkitektur ochsamh¨allsbyggnad, Uppsala, Sweden.

30. Tyagi, B., C.D. Chudasama and R.V. Jasra, 2006. Determination of structural modification in acid activated montmorillonite clay by FT-IR spectroscopy. Spectrochimica Acta, Part A 64: 273–278.

31. Bhattacharyya, K.G. and S.S. Gupta, 2008. Adsorption of a few heavy metals on natural and modified kaolinite and montmorillonite: A review. Advances in Colloid and Interface Science, 140: 114–131.

32. Van Olphen, H., 1977. An introduction to clay colloid chemistry, Wiley Interscience, 187.

33. Adhikari, T. and M.V. Singh, 2003. Sorption characteristics of lead and cadmium in some soils of India. Geoderma, 114: 81-92.

34. Serrano, S., F. Garrido, C.G. Campbel and M.T. Garcia-Gonzalez, 2005. Competitive sorption of cadmium and lead in acid soils of Central Spain. Geoderma, 124: 91-104.

35. Mcbride, M.B., 1994. Environmental Chemistry of Soils, Oxford University Press, New York.

36. da Fonseca, M.G., M.M. de Oliveira and L.N.H. Arakaki, 2006. Removal of cadmium, zinc, manganese and chromium cations from aqueous solution by a clay mineral. Journal of Hazardous materials, 137: 288–292.

37. Coles, C.A. and R.N. Yong, 2000. Aspects of kaolinite characterization and retention of Pb and Cd. Applied Clay Science, 22: 39-45.

38. Auboiroux, M., P. Baillif, J.C. Touray and F. Bergaya, 1996. Fixation of Zn2+ and Pb2+ by a Ca-montmorillonite in brines and dilute solutions: preliminary results. Applied Clay Science, 11: 117-126.

39. Breen, C., C.M. Bejarano-Bravo, L. Madrid, G. Thompson and B.E. Mann, 1999. Na/Pb, Na/Cd and Pb/Cd exchange on a low iron Texas bentonite in the presence of competing H+ ion. Colloids and Surfaces A: Physicochemical Engineering Aspects, 155: 211-219.

40. Echeverria, J.C., I. Zerranz, J. Estella and J.J. Garrido, 2005. Simultaneous effect of pH, temperature, ionic strength, and initial concentration on the retention of lead on illite.

Applied Clay Science, 30: 103-115.

41. Bektas, N., B.A. Agim and S. Kara, 2004. Kinetic and equilibrium studies in removing lead ions from aqueous solutions by natural sepiolite. Journal of Hazardous materials, 112: 115-122.

42. Gu, X., L.J. Evans and S.J. Barabash, 2010. Modeling the adsorption of Cd(II), Cu(II), Ni(II), Pb(II) and Zn (II) onto montmorillonite. Geochimica et Cosmochimica Acta, 74: 5718–5728.

43. Sajidu, S.M.I., I. Persson, W.R.L. Masamb and E.M.T. Henrym, 2008. Mechanisms of heavy metal sorption on alkaline clays from Tundulu in Malawi as determined by EXAFS. Journal of Hazardous materials, 158: 401–409.

44. Brigatti, M.F., S. Colonna, D. Malferrari, L. Medici, L. Poppi, 2005. Mercury adsorption by montmorillonite and vermiculite: a combined XRD, TG-MS, and EXAFS study. Applied Clay Science, 28: 1-8.

45. Hardy, R.G. and M.E. Tucker, 1988. X-ray powder diffraction of sediments. M.E. Tucker (Eds.), Techniques in Sedimentology. Oxford, Blackwell Scientific publishing Inc. New York, pp: 191-228.

46. Ravichandran, J. and B. Sivasankar, 1997. Properties and catalytic activity of acid-modified montmorillonite and vermiculite. Clays Clay Minerals, 45: 854–858.

47. Trombka, J.I., J. Schweitzer, C. Selavka, M. Dale, N. Gahn and S. Floyd, 2002. Crime scene investigations using portable, non-destructive space exploration technology. Forensic Science International, 129: 1–9.

48. Huang, W., D.E. Day, K. Kittiratanapiboon and M.N. Rahaman, 2006. Kinetics and mechanisms of the conversion of silicate (45S5), borate, and borosilicate glasses to hydroxyapatite in dilute phosphate solutions. Journal of Materials Science: Materials in Medicine, 17: 583–596.

49. Huang, L.H., Y.Y. Sun, T. Yang and L. Li, 2011. Adsorption behavior of Ni (II) on lotus stalks derived active carbon by phosphoric acid activation. Desalination, 268, 12–19.

50. Schweitzer, J., J.I. Trombka, S. Floyd, C. Selavka, G. Zeosky and N. Gahn, 2005. Portable generator-based XRF instrument for non-destructive analysis at crime scenes. Nuclear Instruments and Methods in Physics Research Section B, 241: 816–9.

51. Krekeler, M.P.S., J. Morton, J. Lepp, C.M. Tselepis, M. Samsonov and L.E. Kearns, 2008. Mineralogical and geochemical investigation of clay-rich mine tailings from a closed phosphate mine, Bartow Florida, USA. Environmental Geology, 55: 123–147.

52. van Olphen, H. and J.J. Fripiat, 1979. Data Handbook for Clay Materials and other Nonmetallic Minerals, Pergamon Press, New York.

53. Meier, L.P. and R. Kahr, 1999. Determination of the cation exchange capacity (CEC) of clay minerals using the complexes of copper(II) ion with triethylenetetramine and tetraethylenepentamine. Clays and Clay Minerals, 47: 386–388.

54. Amman, L., F. Bergaya and G. Lagaly, 2005. Determination of the cation exchange capacity of clays with copper complexes revisited. Clay Minerals, 40: 441–453.

55. Grim, R.E., 1968. Clay Mineralogy, McGraw-Hill, New York.

56. Eren, E. and B. Afsin, 2008. An investigation of Cu(II) adsorption by raw and acid-activated bentonite: A combined potentiometric, thermodynamic, XRD, IR, DTA study. Journal of Hazardous materials, 15: 682–691.

57. Adebowale, K.O., I.E. Unuabonah and B.I. Olu-Owolabi, 2006. The effect of some operating variables on the adsorption of lead and cadmium ions on kaolinite clay. Journal of Hazardous materials B, 134: 130–139.

58. Steudel, A., L.F. Batenburg, H.R. Fischer, P.G. Weidler and K. Emmerich, 2009. Alteration of swelling clay minerals by acid activation. Applied Clay Science, 44: 105–115.

59. Subrahmanyam, P., B.K. Priya, B. Jayaraj and P. Chiranjeevi, 2007. Determination of Cd, Cr, Cu, Pb and Zn fromvariouswater samples with use of FAAS techniques after the solid phase extraction on rice bran. Environmental Toxicology and Chemistry, 90: 97–106.

60. Ghanemi, K., Y. Nikpour, O. Omidvar and A. Maryamabadi, 2011. Sulfur-nanoparticle-based method for separation and preconcentration of some heavy metals in marine samples prior to flame atomic absorption spectrometry determination. Talanta, 85: 763–769.

61. Karimipour, G.R., M. Ghaedi, R. Sahraei, A. Daneshfar and M.N. Biyareh, 2012. Modification of Gold Nanoparticle Loaded on Activated Carbon with Bis(4-methoxysalicylaldehyde)-1,2-Phenylenediamine as New Sorbent for Enrichment of Some Metal Ions. Biological Trace Element Research, 145: 109–117.

62. Vieiraa, E.G., I.V. Soaresa, N.L.D. Filhoa, N.C. da Silva, S.D. Perujoa, A.C. Bastosa, E.F. Garciaa, T.T. Ferreiraa, L.F. Fracetob and A.H. Rosab, 2012. Study on soluble heavy metals with preconcentration by using a new modified oligosilsesquioxane sorbent. Journal of Hazardous materials, 237–238: 215–222.

63. He, Q., Z. Hu, Y. Jiang, X. Chang, Z. Tu and L. Zhang, 2010. Preconcentration of Cu(II), Fe(III) and Pb(II) with 2-((2-aminoethylamino)methyl)phenol-functionalized activated carbon followed by ICP-OES determination. Journal of Hazardous materials, 175: 710-714.

64. Xie, F., X. Lin, X. Wu and Z. Xie, 2008. Solid phase extraction of lead (II), copper (II), cadmium (II) and nickel(II) using gallic acid-modified silica gel prior to determination by flame atomic absorption spectrometry. Talanta, 74: 836-843.

65. Puzio, B., B. Mikula and B. Feist, 2009. Preconcentration of Cd (II), Pb (II), Co(II), Ni(II), and Cu(II) by Solid-Phase Extraction Method Using 1,10-Phenanthroline. Journal of Analytical Chemistry, 64: 786-7

8

Electrical Behaviour of Chitosan-Silver Nanocomposite in Presence of Water Vapour

Bal Chandra Yadav[1]*, Ritesh Kumar[2], Ravindra Kumar[1], Subhasis Chaudhuri[3], Panchanan Pramanik[3]

[1]*Department of Applied Physics, School of Physical Sciences, Babasaheb Bhimrao Ambedkar University, Lucknow-226025, U.P., India*
[2]*Department of Physics, University of Lucknow, Lucknow-226007, U.P., India*
[3]*Department of Chemistry, IIT Kharagpur, W.B., India*

ABSTRACT

This paper presents the synthesis, characterization of the nanocomposite of silver and chitosan polymer composite reinforced by cellulose fibre and its electrical behaviour in presence of water vapour. The coated paper has been characterized by XRD, IR, SEM and EDX techniques. The size of silver nanoparticles is found to be around 9 nm and deposited uniformly. Chitosan, as well as cellulose, contain a hydrogen attached to electronegative nitrogen and oxygen. This gives a favourable environment for the formation of hydrogen bonds. IR peaks of the composite infer the intermolecular hydrogen bonding between the two constituents. The SEM pictures show that the coating of the fibres with nanoparticles is quite uniform. EDX analysis shows that the coated filter paper has sufficient amount of silver along with carbon and oxygen. The coated paper shows good sensitivity towards humidity. It gives excellent linearity in response with a concentration of water vapour after heat treatment of composite at 130 °C. The sensitivity of the sensor is 0.8 MΩ per unit of relative humidity. Sensing properties originate from protonic conductivity from adsorbed water molecule.

Keywords: *Chitosan, Humidity Sensor, Nanocomposite, Silver Nanoparticles*

INTRODUCTION

Several different kinds and varieties of polymers have been used as different kinds of chemical sensors. Polymers show chemical sensing properties [1-9]. The polymer having the reasonable backbone for hydrogen bonding has the promise to act as a humidity sensor. Thus humidity sensing using polymers has been a major part of the research for a long time. Humidity sensors have attracted increasing attention recently due to its applications in food quality and storage, meteorological studies, environmental humidity for air conditioning systems and also for feedback control in household electric appliances such as drying machines and microwave ovens [10-14]. Humidity sensors based on polymers offer many advantages such as long-term stability, reliability, ease of processing and low fabrication cost. Commonly there are two categories for the humidity sensors: resistive-type [15-17] and capacitive-type [18-19]. However, the poor specific conductivity of polymer sometimes poses great impedance to use a sensor. This problem may be solved by embedding metal particles in the polymer matrix.

* Corresponding Author Email:
  balchandra_yadav@rediffmail.com

The properties of silver nanoparticles have been extensively studied in the past few years for their unique optical, sensing and anti-microbial properties. In fact, silver nanoparticles have been one of the most widely studied of all the nanoparticles. Over the years various techniques have been used to synthesized silver nanoparticles [20]. Chitosan, on the other hand, is a polymer which has found widespread application in recent years. Chitosan is a copolymer of glucosamine and N-acetylglucosamine derived from the natural polymer chitin. This biopolymer can be physically modified to give different forms (e.g., powder, nanoparticles, gel, film, and beads), crosslinked with different substances (e.g., glutaraldehyde, carbodiimide, epichlorohydrin, and tripolyphosphate) and used in various fields of application [21-22].

The present paper reports a unique and very simple technique for making nanocomposite of nano-sized silver and chitosan to use as a humidity sensor. Silver has a high value of conductivity. The addition of silver to the chitosan has ensured that the background resistance value of the sensor material is well within measurable range. Moreover, silver is chemically stable and does not react with the polymer. The deposition of composite material would be carried out on filter paper made of cellulose fibers consisting ether or esters of cellulose. Thus the sensor developed using chitosan and nano-sized silver has a background resistance well within the measurable range.

The success of humidity sensor lies in linearity of the response curve. This composite shows reasonable linearity in response after the proper heat treatment.

EXPERIMENTAL
Preparation of Sensing Material and Its Characterization

0.2% chitosan purchased from Sigma-Aldrich (weight by volume) in 1 % acetic acid (volume by volume) has been prepared. This solution is mixed with 0.2 (M) silver nitrate in a ratio of 1:1 by volume. Strips of filter paper (Whatman 42) have been taken and soaked in this solution followed by soaking in N/10 ammonium hydroxide for 15 mins. The strips are then exposed to hydrazine vapours for 10 mins which produced Ag nanoparticles. On exposure to ammonia gas, chitosan is precipitated on the filter paper along with silver nanoparticles. Thus on the strips of filter paper chitosan and silver

nanoparticles composite are deposited. The strips are washed with water to remove soluble inorganic compounds. The washing is followed by drying. These strips having a smooth deposition are used as a humidity sensor. The amount of silver added to the sample has been optimised. The various different concentration of Ag^+ has been tried ranging from 0.05(M) to 0.4(M). But it is found that on using 0.2(M) Ag^+ the background resistance comes in a measurable range. Too much of silver leads to a decrease in background resistance whereas too little leads to a drastic increase. When 0.05(M) Ag^+ has used the background resistance increases to around 600 MΩ but when 0.4(M) Ag^+ is used the background resistance drops down to around 0.5 MΩ. The concentration of Ag^+ in the range of 0.2(M) produces the background resistance within the range of 100 MΩ to 120 MΩ. This value is comfortable for designing the hardware of the sensor. This value is chosen since it is neither too high nor too low, thus, ensuring that on absorbing moisture there is a sufficient and detectable change in resistance. After deposition of the composite, the thick films have been backed in the oven at 130 C which makes the thick film mechanically stable. The thermal treatment is necessary to eliminate any hysteresis during measurement. The strips having length 5 cm and breadth 1cm are cut from the sheet of thick film for measurement. The contacts of the strips are made with silver paint.

Device assembly for Humidity Detection

A controlled humidity chamber has been designed [23]. A saturated solution of potassium hydroxide in distilled water is used as a dehumidifier and saturated solution of potassium sulphate as a humidifier. A variation in resistance has been noted by using a digital multimeter (Keithley 6541A). Relative humidity is measured using standard hygrometer associated with a thermometer (Huger, Germany) as illustrated in Fig. 1(a). Schematic of proposed sensing structure is shown in Fig. 1(b).

The humidifier/dehumidifier is kept in a dish on a stand. In this process, the temperature of the chamber remains the same throughout the experiment which is being constantly monitored. The ends of the sensor were connected to the Keithley Electrometer which recorded the changes in resistance values. The sensor has been investigated by exposing humidity inside a specially designed controlled humidity chamber. The sensitivity of humidity sensor has been defined as the change

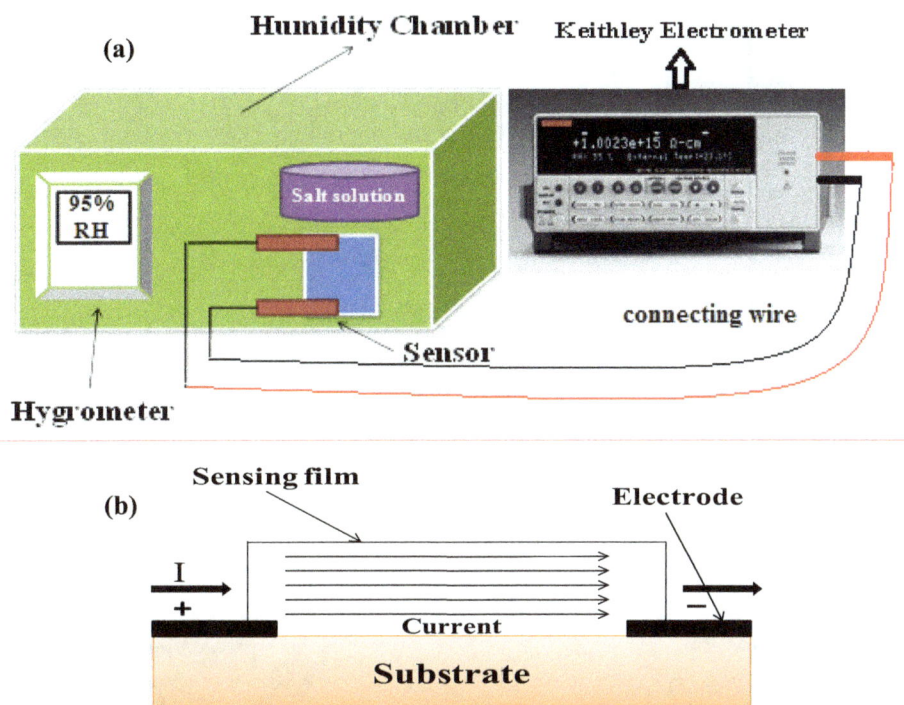

Fig. 1. (a) Device assembly for humidity detection and (b) Schematic of sensing element

in resistance of sensor per unit change in relative humidity (%RH) [24].

Variations in resistance with the variations in %RH for the sensor at room temperature which is around 30°C have been studied.

RESULTS AND DISCUSSION

XRD Studies

XRD studies were done using Philips PW1710 diffractometer with Cu Kα as the target material in which 3 major peaks were obtained apart from some other minor peaks as shown in Fig. 2. The peak at $2\theta = 38.13°$ is due to silver nanoparticle formation.

Using the Scherrer's equation given as under:

$$\beta_{hkl} = \frac{K\lambda}{L_{hkl}\cos\theta_{hkl}}$$

β is the breadth of the peak of a specific phase (hkl).

K is a constant that varies with the method of taking the breadth ($0.89 < K < 1$ in this case taking $K = 0.89$).

λ is the wavelength of incident X-rays. θ is the center angle of the peak. L is FWHM of the peak.

The crystallite size of the silver nanoparticles

is calculated using this equation. The crystallite size thus obtained is around 9 nm. The peak corresponding to (111) is the plane of the silver nanoparticle. However, the largest peak obtained is at $2\theta = 22.67°$ which is due to the chitosan. Any XRD of the chitosan-based sample gives a peak at around 22°. This peak is large owing to the crystallinity of chitosan which is very well known. The other small peak may originate from the AgCl or Ag_2O crystals in the sample.

IR Studies

FTIR spectra of the prepared samples (finely chopped up into very tiny pieces) were done in KBr medium with Thermo Nicolet Nexus FTIR spectrometer (model 870). The IR data was taken for the coated filter paper (sensor) and the data was compared with that of an ordinary filter paper and of chitosan as shown in Fig. 3. Filter paper contains cellulose as its constituent. Chitosan and cellulose have very similar functional groups in its structure. Some of the hydroxyl groups of cellulose have replaced by $-NH_2$ and $-NHCOCH_3$. So this makes the IR data quite similar to both these polymers except some additional peaks in chitosan. When the data were compared, it was found that the coated filter paper (sensor) showed almost all the

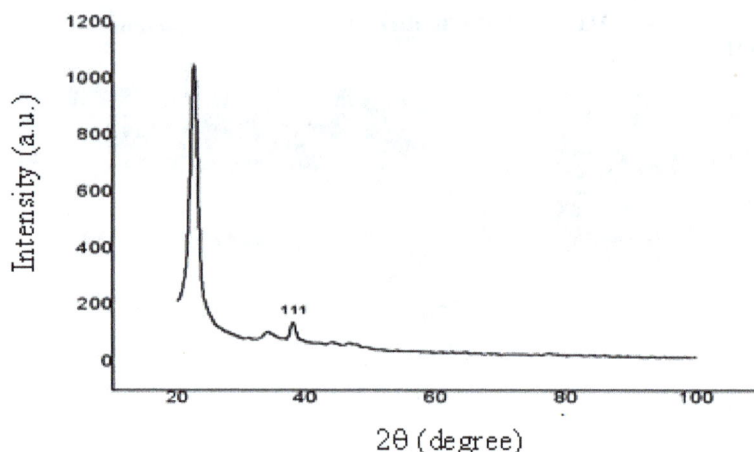

Fig. 2. XRD data of the coated filter paper.

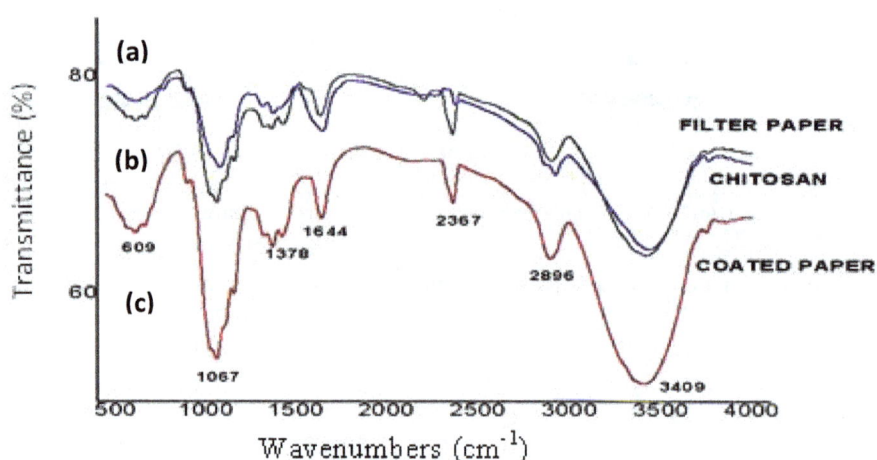

Fig. 3. IR data for ordinary filter paper, chitosan and coated filter paper.

peaks corresponding to both the species. But there is slight shifting and sharpening of peaks in some cases. This can be attributed to the formation of hydrogen bonds between chitosan and cellulose. Chitosan, as well as cellulose, contain a hydrogen attached to electronegative nitrogen and oxygen. This gives a favourable environment for the formation of hydrogen bonds. Thus the changes in IR peaks of the composite infer the intermolecular hydrogen bonding between the two constituents.

The silver nanoparticles loaded chitosan film (Fig. 3c) shows the characteristic peaks with a slight shift of the peak 1350 to 1378 cm^{-1} corresponding to amide III band. In addition, the stretching vibration at 2896 cm^{-1} conforming to OH/NH$_2$ groups has shifted to 3409 cm^{-1}, which indicates that the silver particles are bonded to the functional groups present both in chitosan. The shifting of the peak is due to the formation of coordination bond between the silver atom and the electron rich groups (oxygen/nitrogen) present in chitosan. This causes an increase in bond length and frequency. Binding of silver with N of the amine and amide group results in decreasing of the intensity of amine and amide peaks at 1644 cm^{-1}. Division of combined peak of amine and amide at 1644 cm^{-1}, also indicates the binding of Ag with O and N of those groups. The peak intensities in the range 1000 cm^{-1} and 1350 cm^{-1} due to C–N stretching and bending is very less in silver loaded chitosan nanoparticles because of the complexation of chitosan with silver.

SEM Studies

SEM analysis reveals that fibrous structure of composite due to deposition of composite on cellulose fiber which is depicted in Fig. 4(a) and Fig. 4(b). The SEM of uncoated filter paper is also

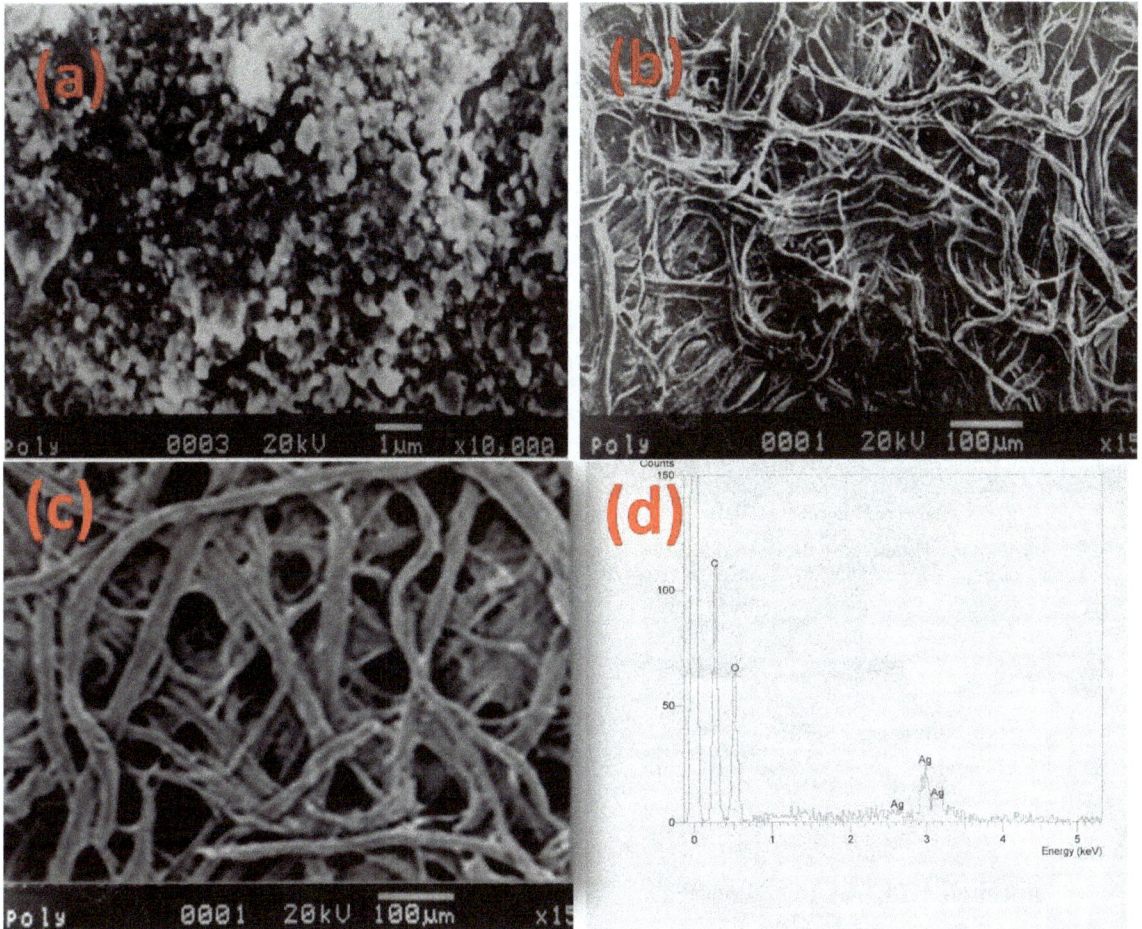

Fig. 4. (a) SEM pictures of the coated filter paper on scale 1μm, (b) SEM pictures of the coated filter paper on scale100 μm, (c) SEM of the uncoated filter paper, (d) EDX data of the coated filter paper.

shown in Fig. 4(c). A filter paper has a high fibrous network. Such a highly fibrous network with nanoparticle deposition gives a good opportunity for the water molecules to get trapped in this material. As seen in the SEM pictures nanoparticle deposition is formed along the fibres. The coating of the fibres with nanoparticles is quite uniform. Such a uniform deposition enables the sample to be used as a good sensor material.

EDX Analysis

EDX analysis (Fig. 4 d) shows that the coated filter paper has sufficient amount of silver along with carbon and oxygen. The filter paper contains cellulose. So the carbon and oxygen come from the filter paper as well as chitosan. Mapping of silver particle shows that the distribution is very uniform. After mapping at different positions along the coated filter paper, almost same values were obtained.

Moisture Sensing

Sensitivity of a humidity sensor can be defined as the change in resistance (ΔR) of sensing element per unit change in relative humidity (%RH) [25] i.e.

$$S = \frac{\Delta R}{\Delta \%RH} \, M\Omega/\%RH$$

The average sensitivity is calculated by taking the average of all sensitivities ranging from 10%RH to 90%RH.

The variation of resistance with a variation of %RH prepared by coating on the paper strip has been shown in Fig. 5(a). The curve for the sensing element shows that resistance decreases slowly with the increase in %RH. The average sensitivity is found to be 0.80 MΩ/%RH. In this case, it is found that the curve of resistance versus relative humidity is nearly linear as supported by the Fig. 5(a). The linearity of the sensing element arises early from 10% relative humidity onwards.

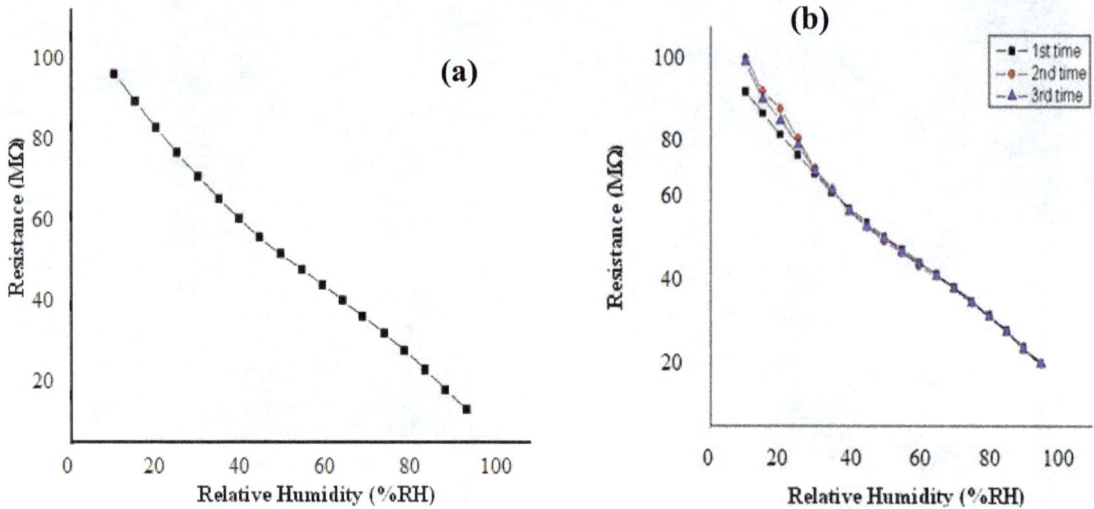

Fig. 5. (a) Variations in resistance with the variations in relative humidity for the sensing element (b) Variations in resistance with the variations in relative humidity after 3 and 6 days respectively.

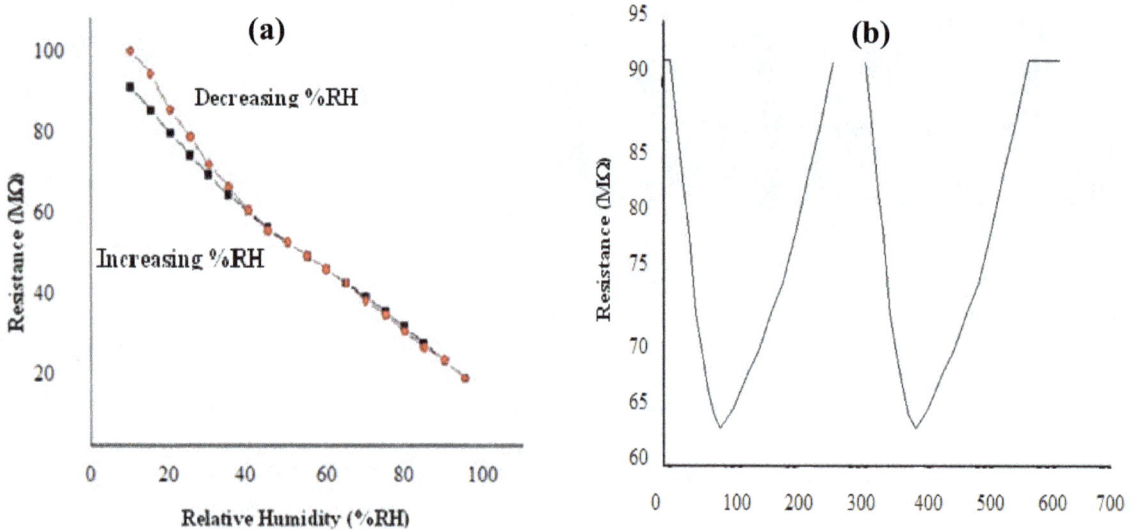

Fig. 6. (a) Hysteresis curve, (b) Response curve at a constant humidity.

The repeatability of the results has also been studied for the sensing element and the characteristics are shown in Fig. 5(b). The results are reproducible in multiple measurements. The hysteresis of the sensor is also tested as shown in Fig. 6(a). Hysteresis is observed especially at low humidity region. With the increase in humidity, the hysteresis loop becomes thinner. It takes in water vapour but during its release does not do so completely. There might even be mild structural changes occurring due to which this phenomenon is being observed. The

sensor is backed in the oven at $130^{\circ}C$ which makes the thick film mechanically stable. The thermal treatment is necessary to eliminate hysteresis during measurement. The response time has also been measured by the sensor. The response curve has been shown in Fig. 6(b) at a constant humidity of around 35%. As shown in the figure, the sensor takes about 80 seconds to reach a stable value of resistance under the given humidity conditions. The recovery time of the sensor is about 3 mins. Thus response curve shows the sensing behaviour is excellent as a humidity sensor.

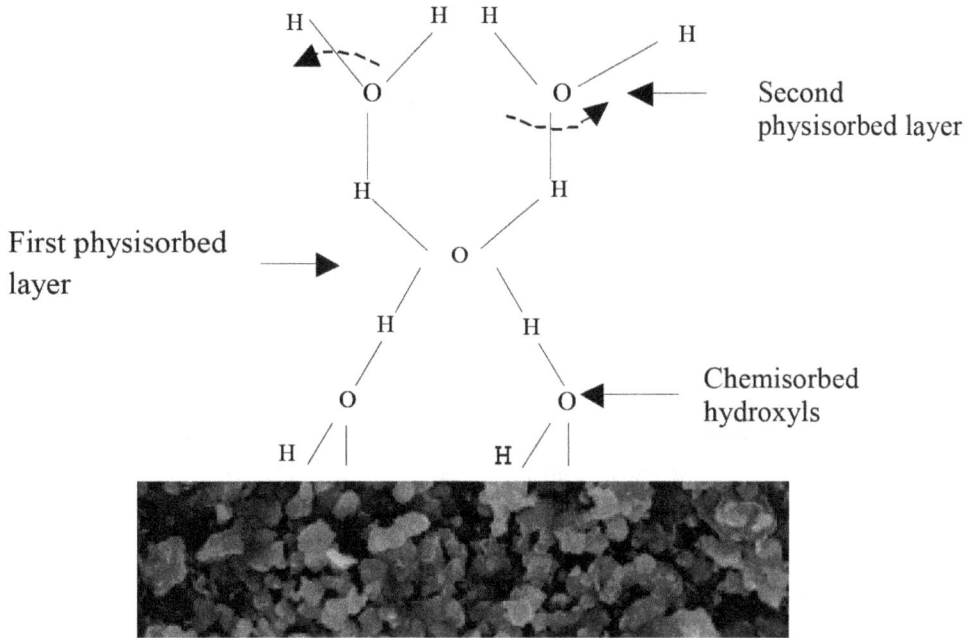

Fig. 7. Schematic of humidity sensing mechanism through a porous film.

Moisture Sensing Mechanism

The sensor material is a highly fibrous network as evident from the SEM analysis. Along these fibers silver nanoparticles are deposited. These silver nanoparticles are distributed all along the fibers. Thus these nanoparticles form small islands along these fibers. There is no long continuous chain between these nanoparticles and hence they are like islands over this fibrous network. Due to the absence of the continuous connectivity, the resistance of the sensor material in the absence of humidity is quite high. The discontinuity has been played upon the action of the sensor. If a higher concentration of silver is used in the fabrication of the sensor then the continuity of the background increases and thus sensitivity of the sensor becomes poor.

When this sensor is exposed to moisture water molecules are adsorbed on the surface of the sensor where highly fibrous networks exist. These water molecules get trapped in between the islands formed by the silver nanoparticles and introduce reasonable conductivity from H^+ generated there. Overall the conduction is due to proton generation through dissociation of H_2O through surface hydrolysis [26-27]. It may be facilitated by proton dissociated from OH group of chitosan. When the film is coated with any nonpolar low boiling organic solvent vapour after adsorption of moisture, the conductivity is drastically reduced and gain to the original value after some time in the open atmosphere. This indicates that surface conductivity is probably due to protonic conductivity. Thus the mechanism of sensing action is simply due to proton conduction as per Grotthus Chain Reaction [25].

Schematic of sensing mechanism is shown in Fig.7. At low humidities, conduction is due to proton hopping between hydroxyl ions on the first layer of chemisorbed water vapour. The chemisorbed hydroxyl ions enhance the electrical conductivity of the sensor either by donating electrons to the conduction band of the base material or through proton hopping between adjacent hydroxyl groups upon application of an electric field. The process of chemisorptions occurs at very low humidity levels and is unaffected by further changes in humidity. However, an increase in humidity makes the water molecules physisorbed onto this hydroxyl layer. The effectiveness of physisorption depends upon the cation charge complexes from the material or the impurities and the water molecules present at the surface of the base material i.e. the hydroxyl ions. During formation of the first physisorbed layer, a water molecule attaches to two neighbouring hydroxyl groups through hydrogen double bonds and a proton may be transferred from a hydroxyl group to the water molecule. At higher humidity levels, the number of physisorbed layers increases, allowing each water molecule to be singly bonded to a hydroxyl group, and proton hopping between adjacent water molecules in the continuous water layer takes place.

CONCLUSION

The sensing element prepared by the nanocomposite of silver and chitosan shows average sensitivity 0.8 MΩ/%RH for the entire range of humidity. The graph of resistance versus relative humidity is found to be linear in all the samples prepared by the same process. The results are also reproducible. The deposition of the silver in the composite is uniform as observed from EDX. The optimization of silver concentration is crucial to maximizing the sensitivity. The measured resistance of the strips (5 cm x 1 cm) ranges from 100 MΩ to 120 MΩ which is a very comfortable range of measurement by the simple electronic gadget. Instruments which can measure resistance up to about 100-150 MΩ will easily do the job. The sensor can be used over and over again due to its reproducibility after its heat treatment. Thus the sensor reported here is user-friendly, cost-effective, easy to fabricate and its operation range is quite large.

CONFLICT OF INTEREST

The authors declare that there are no conflicts of interest regarding the publication of this manuscript.

REFERENCES

1. Kozłowski M, Frąckowiak S. Chemical sensors based on polymer composites. Sensors and Actuators B: Chemical. 2005;109(1):141-5.

2. Kumar R, Singh S. Conducting Polymers: Synthesis, Properties and Applications. International Advanced Research Journal in Science, Engineering and Technology. 2015;2(11):110-24.

3. Singh S, Singh A, Yadav BC, Tandon P, Kumar S, Yadav RR, et al. Frontal polymerization of acrylamide complex with nanostructured ZnS and PbS: Their characterizations and sensing applications. Sensors and Actuators B: Chemical. 2015;207, Part A:460-9.

4. Pomogailo DA, Singh S, Singh M, Yadav BC, Tandon P, Pomogailo SI, et al. Polymer-matrix nanocomposite gas-sensing materials. Inorganic Materials. 2014;50(3):296-305.

5. Meanna Pérez JM, Freyre C. A poly(ethyleneterephthalate)-based humidity sensor. Sensors and Actuators B: Chemical. 1997;42(1):27-30.

6. Lee C-W, Nam D-H, Han Y-S, Chung K-C, Gong M-S. Humidity sensors fabricated with polyelectrolyte membrane using an ink-jet printing technique and their electrical properties. Sensors and Actuators B: Chemical. 2005;109(2):334-40.

7. Singh S, Singh A, Yadav BC, Tandon P, Shukla A, Shershnev VA, et al. Synthesis, characterization and liquefied petroleum gas sensing of cobalt acetylenedicarboxylate and its polymer. Sensors and Actuators B: Chemical. 2014;192:503-11.

8. Lee C-W, Park H-S, Kim J-G, Choi B-K, Joo S-W, Gong M-S. Polymeric humidity sensor using organic/inorganic hybrid polyelectrolytes. Sensors and Actuators B: Chemical. 2005;109(2):315-22.

9. Saxena V, Choudhury S, Gadkari SC, Gupta SK, Yakhmi JV. Room temperature operated ammonia gas sensor using polycarbazole Langmuir–Blodgett film. Sensors and Actuators B: Chemical. 2005;107(1):277-82.

10. Kulwicki BM. Humidity Sensors. Journal of the American Ceramic Society. 1991;74(4):697-708.

11. Arai H, Seiyama T, W. Göpel, Hesse J, Zemel JN. Humidity Sensors, in Sensors Set: A Comprehensive Survey: Wiley-VCH Verlag GmbH, Weinheim,Germany; 1995.

12. Traversa E. Ceramic sensors for humidity detection: the state-of-the-art and future developments. Sensors and Actuators B: Chemical. 1995;23(2):135-56.

13. Sikarwar S, Yadav BC. Opto-electronic humidity sensor: A review. Sensors and Actuators A: Physical. 2015;233:54-70.

14. Li Y, Yang MJ. Humidity sensitive properties of a novel soluble conjugated copolymer: ethynylbenzene-co-propargyl alcohol. Sensors and Actuators B: Chemical. 2002;85(1–2):73-8.

15. Sakai Y, Sadaoka Y, Matsuguchi M. Humidity sensors based on polymer thin films. Sensors and Actuators B: Chemical. 1996;35(1):85-90.

16. Dey KK, Bhatnagar D, Srivastava AK, Wan M, Singh S, Yadav RR, et al. VO₂ nanorods for efficient performance in thermal fluids and sensors. Nanoscale. 2015;7(14):6159-72.

17. Kumar R, Yadav BC. Humidity sensing investigation on nanostructured polyaniline synthesized via chemical polymerization method. Materials Letters. 2016;167:300-2.

18. Yang MJ, Casalbore-Miceli G, Camaioni N, Mari CM, Sun H, Li Y, et al. Characterization of capacitive humidity sensors based on doped poly(propargyl-alcohol). Journal of Applied Electrochemistry. 2000;30(6):753-6.

19. Li D, Jiang Y, Li Y, Yang X, Lu L, Wang X. Fabrication of a prototype humidity-sensitive capacitor via layer-by-layer self-assembling technique. Materials Science and Engineering: C. 2000;11(2):117-9.

20. Prabhu S, Poulose EK. Silver nanoparticles: mechanism of antimicrobial action, synthesis, medical applications, and toxicity effects. International Nano Letters. 2012;2(1):32.

21. Krajewska B. Application of chitin- and chitosan-based materials for enzyme immobilizations: a review. Enzyme and Microbial Technology. 2004;35(2–3):126-39.

22. Kumar MNV. A review of chitin and chitosan applications. Reactive and Functional Polymers. 2000;46(1):1-27.

23. Yadav BC, Srivatava R, Dwivedi CD. Synthesis and Characterization of ZnO Nanorods by the Hydroxide Route and Their Application as Humidity Sensors. Synthesis and Reactivity in Inorganic, Metal-Organic, and Nano-Metal Chemistry. 2007;37(6):417-23.

24. Srivastava R, Yadav C. B. Nanaostructured ZnO, ZnO-TiO₂ and ZnO-Nb₂O₅ as solid state humidity sensor. Advanced Materials Letters. 2012;3(3):197-203.

25. Yadav BC, Singh R, Singh S. Investigations on humidity sensing of nanostructured tin oxide synthesised via mechanochemical method. Journal of Experimental Nanoscience. 2013;8(5):670-83.

26. Kumar R, Yadav BC. Fabrication of Polyaniline (PANI)— Tungsten oxide (WO$_3$) Composite for Humidity Sensing Application. Journal of Inorganic and Organometallic Polymers and Materials. 2016;26(6):1421-7.

27. Fleming WJ. A Physical Understanding of Solid State Humidity Sensors. SAE International; 1981.

9

Novel Adsorptive Mixed Matrix Membrane by Incorporating Modified Nanoclay with Amino Acid for Removal of Arsenic from Water

Elham Shokri, Reza Yegani*

Faculty of Chemical Engineering, Sahand University of Technology, Tabriz, Iran

ABSTRACT

In this work, polysulfone (PSf) mixed matrix membranes were prepared by incorporating modified montmorillonite with lysine amino acid (MMT-Lys) for arsenic removal from water. Different tests including XRD, zeta potential, FE-SEM, contact angle, and pure water flux (PWF) were carried out to characterize modified MMT and fabricated mixed matrix membranes. XRD analysis showed that MMT was successfully modified with Lys and its zeta potentials transferred from negative to positive after modification. Positive charge of MMT-Lys made it proper for anionic arsenic removal from water. The obtained results showed that pure water flux and surface hydrophilicity of the membranes improved as MMT-Lys contents increased from 0 to 1.5 wt.%. The batch adsorption of fabricated membranes as a function of arsenic initial concentration and solution pH was investigated. The removal efficiency was increased with increasing the arsenic initial concentration; however it was decreased with increasing pH of solution. The results also revealed that the arsenic adsorption was most favorable in the neutral pH. Moreover, membrane reusability of the PSf/MMT-Lys (1.5 wt.%) membrane was assessed by conducting five cycles of adsorption-desorption experiments in dead-end filtration. The obtained results showed the applicability of the prepared membrane for multiple cycles.

KEYWORDS: *Amino Acid, Arsenic, Mixed Matrix Membrane, Montmorillonite, Water*

INTRODUCTION

Arsenic contamination of drinking water is one of the most serious concerns in the world. Arsenic can be found in water in inorganic form as oxyanions of arsenite, As(III), or arsenate, As(V) [1]. The major health concern of arsenic exposure through drinking water is the risk of skin, lung, liver and lymphatic cancer. Due to these health hazards associated with arsenic contaminated water, the World Health Organization (WHO) has set 10 ppb as the maximum contaminant level (MCL) of arsenic in drinking water [2]. Various treatment technologies including precipitation [3],

coagulation [4-6], ion exchange, adsorption [7-10] and membrane filtration through nanofiltration (NF) [7, 11-14] have been extensively used for arsenic removal. Among these methods, membrane processes have got much attention and demonstrated to be effective. However, NF membranes require high operational pressure and costly membrane. Consequently, researchers have focused on the development of new membrane based processes for arsenic removal from water. In this practice, adsorptive mixed matrix membrane was introduced for an efficient removal of small pollutants from water at low pressure [15-20].

* Corresponding Author Email: ryegani@sut.ac.ir

Table 1. Casting solution composition of neat and adsorptive mixed matrix PSf membranes.

Membrane	PSf (wt. %)	PEG (wt. %)	NMP (wt. %)	MMT-Lys (wt. %)
PSf	15	10	75	-
PSf/MMT-Lys (0.5 wt.%)	14.925	10	75	0.075
PSf/MMT-Lys (1.0 wt.%)	14.85	10	75	0.15
PSf/MMT-Lys (1.5 wt.%)	14.775	10	75	0.225

Adsorptive mixed matrix membranes are a kind of MF/UF membranes developed by embedding the inorganic materials into porous polymeric matrix [20, 21]. The adsorptive membranes combine the selectivity of adsorption method with the flow behavior of MF/UF membranes [22]. In terms of adsorptive membranes, reversible interaction is very important to have reusable membranes. Since electrostatic interaction is reversible, positive charged inorganic materials possess wider practical in anionic pollutant removal [23]. To prepare highly efficient adsorptive membranes for arsenic removal, investigators have used several adsorbents incorporating in polymer matrix. One of the most common ligand in positive charged adsorbents is amine group. The amino acids with amine and carboxylic groups are zwitterionic compounds that contain both positively and negatively charged chemical groups [24, 25]. Depending on the pH of their environment, amino acids may carry a net positive charge, a net negative charge, or no charge [26]. By taking this point into consideration, Lys as positively charged amino acid was used to modify MMT by a simple cation exchange method. Modified MMT with Lys was embedded in porous polymeric matrix. As for the polymer matrix, polysulfone (PSf) has been widely used in MF/UF membranes due to its outstanding properties such as low cost, availability, high mechanical strength, thermal and chemical stabilities, resistance over wide range of pH, as well as easy processability and variety of active functional groups [27-31]. In this study, modification of MMT and fabrication of PSf/MMT-Lys membranes were done successfully. The key membrane properties, such as hydrophilicity, PWF and morphology were evaluated. The adsorption isotherms of all membranes, dynamic adsorption and regeneration of membrane with high adsorption capacity were investigated and discussed.

MATERIALS AND METHODS
Materials
Commercial grade of polysulfone (ρ =1250 (kg/m^3), Tg=190°C) purchased from Slovay and used as polymer. Natural montmorillonite as nanoclay with cation exchange capacity (CEC) of 92.6 meq/100 g was obtained from Southern Clay Products Inc., Texas, USA. N-Methyl-2-Pyrrolidon (NMP) (ρ=1030 kg/m^3) as solvent and polyethylene glycol (PEG) (Mw = 20000(g/g mol)) as pore former were provided from Merck and used to fabricate membrane. Disodium hydrogen arsenate heptahydrate Na2HAsO4.7H2O to prepare arsenic solutions, lysine as surfactant and finally, NaOH and HCl were purchased from Merck, respectively.

Synthesis and Characterization of MMT-Lys
The modified MMT with Lys was prepared by displacement of the sodium cations of MMT with the protonated Lys. Typically, 0.2 g of the MMT was dispersed in 25ml of distilled water and suspension was continuously stirred for 24 h to swell the layered silicates. The Lys solution prepared separately by dissolving 0.1 g in 25 ml deionized water at 30 ^{0}C following by incorporating of HCl while pH of solution adjusted to 3. Then solution added slowly to clay and mixture was subjected to mechanical stirring for 12 h in water bath at 70 ^{0}C. After precipitation, the final modified MMT was separated by centrifugation and washed with water. Finally, the modified Mt was dried at 60°C for 12 h. X-ray diffraction (XRD) patterns of the resulting clays were obtained on (D500 Siemens, Germany) diffractometer using Cu-Kα radiation (λ=0.154 nm) under a voltage of 35 kV and a current of 30 mA between 2θ of of 2-50°.

The zeta potentials of unmodified and modified MMT were measured on a Nano ZS (red badge) ZEN 3600. The pH dependences of zeta potentials for MMT and MMT-Lys were measured in the aqueous solutions at different pH values, adjusted by NaOH and HCl.

Preparation of Membranes
All membranes were prepared using non solvent induced phase separation method. Modified MMT in various weight percentage (0.5, 1.0 and 1.5) was dispersed into 40 g NMP using sonication by probe

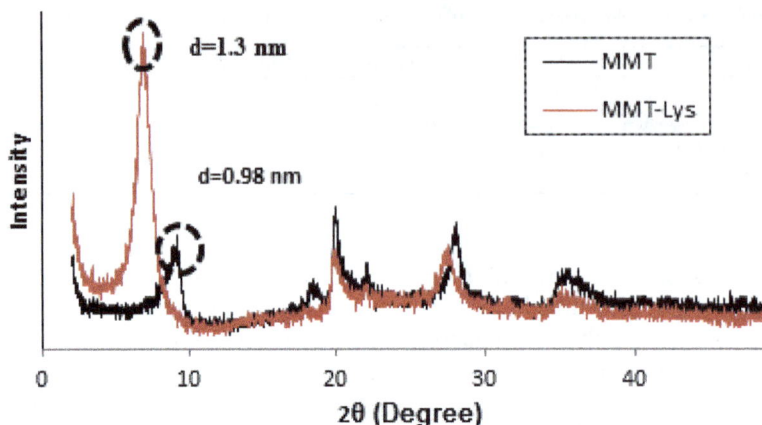

Fig. 1. XRD patterns of MMT and MMT-Lys.

system (Sonopuls HD 3200, Bandelin) for 30 min. Then, 7.8 g PSf and 2.6 g PEG were added to particles solution and mixture was stirred at 60 ˚C for 8h. After that, resulted homogenous solution sonicated for 15 min and then enough time was given for bubbles to be completely released. At least a portion of the achieved solution was poured onto flat plate glass and spread out using an automatic casting knife at the speed of 10 mm/sec. The thickness of all membranes was kept constant at 150 micrometers. Immediately after casting, the film was immersed in a water bath to initiate phase inversion. The obtained membranes were thoroughly rinsed with deionized water and dried at 70 ⁰C to remove the residual solvent. The composition of each casting solution is given in Table 1.

Membrane Characterization

The morphology of the membranes was characterized by FE-SEM (MIRA3 FEG-SEM, Tescan). Cross-section samples were prepared by fracturing the membranes in liquid nitrogen. All samples were coated with gold by sputtering before observation to make them conductive.

The hydrophilicity of membranes was evaluated by measuring contact angle between membrane surface and water droplet using a contact angle goniometer (PGX, Thwing-Albert Instrument Co.). The average of 3 measurements was reported.

Pure water flux of membranes was determined using an in-house fabricated dead-end filtration system having 5cm² of membrane area. To minimize compaction effects, the pre-wetted membranes were compacted for 30 min at 2 bar. Then the pressure was reduced to 1.5.

Adsorption Capacity of Membrane Adsorbers

The adsorption capacity of membrane adsorbers in arsenic removal were investigated in batch adsorption. All adsorption isotherm experiments were performed in a series of sealed volumetric flasks containing 0.1g of membranes and 100 ml of As(V) solutions in the appropriate concentrations (5-15 ppm). The flasks were continuously shaken for 24 h at 25 ⁰C while pH of the solution was adjusted to 7 at 200 rpm. The equilibrium adsorption amount and removal efficiency of As(V) by the membranes were calculated as follows:

$$q_e = \frac{(C_0 - C_e)V}{M_m} \tag{1}$$

$$As\ (V)\ removal\ efficiency\ (\%) = \frac{(C_0 - C_t)}{C_0} \times 100\% \tag{2}$$

where Co (mg/L), Ce (mg/L) and Ct (mg/L) are concentrations at the initial, equilibrium and time t in the solution, respectively, V is the total volume (L) of the arsenic solution and Mm is the mass (g) of dry membrane used in the adsorption study. The equilibrium concentrations of arsenic in the solutions were analyzed by atomic absorption Varian 220-Graphite Furnace spectroscopy.

In order to understand the adsorption in more detail, Langmuir and Freundlich isotherms as two common models were used. The Langmuir isotherm equation which characterizes homogeneous system with single layer adsorption is given by Eq. (3):

$$\frac{1}{q} = \frac{1}{(K_L * C_e * q_{max})} + \frac{1}{q_{m}} \tag{3}$$

where q is defined before, qmax is the maximum

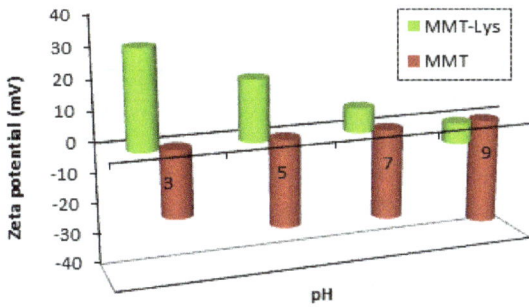

Fig. 2. Zeta potentials of MMT and MMT-Lys at different pH

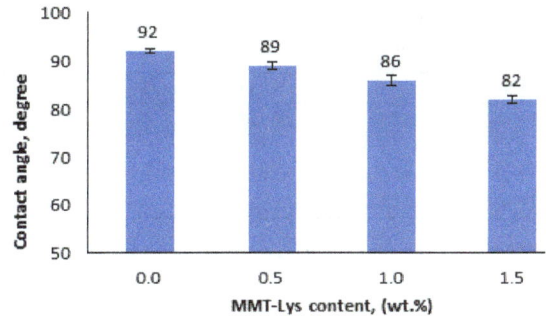

Fig. 3. Variation of contact angle of neat and adsorptive mixed matrix PSf membranes in different MMT-Lys contents

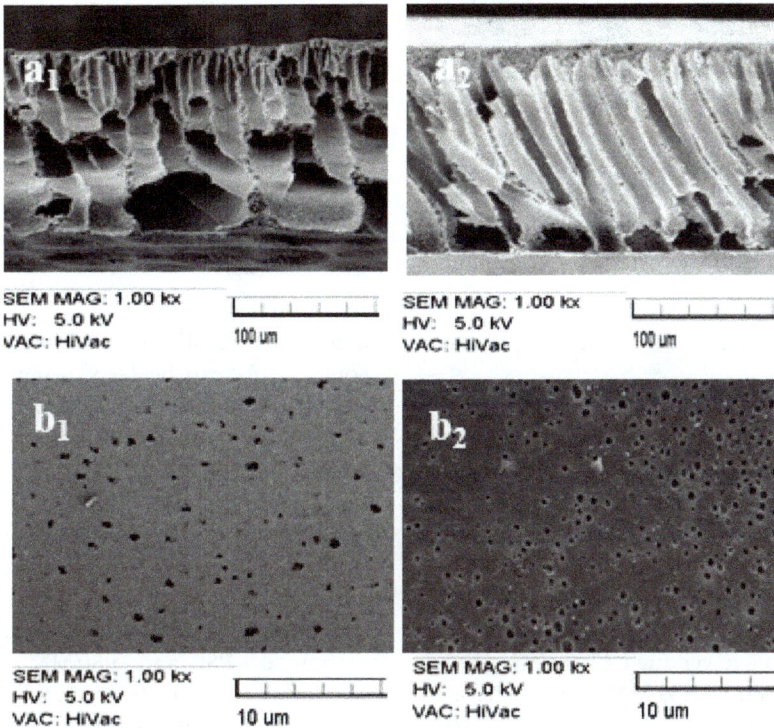

Fig. 4. FE-SEM images of cross-section (1) and top surface (2) of prepared membranes, (a) neat PSf and (b) PSf/MMT-Lys

adsorption capacity (mg/g), KL is the Langmuir adsorption capacity (L/mg), Ce is the arsenic equilibrium concentration in solution (mg/L).

The Freundlich isotherm assumes that different sites with several adsorption energies are involved on a heterogeneous surface. The Freundlich isotherm equation is given by the following equation:

$$\ln(q) = \ln(K_F) + \left(\frac{1}{n}\right)\ln(c_{eq}) \qquad (4)$$

where, K_F is the Freundlich constant and n is the heterogeneity factor.

Dynamic Adsorption and Regeneration Studies

Filtration experiments were carried out in a dead-end filtration setup filled with 100 ppb As(V) solution. The system consisted of a cup connected to a pressure balloon and the filtration experiments were conducted at trans-membrane pressure of 1 bar and feed volume was 300 ml. Each cycle filtration followed by a regeneration step where 50 ml water with pH=9 filtered through the membrane. After regeneration, the second cycle of

Table 2. Langmuir and Freundlich isotherm parameters for As(V) removal using adsorptive mixed matrix membranes with different MMT-Lys contents at pH= 7.

Membrane	Langmuir model			Freundlich model		
	K_L(L/mg)	q_{max}(mg/g)	R^2	K_f (mg/g)	1/n	R^2
PSf/MMT-Lys (0.5 wt.%)	0.27	8.59	0.999	2.21	0.469	0.980
PSf/MMT-Lys (1.0 wt.%)	0.49	11.17	0.999	3.38	0.445	0.984
PSf/MMT-Lys (1.5 wt.%)	0.53	14.12	0.999	4.81	0.501	0.988

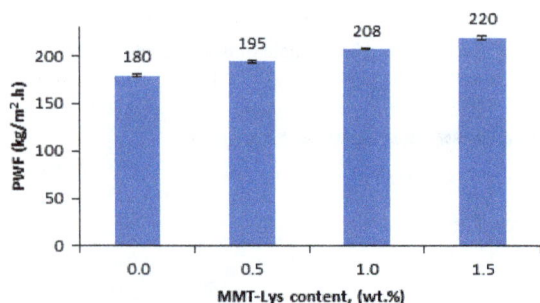

Fig. 5. Variation of pure water flux of neat and adsorptive mixed matrix PSf membranes in different MMT-Lys contents

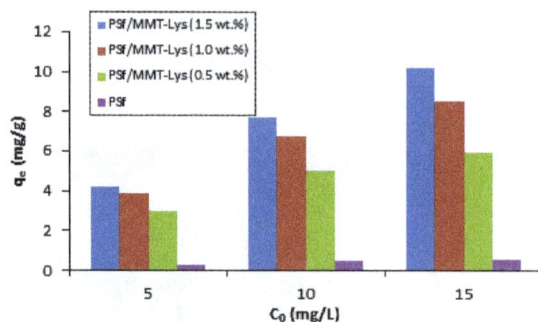

Fig. 6. Equilibrium adsorption curves of As(V) onto the neat and adsorptive mixed matrix membranes, Conditions: m = 1.0 g/L, T = 25 °C, pH= 7

experiments was initiated and this trend continued until 5 cycles.

RESULTS AND DISCUSSION

Characteristics of MMT-Lys

The X-ray diffraction patterns of MMT and MMT-Lys were depicted in Fig. 1. The basal spacing of the MMT is 0.98 nm which is calculated from the peak position at $2\theta = 8.84°$ using Bragg's equation. Diffraction peak of the MMT-Lys after the ion exchange shifted to a lower angle (2θ =6.95, d = 1.3 nm) compared to unmodified MMT. The increase in basal spacing of modified clays confirmed successful modification of MMT with Lys.

Zeta potentials of MMT and MMT-Lys in different pH were shown in Fig. 2. For MMT-Lys, the zeta potential was positive in wide range of pH while it was negative for unmodified MMT. Positive zeta potential means positive surface charge which increases the electrostatic attraction between adsorbent surface and the arsenic anions. This provided another evidence to demonstrate that Lys has been intercalated between MMT.

Characterization of Membranes

The surface hydrophilicity is one of the significant properties of membranes and has significantly effect on the water flux of membranes. Fig. 3 shows the surface contact angles of the

membrane adsorbers as a function of MMT-Lys contents. As can be seen, water contact angle of PSf membrane was about 92° and incorporating of MMT-Lys resulted in reduction of water contact angle to 82° for PSf/MMT-Lys (1.5 wt.%). It can be seen that hydrophilicity of the membrane has been relatively improved. The increasing of membrane hydrophilicity is most due to the hydrophilic polar amine functional groups of modified clay.

FE-SEM images of the top surface and cross-section of PSf and PSf/MMT-Lys (1.5 wt.%) are shown in Fig. 4. As can be seen from Fig. 4(a) images, the membranes exhibit typical asymmetric structure consists of finger-like pores and macrovoids. Some differences can be seen between the PSf and the mixed matrix membrane. The cross sectional images of PSf membrane reveal the presence of macrovoids. It is also observed that macrovoids tend to decrease in size and become narrower, however finger-like pores become longer by incorporating MMT-Lys. Incorporating of MMT-Lys made the dope solution thermodynamically less stable and brought rapid nucleation from the polymer lean phase and promoted macrovoid formation.

The top surface images of membranes (Fig. 4(b)) show that the number of pores increases by incorporating MMT-Lys.

Results of pure water flux of PSf and membrane

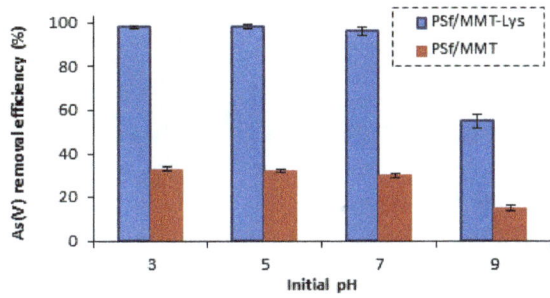

Fig. 7. Effect of pH on the adsorption of As(V) on the PSf/ MMT-Lys (1.5 wt.%), Conditions: As(V) concentration= 100 ppb, m = 1.0 g/L, T = 25 °C

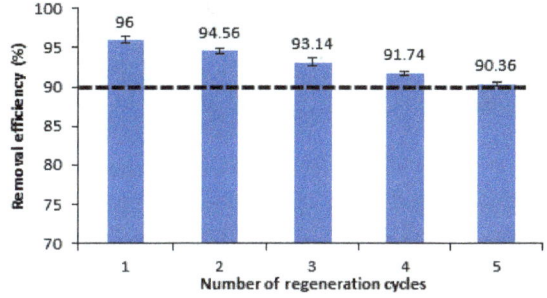

Fig. 8. As(V) removal efficiency in 5 consecutive cycles in dead-end filtration setup using PSf/MMT-Lys (1.5 wt.%)

Table 3. Equilibrium As(V) adsorption capacities and number of regeneration cycles of different mixed matrix membranes.

Adsorptive particles	Polymer	C_0 of arsenate (ppm)	q_e (mg/g)	Number of regeneration cycles	Reference
Zirconia	PVDF	100	20	2	[34]
TiO$_2$	PES	200	125	1	[17]
Laterite	Polyacrylonitrile	100	2.5	4	[35]
MMT-Lys	PSf	20	10.2	5	This study

adsorbers were shown in Fig. 5. According to the effect of MMT-Lys on pure water flux of membranes, it is obvious that addition of MMT-Lys in the casting solution increased pure water flux of membranes. However, higher content of MMT-Lys, i.e. 2 wt. %, resulted in decrease in water flux which has not shown in results. Consequently, 1.5 wt.% loading of MMT-Lys was selected as maximum amount for incorporation into PSf matrix. The main reason to justify the increase in the PWF of the membrane could be attributed to the bigger pore size as well as the extended macrovoids of the mixed matrix membrane[32].

Equilibrium Study

The relationships of the equilibrium adsorption capacities with the initial concentrations of the arsenic are shown in Fig. 6. Fabricated membrane adsorbers demonstrated increasing trend in adsorption capacity with increasing initial concentrations of arsenic, due to the increasing chance of adsorption interaction between arsenic and MMT-Lys in solution with higher As(V) concentrations. Comparison between adsorption capacities of mixed matrix membranes showed that the adsorption capacity increased by increasing the MMT-Lys content. The maximum adsorption capacity was seen in PSf/MMT-Lys (1.5 wt.%) membrane and it was used for filtration studies. The constants for the Langmuir and Freundlich

isotherms were calculated by linear regression of the linear forms of the isotherms and the obtained results were shown in Table 2. From R^2 values, it was found that all mixed matrix membranes favor the Langmuir adsorption isotherm and the adsorption processes were monolayer adsorption.

Effect of Solution pH

The pH is an important parameter that controls the adsorption of arsenic. The removal efficiency of arsenic at different pH is studied in batch experiments and the obtained results are shown in Fig. 7. At initial pH ranging from 3 to 7, 90% of arsenic can be removed successfully. At pH lower than isoelectric point of MMT-Lys the surface of modified MMT in the membrane is in protonation state and causes to increase the positively charged sites. Consequently, the adsorption of arsenic increases through electrostatic attraction between arsenic anions and positive amino chain of MMT-Lys. However, in high pH, the removal efficiency of PSf/MMT-Lys (1.5 wt.%) declined to 50%. The main reason for decreasing the arsenic adsorption at high pH is repulsion between the negatively charged surface sites and arsenic anions.

Dynamic Adsorption Studies

For filtration study, PSf/MMT-Lys (1.5 wt.%) with high batch adsorption capacity was placed in a dead-end filtration cell. As(V) removal efficiency throughout five consecutive cycles was examined at As(V) concentration of 100 ppb. In each cycle 300 ml of As(V) solution was filtered and followed by regeneration step where 50 ml of water with pH=9 was filtered through the membrane. As shown in Fig. 8, removal efficiency was 96% at the first step and decreased with increasing the regeneration cycles. Based on removal efficiency of cycles, it could be concluded that PSf/MMT-Lys (1.5 wt.%) was capable of As(V) removal for 5 cycles with simple regeneration method. This indicates that the adsorptive membrane can be used for multiple cycles.

In Table 3, equilibrium adsorption capacity and the number of regeneration cycles of prepared mixed matrix membrane were compared with other membranes reported in the literature for arsenic removal. The results reported in Table 3 confirm the acceptable adsorption capacity of developed PSf/MMT-Lys in comparison with other mixed matrix membranes with regard to their high As(V) initial concentration. Furthermore, incorporating MMT-Lys into membrane matrix can overcome the drawback of Laterite/polyacrylonitrile membrane which exhibits lower adsorption capacity. The main advantage of developed PSf/MMT-Lys membrane is its high regeneration capability due to the electrostatic interaction between arsenate oxyanions and positive groups of MMT-Lys in neutral pH. However, the adsorption of oxyanions on metal oxides has been regarded as a typical ligand-exchange process which always involves a highly specific chemisorption [33].

CONCLUSION

The modified montmorillonite with lysine amino acid (MMT-Lys) incorporated polysulfone (PSf) membranes were prepared by non-solvent induced phase separation method. The performance of PSf/MMT-Lys adsorptive mixed matrix membranes for arsenic removal from water was investigated. The fabricated membranes exhibited higher pure water flux and hydrophilicity in compare of neat PSf membrane. The FE-SEM studies demonstrated that incorporating MMT-Lys leads to an increase in surface porosity and extended figure like voids. The batch adsorption studies exhibited that the arsenic adsorption was most favorable in the neutral pH. In addition, the obtained results revealed that PSf/MMT-Lys (1.5 wt.%) exposed high adsorption capacity and it was used for filtration studies. The dynamic adsorption experiment showed that

fabricated mixed matrix membrane can be used successfully for 5 multiple adsorption-desorption cycles.

CONFLICT OF INTEREST

The authors declare that there are no conflicts of interest regarding the publication of this manuscript.

REFERENCES

1. Shih M-C. An overview of arsenic removal by pressure-drivenmembrane processes. Desalination. 2005;172(1):85-97.

2. Smedley PL, Kinniburgh DG. A review of the source, behaviour and distribution of arsenic in natural waters. Applied Geochemistry. 2002;17(5):517-68.

3. Harper TR, Kingham NW. Removal of arsenic from wastewater using chemical precipitation methods. Water Environment Research. 1992;64(3):200-3.

4. Parga JR, Cocke DL, Valverde V, Gomes JA, Kesmez M, Moreno H, et al. Characterization of electrocoagulation for removal of chromium and arsenic. Chemical Engineering & Technology. 2005;28(5):605-12.

5. Balasubramanian N, Madhavan K. Arsenic removal from industrial effluent through electrocoagulation. Chemical Engineering & Technology. 2001;24(5):519-21.

6. Bilici Baskan M, Pala A. A statistical experiment design approach for arsenic removal by coagulation process using aluminum sulfate. Desalination. 2010;254(1-3):42-8.

7. Banerjee K, Amy GL, Prevost M, Nour S, Jekel M, Gallagher PM, et al. Kinetic and thermodynamic aspects of adsorption of arsenic onto granular ferric hydroxide (GFH). Water Research. 2008;42(13):3371-8.

8. Camacho LM, Parra RR, Deng S. Arsenic removal from groundwater by MnO2-modified natural clinoptilolite zeolite: Effects of pH and initial feed concentration. Journal of Hazardous Materials. 2011;189(1-2):286-93.

9. Elizalde-González MP, Mattusch J, Einicke WD, Wennrich R. Sorption on natural solids for arsenic removal. Chemical Engineering Journal. 2001;81(1-3):187-95.

10. Giles DE, Mohapatra M, Issa TB, Anand S, Singh P. Iron and aluminium based adsorption strategies for removing arsenic from water. Journal of Environmental Management. 2011;92(12):3011-22.

11. Guan X-H, Su T, Wang J. Quantifying effects of pH and surface loading on arsenic adsorption on NanoActive alumina using a speciation-based model. Journal of Hazardous Materials. 2009;166(1):39-45.

12. Chutia P, Kato S, Kojima T, Satokawa S. Arsenic adsorption from aqueous solution on synthetic zeolites. Journal of Hazardous Materials. 2009;162(1):440-7.

13. Manna B, Ghosh UC. Adsorption of arsenic from aqueous solution on synthetic hydrous stannic oxide. Journal of Hazardous Materials. 2007;144(1-2):522-31.

14. Mandal S, Padhi T, Patel RK. Studies on the removal of arsenic (III) from water by a novel hybrid material. Journal of Hazardous Materials. 2011;192(2):899-908.

15. Casadellà A, Kuntke P, Schaetzle O, Loos K. Clinoptilolite-based mixed matrix membranes for the selective recovery of

potassium and ammonium. Water Research. 2016;90:62-70.

16. Chatterjee S, De S. Adsorptive removal of fluoride by activated alumina doped cellulose acetate phthalate (CAP) mixed matrix membrane. Separation and Purification Technology. 2014;125:223-38.

17. Gohari RJ, Lau WJ, Halakoo E, Ismail AF, Korminouri F, Matsuura T, et al. Arsenate removal from contaminated water by a highly adsorptive nanocomposite ultrafiltration membrane. New Journal of Chemistry. 2015;39(11):8263-72.

18. Jamshidi Gohari R, Lau WJ, Matsuura T, Halakoo E, Ismail AF. Adsorptive removal of Pb(II) from aqueous solution by novel PES/HMO ultrafiltration mixed matrix membrane. Separation and Purification Technology. 2013;120:59-68.

19. Lin L, Zhang L, Zhang C, Dong M, Liu C, Wang A, et al. Membrane adsorber with metal organic frameworks for sulphur removal. RSC Advances. 2013;3(25):9889-96.

20. Niedergall K, Bach M, Schiestel T, Tovar GEM. Nanostructured Composite Adsorber Membranes for the Reduction of Trace Substances in Water: The Example of Bisphenol A. Industrial & Engineering Chemistry Research. 2013;52(39):14011-8.

21. Yin J, Deng B. Polymer-matrix nanocomposite membranes for water treatment. Journal of Membrane Science. 2015;479:256-75.

22. Ladhe AR, Frailie P, Hua D, Darsillo M, Bhattacharyya D. Thiol-functionalized silica–mixed matrix membranes for silver capture from aqueous solutions: Experimental results and modeling. Journal of Membrane Science. 2009;326(2):460-71.

23. Katti KS, Ambre AH, Peterka N, Katti DR. Use of unnatural amino acids for design of novel organomodified clays as components of nanocomposite biomaterials. Philosophical Transactions of the Royal Society A: Mathematical, Physical and Engineering Sciences. 2010;368(1917):1963-80.

24. Ghadiri M, Chrzanowski W, Lee WH, Rohanizadeh R. Layered silicate clay functionalized with amino acids: wound healing application. RSC Advances. 2014;4(67):35332-43.

25. Parbhakar A, Cuadros J, Sephton MA, Dubbin W, Coles BJ, Weiss D. Adsorption of l-lysine on montmorillonite. Colloids and Surfaces A: Physicochemical and Engineering Aspects. 2007;307(1–3):142-9.

26. Ganesh BM, Isloor AM, Ismail AF. Enhanced hydrophilicity and salt rejection study of graphene oxide-polysulfone mixed matrix membrane. Desalination. 2013;313:199-207.

27. Choi J-H, Jegal J, Kim W-N. Fabrication and characterization of multi-walled carbon nanotubes/polymer blend membranes. Journal of Membrane Science. 2006;284(1–2):406-15.

28. Anadão P, Sato LF, Wiebeck H, Valenzuela-Díaz FR. Montmorillonite as a component of polysulfone nanocomposite membranes. Applied Clay Science. 2010;48(1–2):127-32.

29. Yang Y, Zhang H, Wang P, Zheng Q, Li J. The influence of nano-sized TiO2 fillers on the morphologies and properties of PSF UF membrane. Journal of Membrane Science. 2007;288(1–2):231-8.

30. Fan Z, Wang Z, Sun N, Wang J, Wang S. Performance improvement of polysulfone ultrafiltration membrane by blending with polyaniline nanofibers. Journal of Membrane Science. 2008;320(1–2):363-71.

31. Shokri E, Yegani R, Pourabbas B, Kazemian N. Preparation and characterization of polysulfone/organoclay adsorptive nanocomposite membrane for arsenic removal from contaminated water. Applied Clay Science. 2016;132–133:611-20.

32. Zhu R, Chen Q, Zhou Q, Xi Y, Zhu J, He H. Adsorbents based on montmorillonite for contaminant removal from water: A review. Applied Clay Science. 2016;123:239-58.

33. Zheng Y-M, Zou S-W, Nanayakkara KGN, Matsuura T, Chen JP. Adsorptive removal of arsenic from aqueous solution by a PVDF/zirconia blend flat sheet membrane. Journal of Membrane Science. 2011;374(1–2):1-11.

34. Chatterjee S, De S. Adsorptive removal of arsenic from groundwater using a novel high flux polyacrylonitrile (PAN)-laterite mixed matrix ultrafiltration membrane. Environmental Science: Water Research & Technology. 2015;1(2):227-43.

10

Desalination of Kashan City's Water Using PEBA-Based Nanocomposite Membranes via Pervaporation

Soheill Azadikhah Marian[1]*, Morteza Asghari[2], Zahra Amini[2]

[1]*Department of Chemical Engineering, Faculty of Engineering, University of Azad, North Branch, Tehran, Iran*
[2]*Separation Processes Research Group (SPRG), Department of Engineering, University of Kashan, Kashan, Iran*

ABSTRACT

In this work, performance of composite membranes was investigated for desalination of Kashan city's water via pervaporation process. PEBA/PAN/PE, PEBA/PSF/PE and PEBA+NaX/PSF/PE composite membranes that used, was synthesized via a phase inversion route. For all experiments under 45°C, salt rejection was too high and equals to 99.9% that this quantity dropped by increasing the temperature that cause membrane swelling in high temperatures. Water contact angle and water take-up were measured to evaluate the hydrophilicity of the membrane. Also the effect of operating conditions including feed temperature and permeate pressure on permeability and selectivity is discussed. A permeate flux of 4.93 kg/m2h with salt rejection of 99.9% could be achieved at a feed temperature of 50 °C and a vacuum of 0.04 bar. Apparent diffusion coefficients of water at various permeate pressure and feed temperature are calculated. The most effective parameter was feed temperature.

Keywords: *Composite Membrane, Desalination, Operational Conditions, Pervaporation*

INTRODUCTION

In the past few decades, water scarcity has become one of the most serious challenges globally in the society. Over 2.3 billion people on the Earth live in the water-stressed areas, and this number is expected to increase to 3.5 billion by 2025 [1]. In order to maintain the sustainable development of economy and environment, Global Water Partnership (GWP) was established in 1996 to develop Integrated Water Resources Management, focusing on the adjustment, management and development of water, land and related resources [2]. Technologies for water desalination have been developed in two approaches: one is based on distillation, including multi-stage flash distillation and multiple-effect distillation; the other is membrane-based desalination, including nanofiltration, vacuum membrane distillation and reverse osmosis [3]. In recent years, membrane separation processes become more and more popular in desalination because the energy requirements are lower than that of the traditional distillation processes[4]. As a result, Membrane processes are environmental friendly since the membranes are made of relatively simple and non-harmful materials. A large number of polymers can be used to prepare membranes. In general, a high salt rejection and permeation flux are required for desalination with membrane processes. Until now, RO has been one of the most important membrane processes for desalination in industrial scale [5].

* Corresponding Author Email: soheill.azadi.eng@gmail.com

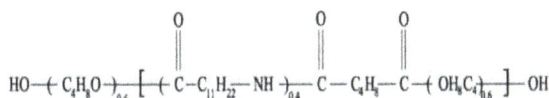

Fig. 1. Chemical structure of PEBAX-1657 [10].

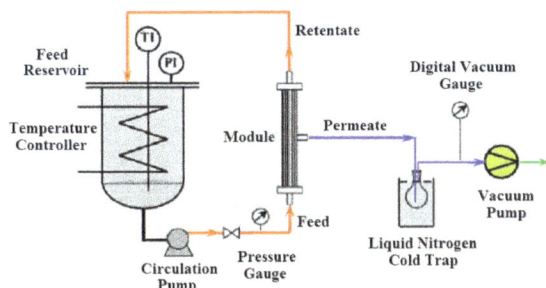

Fig. 2. Schematic drawing of the pervaporation unit.

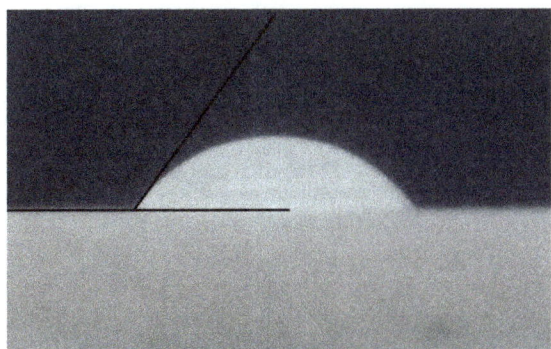

Fig. 3. Water contact angle of hybrid membrane.

However, the wide spread use of RO process is restricted by some operating conditions, high energy cost and easy fouling [6,7]. To deal with high-salinity water, an extremely high operating pressure is needed in RO process. Comparing with the membrane distillation, pervaporation desalination using hydrophilic materials can effectively reduce membrane fouling and maintain membrane separation performance. Currentefforts are focusing on pervaporation with the hope touse this technology for sea water desalination in the future [8,9].

The composite membranes used in this study were made of poly(ether block amide) (Pebax-1657) which is a hydrophilic polymer and have been shown in Fig. 1. Pebax is copolymer with soft and flexible segments, which make it useful in many areas, including medical, textile and membrane applications. The Pebax® polymer used in this work had high sorption of water vapor [10]. However, very little research is done related

to Pebax for desalination applications. Therefore, the performance of PEBA-based nanocomposite membranes for desalination of high-salinity water was studied in this work.

EXPERIMENTAL

Hybrid Membrane Synthesis and Characterization

The PEBA/PAN/PE, PEBA/PSF/PE and PEBA+NaX/PSF/PE composite membranes was synthesized via n phase inversion route. The porous PAN and PSF substrate was prepared by spin coating method. Solution of PAN or PSF was casted by the spin coating method on the PE and immediately submerged into a de-ionized water bath as non-solvent at 40°C. Finally, the prepared membrane was dried at room temperature. Then, Polymer solution containing 8% PEBA granules or PEBA+NaX and 92% de-ionized water and ethanol (30/70) was prepared by stirring for 8 h. the prepared solution was casted on prepared porous PAN or PSF substrate in the last process and then put in oven at 50 °C for 48 h to evaporating of solvents. Pebax-1657 was chosen to enable greater interaction with H2O molecules through H-bonding as this grade of the polymer contains 40% amide groups. Chemical structure of PEBAX-1657 was shown in Fig.1. Pebax-1657 was purchased from Arkema, France. Reagent grade chemicals Polyacrylonitrile (PAN), polyester (PE), Polysulfone (PSF) and NaX were obtained from Sigma-Aldrich and used without further purification.

Pervaporation Testing

The pervaporation experiments were carried out at variable temperature (from 25 to 29 °C) on a laboratory scale pervaporation unit as shown in Fig. 2. The membrane was placed in the middle of a pervaporation cell with an effective surface area of the membrane of 21.237 cm2. During the experiment, the feed solution was preheated in a water bath to a desired temperature and pumped to the pervaporation cell using a pump.

The pressure on the permeate side of the membrane cell was maintained at constant pressure with a vacuum pump. Permeate was collected in a dry-ice cold trap that its temperature was under −196 °C. A thermometer fixed in the feed chamber was used to measure the operating temperature of feed solution and the feed temperature was varied from 30 to 50 °C in this study. Kashan City's water containing 0.076 wt% NaCl (760 TDS) was used as

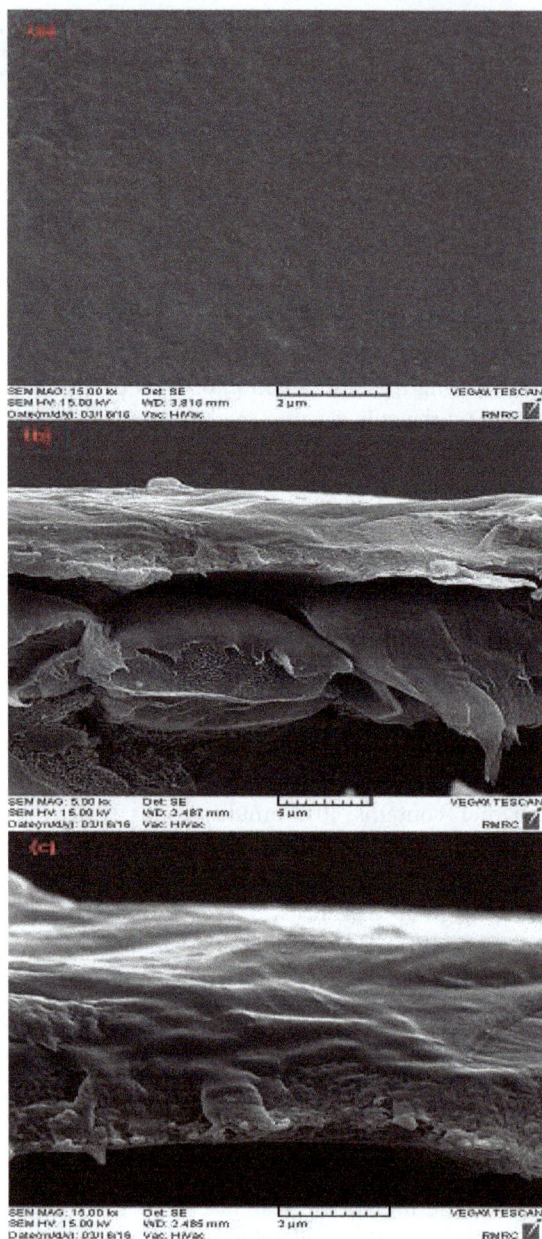

Fig. 4. SEM of the top surface of the PEBA1 composite membrane (a), the cross-sectional SEM image of PEBA/PAN layers (b), and a closer cross-sectional SEM image of PEBA layer of the PEBA1 composite membrane (c)

the feed solution. The pervaporation desalination performance of composite membranes was evaluated by measuring water flux and salt rejection. The water flux (J) was determined from the mass (M) of the permeate collected in the cold trap, the effective membrane area (A) and the experimental time (t) using the following equation:

$$J = \frac{M}{At} \tag{1}$$

The salt concentrations of the feed (C_f) and the permeate (C_p) were determined from the conductivity measured with an AZ® 8361.Cond. & TDS meter. The saltrejection (R) was determined by the following equation:

$$R = \frac{C_f - C_p}{C_f} \times 100\% \tag{2}$$

Diffusion coefficient is an important factor to estimate the diffusion of the penetrants through membranes and permeation flux. Based on Fick's law, the apparent diffusion coefficient can be calculated by the equation [11]:

$$D_i = \frac{J_i \delta}{C_{i,f}} \tag{3}$$

Where D_i is diffusion coefficient (m2/s), J is the permeate flux, is the membrane thickness, and $C_{i,f}$ is the concentration of component i in the feed.

SEM Analysis

Composite membrane was investigated by scanning electron microscopy (SEM) (VEGA \\TSCAN-LMU). SEM images of the membranes reported the cross-sectional and surface morphologies of the prepared composite membrane.

Swelling Properties

PEBA side of composite membrane was in contacted with water at room temperature for 24 h to reach the absorption equilibrium. Wet layer was then dried with filter paper carefully and quickly weighed within 10 s (wet layer) following which the membrane was dried in a vacuum oven at 50 °C for overnight and then weighed (dried layer). Two tests were conducted on each sample. The swelling degree (S) of membrane was calculated according to:

$$S = \frac{W_s - W_d}{W_s} \times 100\% \tag{4}$$

Where W_s and W_d are the weight of wet and dry membrane layers, respectively [12].

Contact Angle

The hydrophilic properties of membrane samples were assessed by capturing of water droplet. Static contact angles were measured by the sessile drop method. A 6 μL water drop was formed on the

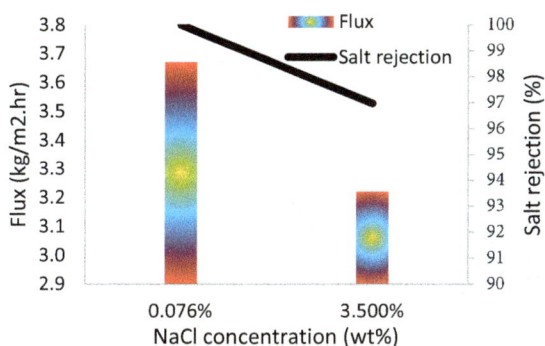

Fig. 5. Effect of feed concentration on water flux and salt rejection (feed temperature 40 °C, vacuum 0.04 bar) for PEBA1 composite membrane.

Table 1. Membranes composition and codes.

Membrane composition	Code
PEBA/PAN/PE	PEBA1
PEBA/PSF/PE	PEBA2
PEBA+NaX/PSF/PE	PEBA2X

Table 2. Contact angle and swelling of composite membranes.

Membrane	Contact angle	Swelling
PEBA1	53	21.8%
PEBA2	53	21.8%
PEBA2X	55	21.2%

levelled surface of the membrane for contact angle measurements. Samples were fully dried before testing and five different locations of each sample were selected for testing.

RESULTS

SEM

SEM images of surface and cross-section of the composite membrane are reported in Fig 4. Fig. 4 (a) exhibits the surface of the PEBA1 composite membrane manufactured from pure PEBA 1657 which is dense, homogeneous, defect free, without any leak, and suitable for gas separation.

Fig. 4 (b,c) represents the cross-section of the composite membrane indicating a PAN ultra-porous substrate with thickness of about 21 μm without any defect or split formed on PE non-woven paper layer. Also a PEBA dense layer with thickness of about 11 μm shaped on PAN substrate.

Swelling and Contact Angle

Table 2 shows the contact angle and swelling of composite membranes. As can be seen, the swelling of the composite membrane has been slightly reduced, indicating suppressed swelling due to crosslinking among PEBA and NaX. The composite membrane remained hydrophilic. This could be due to the contribution of hydrophilic —OH groups from.

Salt Rejection

Each experiment was run for 4 h. At the end of each experiment, the downstream (permeate side) of the membrane cell was flushed with a known amount of de-ionized water and the conductivity of this stream was measured to check the salt leaking. In the study, hybrid membranes remained cleanand there was no evidence of salt precipitation onthe

permeate side of the membrane. The results during pervaporation testing were reproducible, with the variation generally within±0.2 kg/m2 h for water flux and ±1.0% for salt rejection.

Fig. 5 shows the pervaporation desalination performance of PEBA hybrid membranes with same thickness at a feed temperature of 40 °C and a vacuum of 0.04 bar. All prepared membranes had the same amount of PAN (40 wt% with respect to PEBA. Overall, the PEBA based hybrid membranes demonstrated good desalination performance with high flux (>3 kg/m2 h) while maintaining a high salt rejection (>93%). The salt rejection increased with the filler content and achieved >99%. The salt rejection increased with the filler content and achieved >99%. The incorporation of NaX nanoparticles in the polymer chain may disrupt the polymer chain packing and therefore lead to reduced free volume radius and consequently a high salt rejection [13].

Effect of Feed Concentration

Fig. 6 shows the effect of salt concentration in the feed solution on separation performance of aqueous salt solution at various feed temperatures. At room temperature, salt concentration has negligible effect on water flux. At a higher temperatures (50 °C), the water flux decreases with increasing salt concentration. This increase became more significant as the feed temperature was increased further to 60 °C.

Feed concentration is believed to directly affect the sorption of its components at the liquid/membrane interface [14]. That is, the concentration of the components in the membrane tends to increase with its increase in the feed concentration. Since diffusion in the membrane is concentration dependent, the permeate flux generally increases

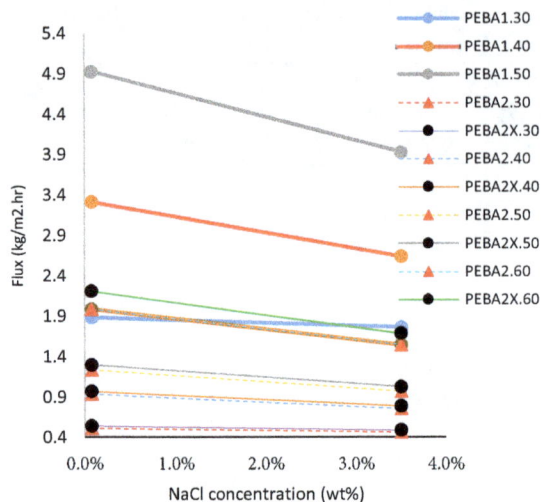

Fig. 6. Effect of feed concentration on water flux (vacuum 0.04 bar).

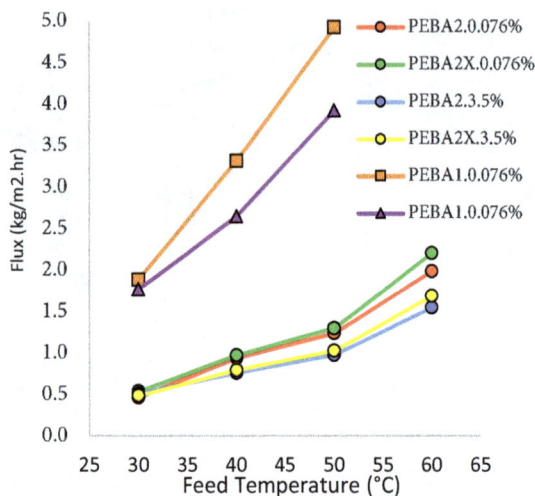

Fig. 7. Effect of feed temperature on water flux (vacuum 0.04 bar).

with the bulk feed concentration [15].

As the salt concentration increased from 0.076 to 3.5 wt%, the water concentration decreased from 99.924 to 96.5 wt%. At room temperature, this decrease in water concentration may not have any effect on diffusion within the membrane as the majority of the feed is water and there is no major difference of water vapor pressure at salt concentration range of 96.5–99.924%. It is therefore expected that the diffusivity of the membrane towards water remained constant at room temperature.

Therefore, there was no or negligible change on the flux. On the otherhand, at the higher temperatures, as the vapor pressure is exponentially related to the temperature, differences in bulk feed water concentration would have pronounced effect on the water concentration in the membrane surface, and consequently affect the diffusivity and flux. Therefore, it is expected that, at the higher temperature, that increasing salt concentration would lead to adecrease in diffusivity in the membrane due to the decreased water concentration.

Effect of Feed Temperature

Fig. 7 shows the effect of feed temperature on the pervaporation desalination performance of hybrid PEBA composite membranes at a vacuum 0.04 bar. For all feed concentrations, there was an exponential increase of water flux when the feed temperature increased from 30 to 60 °C. A high water flux of 7.63 kg/m2 h was achieved at the feed temperature of 60 °C. This is not surprising, as firstly, the driving force

for the pervaporation process is the partial vapor pressure difference of permeant between the feed and permeate conditions. As the feed temperature increased, the water vapor pressure on the feed side increased exponentially. As the vapor pressure on the permeate side was held constant, the increasing vapor pressure in feed led to an increase in the driving force and consequently the water flux. Secondly, an increase in temperature also raises the diffusion coefficient for transport through the membrane, making it easier for the transport of the water molecules.

This is confirmed by the diffusion coefficient results as shown in Table 2. As can be seen, there is an increasing trend in the diffusivity coefficient of water in the hybrid membrane as the feed temperature rose. In addition, the mobility of the polymer chains also increased with the feed temperature, which led to the increase of the free volume of the membranes. According to the free volume theory [16], the thermal motion of polymer chains in the amorphous region creates momentary free volumes. As the temperature increases, the frequency and amplitude of the chain motion increase and the resulting free volumes become larger. Consequently, water molecules which have smaller size can diffuse through these free volumes more easily. Therefore, the water flux increases.

Effect of Permeate Pressure

Permeate pressure is another important operating parameter as a high vacuum is directly related to a high energy cost. Theoretically, the maximum flux is achieved at zero absolute permeate pressure.

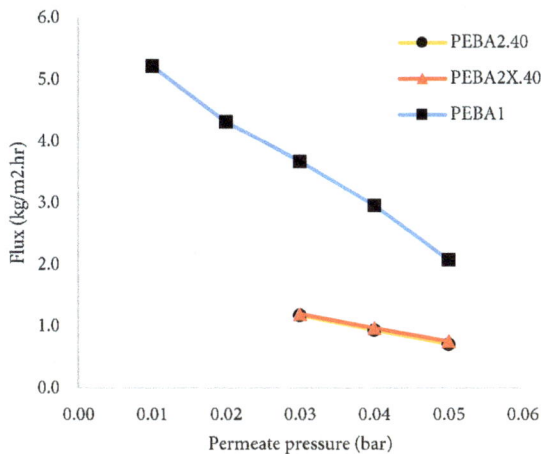

Fig. 8. Effect of vacuum on water flux (feed temperature 40 °C).

Table 3. Apparent diffusion coefficients of water at various feed temperatures.

Feed temperature (C°)	30	40	50
diffusion coefficient (10^{-11} m^2/s)	0.525	0.922	1.370

Table 4. Apparent diffusion coefficients of water at various permeate pressure (feed temperature 40 °C).

Permeate pressure (bar)	0.01	0.02	0.03	0.04	0.05
diffusion coefficient (10^{-11} m^2/s)	1.451	1.201	1.020	0.823	0.578

Fig. 8 shows the effect of permeate pressure on water flux. Generally, the water flux decreased as the permeate pressure is increased since there is a decrease of driving force for mass transport.

For pervaporation processes, the driving force is provided by the vapor pressure difference between the feed and permeate side of the membrane. With increasing permeate pressure (i.e. decreasing vacuum), as the feed side pressure remains unchanged, the transmembrane vapor pressure difference is increased. This leads to a decreased driving force and consequently water flux. It was observed that the water flux dropped down to less than 0.5 kg/m2 h when the permeate pressure increased to 0.07 bar. At temperature 40 °C, the saturation vapor pressure of water is about 0.073 bar [17]. When the permeate pressure is increased 0.07 bar, the driving force for water vaporization approaches zero, leading to near zero net evaporation and consequently the low mass transport of water. Table 3 presents the diffusion coefficient of water at various permeate pressure. Decreasing diffusion coefficient with permeate pressure indicates that the permeation process is mainly controlled by diffusion through the hybrid membrane. As permeate pressure increased above 0.07 bar, the diffusion coefficient dropped significantly, by nearly90%, indicating the diffusion of water has been greatly reduced.

CONCLUSIONS

Pervaporation under various operating conditions was carried out to evaluate the separation performance of aqueous salt solution through the hybrid PEBA membranes. Membrane PEBA1 showed the best performance leading to

interesting fluxes and salt rejection values for feed salt concentrations of 0.076 and 3.5 wt%. A high water flux of 4.93 kg/m2h could be achieved for PEBA1 composite membrane at a feed temperature of 50 °C and a vacuum of 0.04 bar, while for PEBA2 and PEBA2X this quantity were 1.24 and 1.3 kg/m^{2}h, respectively. Under all operating conditions, salt rejection remained high (up to 99.9%), indicating salt rejection performance of hybrid PEBA membranes is independent of the operating conditions due to the non-volatile nature of NaCl. High feed temperature and high vacuum had a significant enhancing effect on the water flux and diffusivity coefficients of water due to the increased driving force and increased free volume of the membrane. The effect of feed concentration had differing impacts depending on the operating temperature. At low feed temperatures, the salt concentration in the feed solution had little or negligible effect on water flux and diffusion coefficients. However, at high feed temperature (50–60 °C), feed flux and diffusivity of water decreased with increasing salt concentration due to the decreased water vapor pressure and consequently water concentration in the membrane surface.

ACKNOWLEDGMENT

The authors would like to acknowledge the Mr. Azadikhah for the financial support of this work.

CONFLICT OF INTEREST

The authors declare that there are no conflicts of interest regarding the publication of this manuscript.

REFERENCES

1. Elimelech M, Phillip WA. The Future of Seawater Desalination: Energy, Technology, and the Environment. Science. 2011;333(6043):712-7.

2. Rogers P, Bhatia R, Huber A. Water as a social and economic good: How to put the principle into practice: Global Water Partnership/Swedish International Development Cooperation Agency Stockholm, Sweden; 1998.

3. Matteo, G.,Seawater desalination: thermal desalination vs membrane, Separation Processes Laboratory, (2015).

4. Wang LK, Chen JP, Hung Y-T, Shammas NK. Membrane and desalination technologies: Humana Press; 2010.

5. Cath TY, Childress AE, Elimelech M. Forward osmosis: Principles, applications, and recent developments. Journal of Membrane Science. 2006;281(1–2):70-87.

6. Kwak SY, Kim SH, Kim SS. Hybrid organic/inorganic reverse osmosis (RO) membrane for bactericidal anti-fouling. 1. Preparation and characterization of TiO2 nanoparticle self-assembled aromatic polyamide thin-film-composite (TFC) membrane. Environ Sci Technol. 2001;35(11):2388-94.

7. Chen KL, Song L, Ong SL, Ng WJ. The development of membrane fouling in full-scale RO processes. Journal of Membrane Science. 2004;232(1–2):63-72.

8. Kuznetsov YP, Kruchinina EV, Baklagina YG, Khripunov AK, Tulupova OA. Deep desalination of water by evaporation through polymeric membranes. Russian Journal of Applied Chemistry. 2007;80(5):790-8.

9. Xie Z, Ng D, Hoang M, Duong T, Gray S. Separation of aqueous salt solution by pervaporation through hybrid organic–inorganic membrane: Effect of operating conditions. Desalination. 2011;273(1):220-5.

10. Hamouda SB, Boubakri A, Nguyen QT, Amor MB. PEBAX membranes for water desalination by pervaporation process. High Performance Polymers. 2011;23(2):170-3.

11. Villaluenga JPG, Godino P, Khayet M, Seoane B, Mengual JI. Pervaporation of Alcohols and Methyl tert-Butyl Ether through a Dense Poly(2,6-dimethyl-1,4-phenylene oxide) Membrane. Industrial & Engineering Chemistry Research. 2004;43(10):2548-55.

12. Wang Q, Lu Y, Li N. Preparation, characterization and performance of sulfonated poly(styrene-ethylene/butylene-styrene) block copolymer membranes for water desalination by pervaporation. Desalination. 2016;390:33-46.

13. Xie Z, Hoang M, Ng D, Doherty C, Hill A, Gray S. Effect of heat treatment on pervaporation separation of aqueous salt solution using hybrid PVA/MA/TEOS membrane. Separation and Purification Technology. 2014;127:10-7.

14. Jiraratananon R, Chanachai A, Huang RYM, Uttapap D. Pervaporation dehydration of ethanol–water mixtures with chitosan/hydroxyethylcellulose (CS/HEC) composite membranes: I. Effect of operating conditions. Journal of Membrane Science. 2002;195(2):143-51.

15. Wang Q, Li N, Bolto B, Hoang M, Xie Z. Desalination by pervaporation: A review. Desalination. 2016;387:46-60.

16. Burshe MC, Sawant SB, Joshi JB, Pangarkar VG. Sorption and permeation of binary water-alcohol systems through PVA membranes crosslinked with multifunctional crosslinking agents. Separation and Purification Technology. 1997;12(2):145-56.

17. Shakhashiri BZ. Chemical demonstrations: A handbook for teachers of chemistry: Univ of Wisconsin Press; 1985.

11

Silica-Supported Copper Oxide Nanoleaf with Antimicrobial Activity Against Escherichia Coli

Mohsen Moghimi, Mohammad Ghorbanpour*, Samaneh Lotfiman

Chemical Engineering Department, University of Mohaghegh Ardabili, Ardabil, Iran

ABSTRACT

In this research, a simple and fast method was employed to synthesize CuO nanoleaves/silica gel nanocomposites (CuO/SGn), which is a cost effective antimicrobial material. $CuSO_4.5H_2O$ is the only raw material used in CuO/SGn production through the molten salt method. The structure and morphology of the nanocomposites were characterized by DRS, XRD, and SEM. The copper size in CuO/SG was found to be dependent on the immersion time in molten salt. SEM images revealed smaller-sized particle leaves, from a sample obtained after longer immersion time.The antimicrobial activity of CuO/SGn was investigated against Escherichia coli. The produced CuO/SGn showed inhibitory effect against E. coli. However, the growth-inhibitory effect depends on the copper content and size. Lower loading of CuO nanoparticles in 10 min immersion time resulted in less antibacterial activity (73.33%) and sample obtained from longer immersion time demonstrated higher antibacterial activity (up to 99.96%). The maximum amount of released Cu ions from nanocomposites produced in 90 min was 12.2 ppm after 6 h. Furthermore, the minimum release of Cu ions was observed by 3 h for 40 min nnanocomposite.

KEYWORDS: *Antimicrobial Activity, Copper Oxide, Nanoleaf, Silica*

INTRODUCTION

Broad applications of metal oxide in technologically-driven countries, particularly in the field of antibacterial materials, have resulted in extensive industrial uses for metal oxide, including metal oxide in water treatment processes and in food production and storage. Metal nanoparticles such as copper, silver, and zinc are well known for their excellent antibacterial activity [1-5]. With their low toxicity, high chemical and thermal stability, high and long-lasting antibacterial activity, metal nanoparticles are suitable as bactericidal agents in applications including water disinfection [1]. On the other hand, typical water disinfection has notable drawbacks, and post-treatment

removal of nanoparticles greatly reduces the processss typical benefits. It is widely acknowledged that removing nanoparticles from water is a time consuming and expensive procedure. The solution has been to immobilize the nanoparticles on an inert support. This helps to eliminate expensive phase separation procedures. Furthermore, several reports have observed antibacterial activity of nanoparticles from immobilized nanoparticles on various supports, such as clay [2,3,5], silica gel [4] and metal foil [5]. In this case, copper oxide serves as an excellent antibacterial material against bacteria, fungus, and algae because of its low price, nontoxicity, and facile preparation. Recently, different coating technologies have been developed

* Corresponding Author Email: ghorbanpour@uma.ac.ir

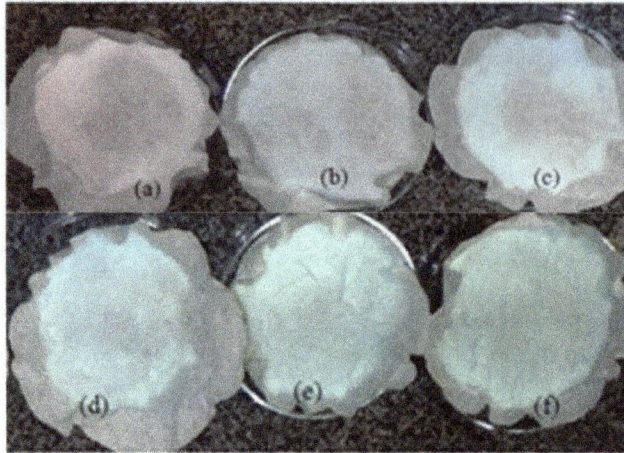

Fig. 1. Appearance and color of (a) silica gel and CuO/SGn nanocomposites
with immersion times of (b) 10, (c) 20, (d) 40, (e) 60, and (f) 90 min.

to produce Cu/nanocomposites as antibacterial materials [7,8].

Various synthesis methods have been researched for the preparation of nanostructures, such as hydrothermal, thermal oxidation, solgel, wet-chemical, electrochemical deposition, sputtering, thermal evaporation, thermal relaxation, alkaline ion exchange and anodizing [3, 9-17]. Different morphological nanostructures of CuO have been reported in recent years, such as ribbon-like CuO [12], spindle-like 314] and 2D leaf-like [14,15]. For example, Zhu et al. [16] synthesized one-dimensional CuO nanowires by thermal transformation of $Cu(OH)_2$ nanowires, made from $Cu_7C_{14}(OH)10H_2O$. Liang et al. [10] successfully synthesized 2D CuO nanoleaves by microwave heating of an aqueous solution containing copper salt and sodium hydroxide. Most reported preparation methods of CuO nanostructure still exhibit several deficiencies, such as time constraints, special instrumentation requirements, complex operation, and high cost. Thus, the development of a simple, rapid, and relatively inexpensive technique to prepare CuO nanostructures is necessary and attractive.

This research introduces a novel method for covering a surface in order to produce antibacterial material via the molten salt method. Synthesis of CuO/SGn through molten salt involves selecting a molten salt like $CuSO_4$ for use as a solvent in a medium for preparing complex oxides from the salts constituent materials. For this reason, $CuSO_4$ and silica gel were heated to the salts estimated melting point. This method can easily be expanded for large-scale commercial production. The characteristics of the produced powder are controlled by selecting the heating time. Then, the reacted mass is cooled to room temperature and washed with water to remove the salt. A complex oxide powder is obtained after drying. In contrast with other common methods, expensive equipment and chemical reagents are unnecessary. Therefore, minimum process time, high purity, and inexpensive substrates are the main advantages of our product, enabling it to be produced on an industrial scale.

EXPERIMENT
Materials

All reagents were of analytical grade and were used as received without further refinement. Also, all aqueous solutions were prepared with distilled water. Silica gel, $CuSO_4.5H_2O$, Mueller-Hinton agar broth, and nutrient agar were purchased from Merck Co., Inc. (Tehran, Iran). The bacterial strain used for the antibacterial activity was gram-negative Escherichia coli (PTCC 1270), received from the Iranian Research Organization for Science and Technology (Tehran, Iran).

CuO/SGn Preparation

Silica gel was immersed in the melted $CuSO_4.5H_2O$ at 550–560 °C for 10, 20, 40, 60, and 90 min. This operation was conducted using 5 g of silica gel and 5 g of $CuSO_4.5H_2O$. After this step, the silica gel was adequately washed with distilled water and sanitation. After dissolving, the product was dried in an oven.

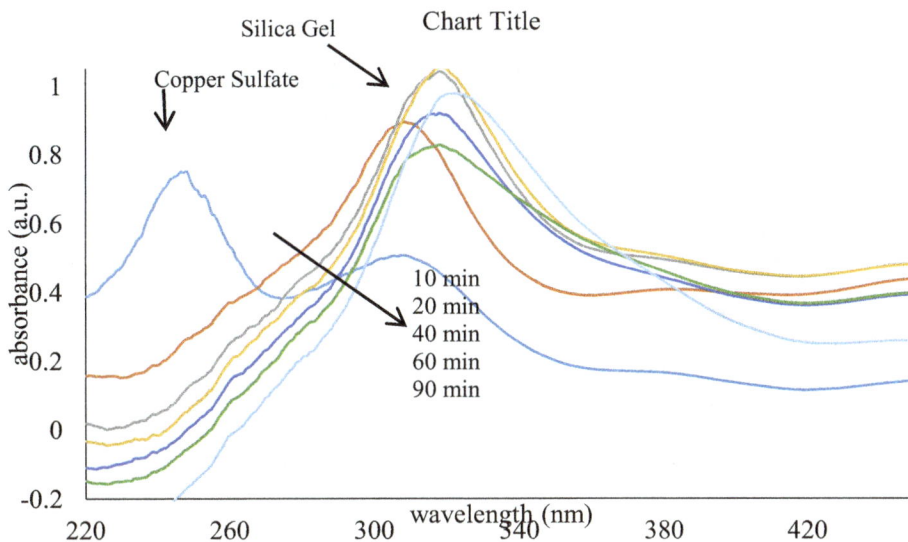

Fig. 2. The DRS graphs of the parent silica gel, copper sulfate, and the final products of CuO/SGn processing.

Characterization

Synthesized samples were analyzed by X-ray diffractometer, XRD (Philips PW 1050, Germany). The patterns were registered in the 2θ range from 10° to 60° with a scanning step size of 0.05°. For this purpose, Silica gel was used as a reference. The morphology of the samples was investigated by scanning electron microscope (LEO 1430VP, Germany). Absorption spectra of nanocomposites was measured by a UV-visible diffuse reflectance spectrophotometer (Sinco Model S4100, Korea), in the wavelength range 200–700 nm.

Antibacterial Activity

The antibacterial activity of the samples was studied by plate count technique of *E. coli*. 50 mg of powder was added to bacterial cultures in 5 ml sterile Mueller-Hinton broth. The cultures were then incubated at 37 °C on a rotary shaker for 24 h. Growth inhibition with time was followed by plating 100 μL of the treated cultures on nutrient agar plates. Bacterial colonies were counted and compared with the control after 24 h incubation at 37 °C. The experiment was repeated twice to confirm the data. The antibacterial effect was calculated using -Eq. 1:

Mortality (%) = (B-C)/B × 100   (1)

Where B is the mean number of bacteria in the control samples (CFU/sample) and C is the mean number of bacteria in the treated samples (CFU/sample) [2,3].

Leaching Test

In order to evaluate the stability of the nanocomposites, leaching tests were performed. For each composite material, 0.2 g was immersed in 10 ml of distilled water and vigorously shaken in a shaking water bath (30 °C, 200 rpm) for 3 and 6 h. Supernatant from each test tube was collected after 3 and 6 h by centrifugation at 4000 rpm for 10 min. Copper ion release from the nanocomposites was qualitatively determined through atomic absorption spectroscopy analysis (Varian/AAS, 20BQ, USA) [2,3].

RESULTS AND DISCUSSION
Characterization

The nanocomposites consisting of leaf-like copper nanoparticles supported by pure silica gel were prepared by immersion of silica gel in molten copper sulfate. The chemical reaction is as follows:

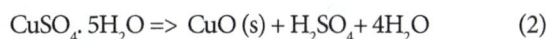

$$CuSO_4 \cdot 5H_2O => CuO (s) + H_2SO_4 + 4H_2O \qquad (2)$$

High temperature prompted the chemical reaction between $CuSO_4$ and water, which produced an acidic solution of solid CuO and H_2SO_4. As mentioned before, this method is much easier than the common synthesizing approaches for CuO nanoparticles production [12-17]. Furthermore, this method generates inexpensive nanoparticles that have been one of the most important parameters, which has widespread applications.

Appearance and color of the nanocomposites are shown in Fig. 1. As shown in this figure, the parent

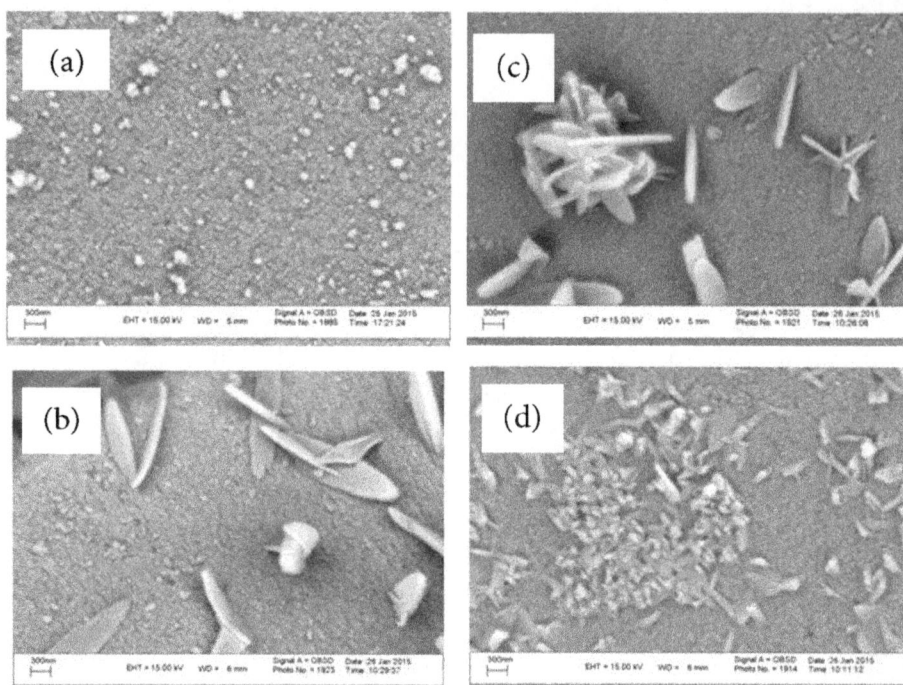

Fig. 3. Typical SEM image of (a) parent silica gel and CuO/SGn produced by immersion times of (b) 40, (c) 60, and (d) 90 min.

Table 1. Copper concentration in water at different contact times between CuO/SGn and the aqueous media.

Ion exchange time (min)		Copper Concentration in Water (ppm)			
		MML	40	60	90
Contact time (h)	3 h	0	4.3	7.7	11.65
	6 h	0	6.1	7.9	12.2

silica gel color was white. After nanocomposite production, the color of composites changed to greenish. Fig. 1 shows that increasing immersion time in molten salt enhances nanocomposites color change. The color changes are a result of the amount of copper oxide loaded onto the silica gel or the aggregation of nanoparticles on the surface of the silica gel [17]. Other studies reported that nanoparticles coated with Ag, Zn, and Cu on bentonite (white) or silica gel (no color) change substrate color [2,4]; for example, coating nanoparticles of ZnO and Ag on bentonite and silica gel changes the parents color to cream [2] and yellow-brown [4], respectively. Therefore, color change is a sign for loading of nanoparticles on silica gel.

Fig. 2 shows that silica gel has a maximum absorption peak around 304 nm and pure CuSO$_4$.5H$_2$O has two maximum absorption peaks around 243 and 304 nm. After composite formation, the absorption band of pure CuSO4.5H2O gradually red shifts from 304 nm to a wavelength of 314 to 320 nm, and the absorption band at 243 nm has disappeared. These changes indicate that copper oxide nanoparticles have formed on the silica gel surface. On the other hand, longer immersion of silica gel in the molten copper sulfate results in the broadening of the peak between 314–320 nm. According to the previous findings, the broadening absorption peak of a nanoparticle corresponds to the wider dispersion of synthesic nanoparticles sizes [18, 4]. Therefore, it can be concluded that the increased immersion time resulted in the formation of nanoparticles of various sizes.

To better characterize CuO/SGn, its morphology was observed by SEM at different immersion

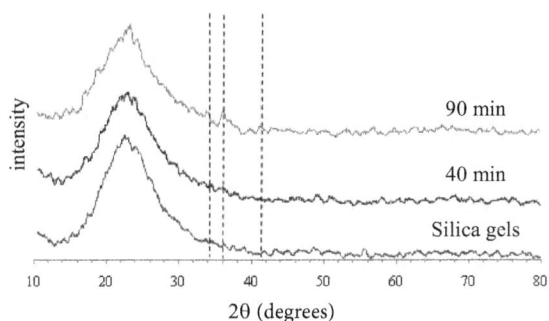

Fig. 4. XRD patterns of silica gel and CuO/SGn

Table 2. Antibacterial activity of CuO/SG nanocomposites Ionexchangetime.

Ion exchange Time (min)	Number of Colonies (CFU)	Mortaliy (%)
Silica gel	7.4×10^{14}	
10	2.0×10^{14}	73.33
20	2.5×10^{12}	99.67
40	1.4×10^{12}	99.81
60	9.0×10^{11}	99.90
90	3.0×10^{11}	99.96

times. Fig. 3 shows SEM images of the CuO/SGn. No leaf-like CuO particles were deposited on the silica surface from samples with a process time of less than 40 min. By increasing immersion time to 40 min, leaf-like shapes appeared (Fig. 3b), and the amount of nanoleaves increases with higher annealing time (Figs. 3c and 3d). The silica surface was relatively covered with CuO leaves at 90 min immersion time (Fig. 3d). Interestingly, the produced particles show leaf-like morphology of similar size on the silica surface. Fig. 3 also shows that the samples obtained from longer immersion time have smaller-sized particle leaves.

The diffraction peaks at approximately 35.5° and 38.7° were indexed to the monoclinic structure of the CuO phase (Fig. 4). Thus, CuO formed on the silica surface, as expected from other studies [12, 15]. The XRD pattern of CuO/SGn shows very weak diffraction peaks corresponding to the crystallites of the copper species, which may result from the high dispersion of copper species on the substrates surface.

Leaching Test

Table 1 shows the amount of copper released into the aqueous media is a function of the contact time between phases. CuO/SGn prepared with immersion time of 90 min releases a maximum amount of copper after 6 h contact with water. It should be mentioned that the condition of this test is very violent compared with the laminar flow of water in water handling operations. The obtained results indicate that these materials pose no risk to drinking water treatment since the leached metals are quite limited and in acceptable concentrations.

Antibacterial Assay

The antibacterial properties of CuO/SGn were studied against gram-negative *E. coli*. The effect of synthesis time on the antibacterial properties of

the CuO/SGn is presented in Table 2. Parent silica gel showed no antibacterial activity. On the other hand, all fabricated composites containing CuO nanoparticles displayed antibacterial properties against *E. coli* with high mortality. The total numbers of bacteria for samples were approximately 7.5×10^{14} CFU/ml at zero contact time. These amounts declined considerably (more than 99% for *E. coli*) for doped silica-treated samples with CuO nanoparticles, but increased for control samples (undoped silica).

As can be seen from Table 2, bacterial growth was increasingly inhibited with enhanced immersion time of the CuO/SGn samples in molten salt. Researchers acknowledge that the antibacterial activity of metallic nanoparticles is strongly correlated with their sizes and loading [2,3,7]. The CuO/SGn with 10 min immersion time showed less antibacterial activity. This may result from lower loading of CuO nanoparticles. The samples prepared with longer immersion time revealed more strongly inhibited bacterial growth (Table 2).

The cell walls of viable bacteria usually are negatively charged due to functional groups such as carboxylates present in lipoproteins at the surface. CuO/SG attracts bacteria by electrostatic forces and immobilizes them on the surface. CuO may also disassociate and directly exert its antimicrobial effect on the bacteria in the dispersion [18]. As mentioned previously, increasing synthesizing time enhances antibacterial properties. The SEM images demonstrate that longer process time results in loading more nanoparticles. On the other hand, release test for the sample produced in 90 min shows greater antibacterial activity. These results support each other and concur with others findings (2-4).

CONCLUSION

In this research, CuO/SG was successfully synthesized by a fast and simple method. The results demonstrated that immersion time affected nanoparticle size, and longer immersion time produced various sizes of CuO nanoparticles. SEM images revealed smaller-sized particle leaves, from a sample obtained after longer immersion time. The maximum amount of released CuO was found after 6 h contact with water, with no risk to drinking water. All produced composites containing CuO nanoparticles displayed antibacterial properties against *E. coli* with high mortality. Lower loading of CuO nanoparticles in 10 min immersion time resulted in less antibacterial activity and sample obtained from longer immersion time demonstrated higher antibacterial activity.

CONFLICT OF INTEREST

The authors declare that there are no conflicts of interest regarding the publication of this manuscript.

REFERENCES

1. Appendini P, Hotchkiss JH. Review of antimicrobial food packaging. Innovative Food Science & Emerging Technologies. 2002;3(2):113-26.

2. Pouraboulghasem H, Ghorbanpour M, Shayegh R, Lotfiman S. Synthesis, characterization and antimicrobial activity of alkaline ion-exchanged ZnO/bentonite nanocomposites. Journal of Central South University. 2016;23(4):787-92.

3. Pourabolghasem H, Ghorbanpour M, Shayegh R. Antibacterial Activity of Copper-doped Montmorillonite Nanocomposites Prepared by Alkaline Ion Exchange Method. Journal of Physical Science. 2016;27(2):1-12.

4. Payami R, Ghorbanpour M, Parchehbaf Jadid A. Antibacterial silver-doped bioactive silica gel production using molten salt method. Journal of Nanostructure in Chemistry. 2016;6(3):215-21.

5. Gilani S, Ghorbanpour M, Parchehbaf Jadid A. Antibacterial activity of ZnO films prepared by anodizing. Journal of Nanostructure in Chemistry. 2016;6(2):183-9.

6. Ghorbanpour M, Lotfiman S. Solid-state immobilisation of titanium dioxide nanoparticles onto nanoclay. Micro & Nano Letters [Internet]. 2016; 11(11):[684-7 pp.]. Available from: http://digital-library.theiet.org/content/journals/10.1049/mnl.2016.0259.

7. Top A, Ülkü S. Silver, zinc, and copper exchange in a Na-clinoptilolite and resulting effect on antibacterial activity. Applied Clay Science. 2004;27(1–2):13-9.

8. Stanić V, Dimitrijević S, Antić-Stanković J, Mitrić M, Jokić B, Plećaš IB, et al. Synthesis, characterization and antimicrobial activity of copper and zinc-doped hydroxyapatite nanopowders. Applied Surface Science. 2010;256(20):6083-9.

9. Ghorbanpour M. Amine Accessibility and Chemical Stability of Silver SPR Chips Silanised with APTES via Vapour Phase Deposition Method. Journal of Physical Science. 2016;27(1):39-51.

10. Zhen-Hua L, Ying-Jie Z. Microwave-assisted Synthesis of Single-crystalline CuO Nanoleaves. Chemistry Letters. 2004;33(10):1314-5.

11. Ghorbanpour M. Fabrication of a New Amine Functionalised Bi-layered Gold/Silver SPR Sensor Chip. Journal of Physical Science. 2015;26(2):1-10.

12. Ke F-S, Huang L, Wei G-Z, Xue L-J, Li J-T, Zhang B, et al. One-step fabrication of CuO nanoribbons array electrode and its excellent lithium storage performance. Electrochimica Acta. 2009;54(24):5825-9.

13. Zhang X, Wang G, Liu X, Wu J, Li M, Gu J, et al. Different CuO Nanostructures: Synthesis, Characterization, and Applications for Glucose Sensors. The Journal of Physical Chemistry C. 2008;112(43):16845-9.

14. Yan Z, Jingzhe Z, Yunling L, Dechong M, Shengnan H, Linzhi L, et al. Room temperature synthesis of 2D CuO nanoleaves in aqueous solution. Nanotechnology. 2011;22(11):115604-11.

15. Huang C-C, Hwu JR, Su W-C, Shieh D-B, Tzeng Y, Yeh C-S. Surfactant-Assisted Hollowing of Cu Nanoparticles Involving Halide-Induced Corrosion–Oxidation Processes. Chemistry – A European Journal. 2006;12(14):3805-10.

16. Zhu L, Chen Y, Zheng Y, Li N, Zhao J, Sun Y. Ultrasound assisted template-free synthesis of Cu(OH)2 and hierarchical CuO nanowires from Cu7Cl4(OH)10·H2O. Materials Letters. 2010;64(8):976-9.

17. Ghorbanpour M, Falamaki C. A novel method for the production of highly adherent Au layers on glass substrates used in surface plasmon resonance analysis: substitution of Cr or Ti intermediate layers with Ag layer followed by an optimal annealing treatment. Journal of Nanostructure in Chemistry. 2013;3(1):66.

18. Zhou Y, Xia M, Ye Y, Hu C. Antimicrobial ability of Cu2+-montmorillonite. Applied Clay Science. 2004;27(3–4):215-8.

12

Removal of Reactive Red 198 by Nano-particle Zero Valent Iron in the Presence of Hydrogen Peroxide

Siroos Shojaei[1], Somaye Khammarnia[2], Saeed Shojaei[3], Mojtaba Sasani[4]*

[1]*Department of Chemistry, University of Sistan and Baluchestan, Zahedan, Iran*
[2]*Department of Chemistry, Payam-e-noor University, Zahedan, Iran*
[3]*Department of Desertification, University of Yazd, Iran*
[4]*Analytical Chemistry, University of Sistan and Baluchestan, Zahedan, Iran*

ABSTRACT

Although dyes are widely used in textile industries, they are carcinogenic, teratogenic and mutagenic. Industries discharge their wastewater containing a variety of colors into water resources and make harmful effect on the environment. The present study aims to Evaluate removal of reactive red 198 by nanoparticle zero valent iron (NZVI) in the presence of hydrogen peroxide from aqueous solution. The effective parameters on the removal of dye such as the hydrogen peroxide concentration of NZVI, contact time, pH and dye concentration were investigated and optimized. According to the results, the combination of NZVI with hydrogen peroxide is more effective than single hydrogen peroxide. At pH = 4, contact time= 40 min, 200 M of hydrogen peroxide, dye concentration= 75 mg/L and concentration of NZVI 2g/L, color removal was achieved 91% approximately. Based on the results of experiments, using hydrogen peroxide- NZVI has high efficiency in removal of azo dye type.

Keywords: *Dye removal, Hydrogen peroxide, Reactive Red 198, Zero valent-iron nanoparticles*

INTRODUCTION

Textile industry is known as an important industry and indicates the rate of development in countries around the world. In addition, industries which produce chromatic wastewater include intiction industry, cosmetic industry, paper and pharmaceutical [1-8]. In intiction industries, about 15 percent of total color enter into the sewage system and color wastewater becomes produced [9]. There are several chromogenic materials which are used in industry and most common colors are azo group. Azo dyes contain a large synthetic group which have one or more azo bands (-N = N-). In order to decolor colored wastewater, various methods have been studied by many researchers, including physio-chemical methods such as ultra-filtration, reverse osmosis, ion exchange and surface adsorption to activated carbon, charcoal, wood chips and silica gel for color removal from effluent, which are relatively successful in applicability of COD [10, 11]. Since they just transfer aqueous pollution to the solid phase, they are not degraded and removed and they are not considered as comprehensive techniques as well. Thus, in recent years, advanced oxidation process, which is based on the production of free and active radicals, especially OH, is widely used by researchers and that is because of high oxidation power [12, 13]. In the process of zero-valent iron nanoparticles and

* Corresponding Author Email: shojaeisiroos@gmail.com

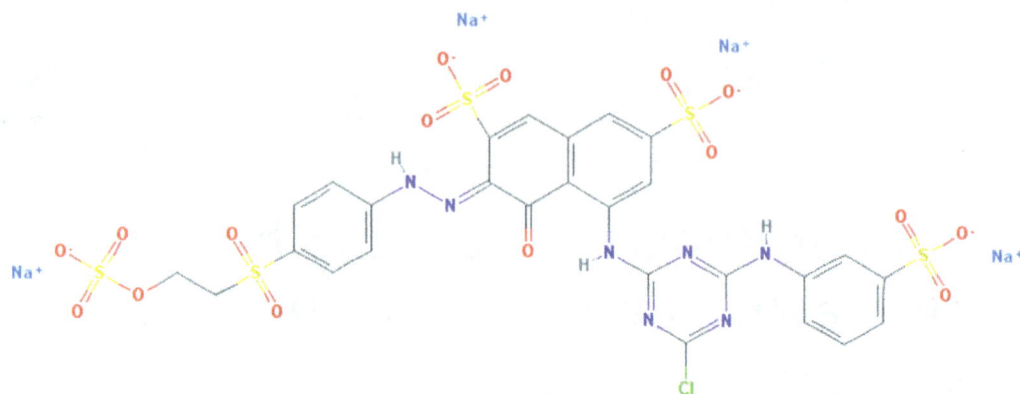

Fig 1. The structure of the chemical composition of Reactive Red 198.

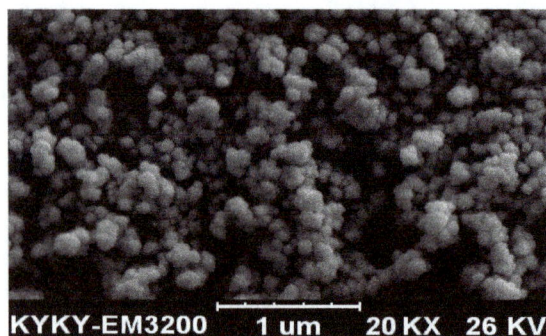

Fig 2. SEM image of synthesized iron nanoparticles.

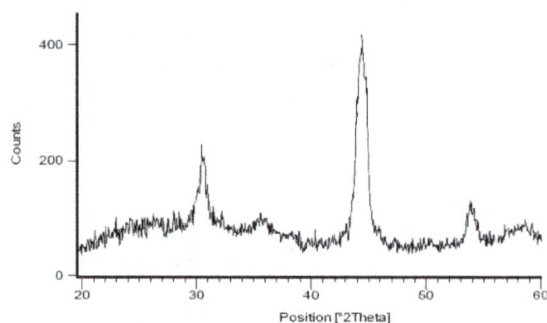

Fig 3. XRD pattern of NZVI particles.

hydrogen peroxide in contrast with conventional Fenton process, hydroxyl radical production occurs in two stages. So, process efficiency is increased. On the other hand, after ferrous ion formed in Fenton processing, the efficiency of the process is reduced and stopped [14]. The most important reactions in the removal of organic material by hydrogen peroxide in the presence of zero-valent iron nanoparticles, are included as follows:

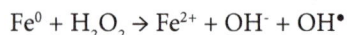

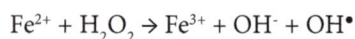

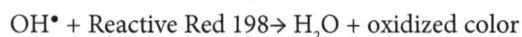

$$Fe^0 + H_2O_2 \rightarrow Fe^{2+} + OH^- + OH^\bullet$$

$$Fe^{2+} + H_2O_2 \rightarrow Fe^{3+} + OH^- + OH^\bullet$$

$$OH^\bullet + Reactive\ Red\ 198 \rightarrow H_2O + oxidized\ color$$

A wide variety of polluters including chlorinated organic compounds, polychlorinated biphenyls, heavy metal ions, oxy-anions and dimethyl phthalate could be decolored with Nanoparticles Zero Valent Iron (NZVI) [15, 16]. Furthermore, inactivation of microorganisms in drinking water, wastewater, surface water and other resources are some applications of NZVI [17]. Colors are kinds of chemicals that may be unstable by this process and recently some azo colors are used for

investigating on degradation [18, 19]. So, according to the introduction, due to the high efficiency of this process for removal of various polluters like chromatic materials, the main purpose of this study is to evaluate the efficiency of hydrogen peroxide with the presence of zero-valent iron nanoparticles for removal of Reactive Red 198 solutions in water.

EXPERIMENTAL

Materials and methods

All the sodium borohydride and ferric chloride (96%w/w) were supplied from Merck and High Media Co. respectively. Synthesis of Nano-particles zero-valent iron Nanoparticles were synthesized by adding the solution of sodium borohydride 0.16 M to the solution of hydrated ferric chloride 0.1 M at ambient temperature. In order to produce the Ferric chloride solution, deionized water that neutralized with N_2 gas is used. In order to produce a solution of sodium borohydride, Sodium hydroxide 0.1 M is applied. After preparation, a solution of sodium borohydride is added to ferric chloride solution drop by drop and in vacuum conditions and

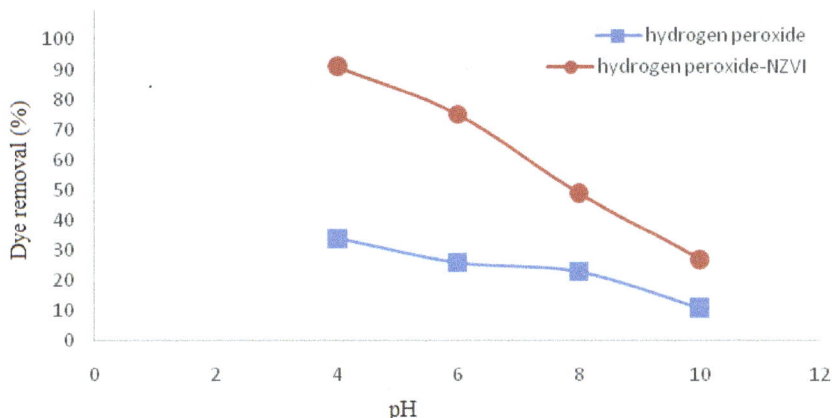

Fig 4. Effect of pH on process efficiency. (Dye concentration: 75 mg/L, contact time: 40 min, concentration of nanoparticles: 2 g/L, concentration of hydrogen peroxide 200 Mmol).

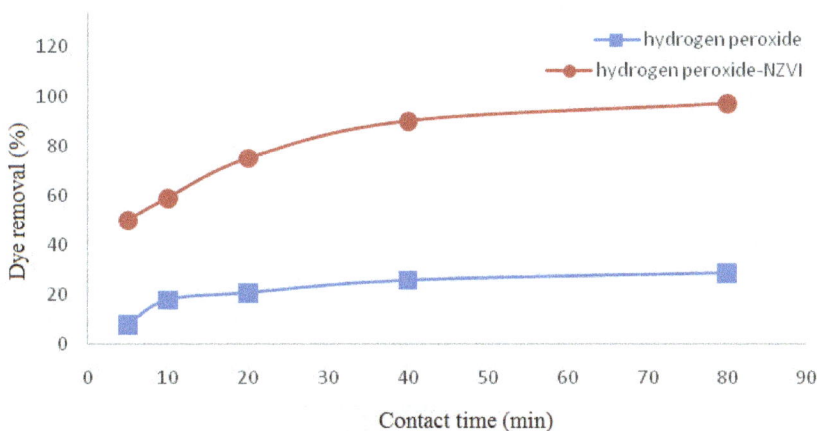

Fig 5. Effect of contact time on efficiency process. (Saturation: 75 mg/L, pH=4, concentration of nanoparticles: 2 g/L, hydrogen peroxide concentrations: 200 Mmol).

intense mixing. This phase takes about 30 min. During the process, the beaker was closed so that the synthesis takes place in a vacuum condition. Sodium borohydride reacted with ferric chloride according to first reaction and finally it leads to Fe^0 revival.

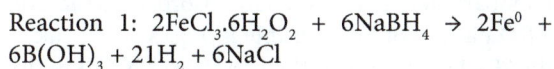

Reaction 1: $2FeCl_3.6H_2O_2 + 6NaBH_4 \rightarrow 2Fe^0 + 6B(OH)_3 + 21H_2 + 6NaCl$

After reaction, iron nanoparticles were deposited in black particles. Reaction was conducted under hood because as a result of chemical reactions, hydrogen gas is produced. In addition, it is necessary to mix reaction vessel with a blast-resistant hopper to reduce possibility of ignition. SEM electron microscopy is used to determine characteristics of iron nanoparticles.

For measurement pH with pH meter model Mettler Toledo and to pH adjustment we use Sodium hydroxide (1.5 mol L^{-1}) 0.1 N and hydrochloric acid (37%w/w) 0.1 N. Reactive Red 198 is a color-molecular-weight anionic equivalent of 968.21 g/mole and absorption peak λ_{max}=530 nm. Reactive Red 198 ($C_{27}H_{18}ClN_7Na_4O_{15}S_5$) which is used in this study is of laboratory type and manufactured by Merck, Germany. Chemical formula of this compound is shown in Fig 1 [20].

Removal of Reactive Red 198 Tests

In this research, studied variables include: exposure time (5, 10, 20, 40, 80 min), initial concentration of nanoparticles (0.5, 1, 2, 3, 4 g/L), pH (4, 6, 8, 10), hydrogen peroxide (25, 100, 150, 200, 300 Mmol) and initial concentration of

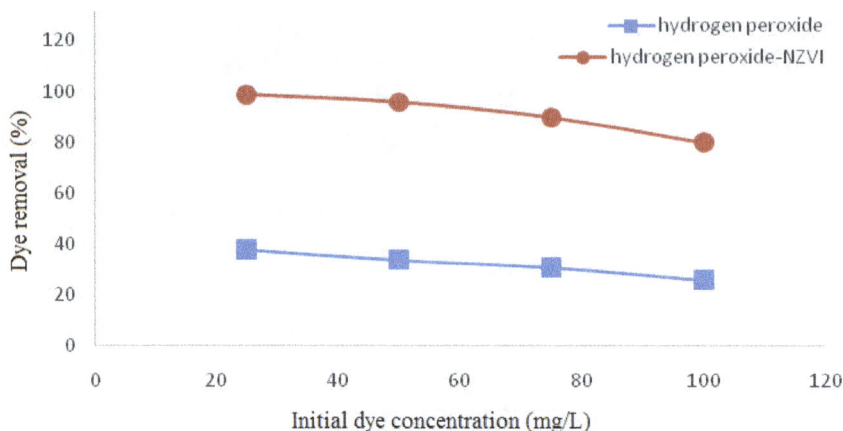

Fig 6. Effect of initial dye concentration on process efficiency. (Contact time: 40 min, pH=4, concentration of iron nanoparticles: 2 g/L, hydrogen peroxide concentrations: 200 Mmol).

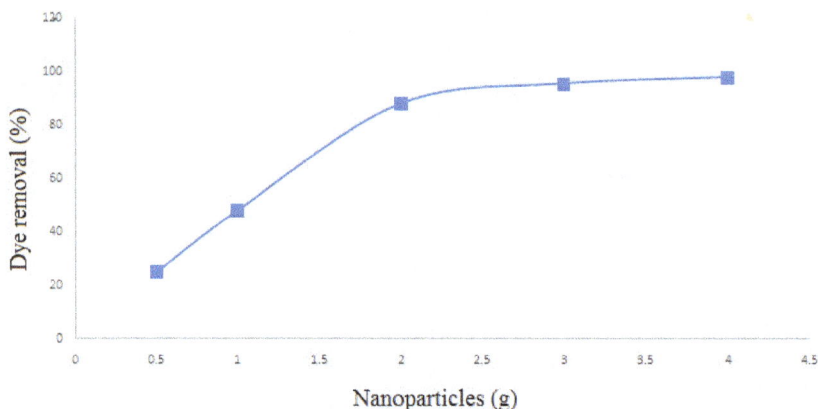

Fig 7. Effect of nanoparticles on process efficiency. (Contact time: 40 min, pH=4, initial dye concentration of 57 mg/L, hydrogen peroxide concentration of 200 Mmol).

reactive red 198 (25, 50, 75, 100 mg/L).

To perform experiments, different concentrations of color was added to volume of 250 ml to 500 ml beakers and pH of color was adjusted in desired range. Then, different concentrations of nanoparticles and hydrogen peroxide were added in 250 ml of color with different initial concentrations and after mixing by jar with around 250 rpm at certain intervals, sampling were done. Remaining concentration was determined by rate of absorbance and using a calibration curve.

RESULTS

Characterizations of Nanoparticles

In Fig 2. image of synthesized nanoparticles is shown. results of SEM particle, shows size range of particles in nano limitation (diameters of <100 nm). XRD pattern of dry NZVI particles was shown in Fig. 3. X-ray diffraction (XRD) analysis (Siemens, D5000 with Cu Kα radiation) was used to determine the presence of zero-valent iron nanoparticles in the studied absorbent structure. XRD pattern of Fe^0 is visible in Fig. 3. The peak at the $2\theta=44.7°$ indicated the presence of NZVI in the synthesized absorbent structure.

Results of Reactive Red 198 removal

Results are shown in Figs. 4 to 8. In these graphs effect of contact time, initial dye concentration, pH, concentration of nanoparticles and hydrogen peroxide concentration on color removal efficiency is shown. Results show that increasing

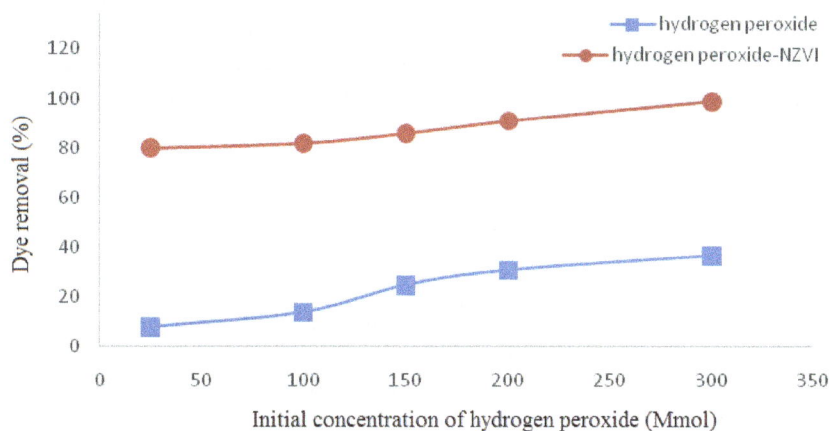

Fig 8. Effect of initial concentration of hydrogen peroxide on process efficiency. (Contact time: 40 min, pH=4, initial dye concentration of 75 mg/L, concentration of iron nanoparticles: 2 g/L).

concentrations of nanoparticles, hydrogen peroxide, contact time, initial dye concentration and pH reduction to a specified level, increases removal efficiency. Optimum experimental results show that exposure time of 40 min, PH=4, dye concentration of 75 mg/L, concentration of nanoparticles 2 g/L and hydrogen peroxide concentration 200 Mmol, hydrogen peroxide dye removal in process with nanoparticles and hydrogen peroxide about 91 and 34% respectively and by increasing pH to 10, color removal rate of about 27 and 10% reduced. Effects of pH on color removal efficiency are clearly shown in Fig. 4. So that by lowering pH of 10 to 4, removal process, hydrogen peroxide and hydrogen peroxide-iron nanoparticles has increased from 27 to 91 and 10 to 34%, respectively. Fig. 5 shows that with increasing contact time, color removal rate is increased so that by increasing time from 5 to 80 min, removal process, hydrogen peroxide and hydrogen peroxide-iron nanoparticles increased from 50 to 97 and 8 to 29%, respectively. Fig. 6 shows that with increasing concentrations of color, its removal rate is reduced so that by increasing the dye concentration of 25 to 100 mg/L, the removal process, hydrogen peroxide and hydrogen peroxide-iron nanoparticles reduced 99 to 80 and 38 to 26%, respectively. Fig. 7 shows clearly different levels of color removal efficiency of NZVI. With increasing NZVI from 0.5 to 4 g/L at pH 4 and 200 Mmol for hydrogen peroxide, in the process of hydrogen peroxide-iron nanoparticles, color removal rate is increased about 25 to 98%. In Fig. 8 The effect of hydrogen peroxide was investigated in doses of

25 to 300 Mmol, in pH=4 and it was found that the removal efficiency for the highest and lowest amounts of H_2O_2 was about 8 to 37%. The impact of increased efficiency of hydrogen peroxide color removal clearly has shown in Fig. 8. Efficiency of color removal increases by increasing amount of hydrogen peroxide in presence of NZVI. So that by increasing amount of hydrogen peroxide at a pH of 4 from about 25 to 300 Mmol, color removal rate increases 80 to 99 percent.

DISCUSSION

Results of this study show that acidic pH is effective in achieving maximum color removal. This is a major problem that requires to acidification environment. At lower pH than 3 to formation of $Fe(OH)^{2+}$ developments with hydrogen peroxide reacts slowly, reduces amount of hydroxyl radicals, resulting in reduced process efficiency. In alkaline pH Fe^{2+} is also converted to Fe^{3+} and in $Fe(OH)_3$ state removed from catalytic cycle [21]. In this study, pH of 4 is optimized condition and it's clear in Fig. 4 that maximum efficiency is achieved in pH=4 and color removal efficiency has decreased with increasing pH. It has been demonstrated in previous studies that solution pH can affect on absorbent surface charges, degree of ionization various contaminants, separating functional groups on active sites attractive and effective dye molecule structures [22, 23] In fact, under acidic pH, linked sites absorbing hydrogen ions works closely with each other as ligands interface (bridge) between absorber and dye molecule.

In these conditions, lower pH values, provides favorite conditions for reactive dyes removal [24, 25]. Gulnaz et al removal of reactive dye by Potamogeton Crispus 198 indicates that decrease pH from 5 to 1, increase color removal efficiency that is compatible with results of present study [20]. As shown in Fig. 8 is color removal efficiency increases by increasing amount of hydrogen peroxide at pH=4. This is because of increasing concentration of hydrogen peroxide, which dues to acidification of environment and amplify generates ions Fe^{2+} and increased production of hydroxyl radicals, and result increasing efficiency of dye removal [26]. Contact time is one of most important parameters for practical applications in adsorption process. Based on results of this study (Fig. 5), removal percent of Reactive Red 198 by adsorbent increased rapidly during early stages of adsorption and then at a slower speed and appropriate time, spend time to reach a state of equilibrium during a period of about 80 min. In other hand, as shown in Fig. 5, dye uptake was very rapid during first 5 min, and then absorbs speed decreased rapidly over time and eventually balance is reached after 80 min, maximum reduction in contact time of 80 min was 97 percent [27]. Generally, rate of removal (adsorbent) is rapid at first, but gradually decreased with time until equilibrium is reached. It is due to fact that at beginning and in early stages of absorption, a large number of empty surface sites are available for absorption, but with spending of time, empty remain site surface, are difficult to absorb pollutants. Which could be due to deterrent forces between molecules adsorb on surface of solid and liquid mass [28]. Similar results have been reported by Gulnaz [20]. In another study, mozia et al have used hybrid membrane system to analyze Acid Red 18 for decomposition in photocalystic process. In this study, color is completely removed in 5 hours had less efficiency compared to presented study [29]. Initial dye concentration, provides a significant driving force to overcome overall resistance substantially phases of mass transfer between liquid and solid colors. With increasing concentrations of color, its removal rate decreases that this could be due to NZVI surface occupation by dye molecules, ions Fe^{+2} less access to NZVI dye molecules and less access to hydrogen peroxide. This topic is similar to other studies [30]. Adsorbent concentration, an important parameter that effects on absorption and hence determining absorption capacity for a given initial concentration of color is

for Reactive Red 198, is important. As is shown in Fig. 7, Removal of Reactive Red 198, is applied by adsorbent dosage and by increasing adsorbent dose increases up to a certain value highly. This result can be explained by fact that absorbtion site, remained saturated during adsorption, while Increasing number of sites available dose increases adsorbents. In another study, mozia et al use photocatalytic process to analyze Acid Red 18. In this study, initial concentration of adsorbents, catalysts dose and reaction temperature were investigated [31]. This process compared to processes of this research is a quite complex process. According that this process iron nanoparticles in presence of hydrogen peroxide, is more accessible, environmentally safer, easy operation with high efficiency rather than other advanced oxidation processes [18]. Widespread use of this method is result of these advantages. Therefore, this process is useful for removal of azo dyes.

CONCLUSION

According to results of research, nanoparticles of hydrogen peroxide has advantages such as high removal efficiency and short reaction time and could be considered as an appropriate option for removal of azo dyes in aqueous environment. But hydrogen peroxide has less efficiency for removal of azo dyes in aqueous the environments. The results confirm, Azo dyes can be removed by this process. When NZVI were used with Fenton process, amount of removing will be increased. 5 factors effects were studied like pH, contact time, initial dye concentration, adsorbent concentration and initial Hydrogen peroxide concentration. Optimized parameters for this process are pH=4, contact time= 40 min, initial dye concentration= 75 mg/L, adsorbent concentration= 2 g/L and initial Hydrogen peroxide concentration= 200 Mmol.

CONFLICT OF INTEREST

The authors declare that there are no conflicts of interest regarding the publication of this manuscript.

REFERENCES

1. Pagga U, Brown D. The degradation of dyestuffs: Part II Behaviour of dyestuffs in aerobic biodegradation tests. Chemosphere. 1986;15(4):479-91.
2. Wang J, Huang CP, Allen HE, Cha DK, Kim D-W. Adsorption Characteristics of Dye onto Sludge Particulates. Journal of Colloid and Interface Science. 1998;208(2):518-28.
3. Oh SY, Cha DK, Chiu PC, Kim BJ. Conceptual comparison

of pink water treatment technologies: granular activated carbon, anaerobic fluidized bed, and zero-valent iron-Fenton process. Water Science and Technology. 2004;49(5-6):129-36.

4. Papić S, Koprivanac N, Lončarić Božić A, Meteš A. Removal of some reactive dyes from synthetic wastewater by combined Al(III) coagulation/carbon adsorption process. Dyes and Pigments. 2004;62(3):291-8.

5. Bayer P, Finkel M. Modelling of sequential groundwater treatment with zero valent iron and granular activated carbon. Journal of Contaminant Hydrology. 2005;78(1–2):129-46.

6. Akcil A, Erust C, Ozdemiroglu S, Fonti V, Beolchini F. A review of approaches and techniques used in aquatic contaminated sediments: metal removal and stabilization by chemical and biotechnological processes. Journal of Cleaner Production. 2015;86:24-36.

7. dos Santos AB, Cervantes FJ, van Lier JB. Review paper on current technologies for decolourisation of textile wastewaters: Perspectives for anaerobic biotechnology. Bioresource Technology. 2007;98(12):2369-85.

8. Lu C-S, Chen C-C, Mai F-D, Li H-K. Identification of the degradation pathways of alkanolamines with TiO2 photocatalysis. Journal of Hazardous Materials. 2009;165(1–3):306-16.

9. Özer A, Dursun G. Removal of methylene blue from aqueous solution by dehydrated wheat bran carbon. Journal of Hazardous Materials. 2007;146(1–2):262-9.

10. Robinson T, McMullan G, Marchant R, Nigam P. Remediation of dyes in textile effluent: a critical review on current treatment technologies with a proposed alternative. Bioresource Technology. 2001;77(3):247-55.

11. Qu S, Huang F, Yu S, Chen G, Kong J. Magnetic removal of dyes from aqueous solution using multi-walled carbon nanotubes filled with Fe2O3 particles. Journal of Hazardous Materials. 2008;160(2–3):643-7.

12. Wang C-B, Zhang W-x. Synthesizing Nanoscale Iron Particles for Rapid and Complete Dechlorination of TCE and PCBs. Environmental Science & Technology. 1997;31(7):2154-6.

13. Bigg T, Judd SJ. Zero-Valent Iron for Water Treatment. Environmental Technology. 2000;21(6):661-70.

14. Kusic H, Koprivanac N, Srsan L. Azo dye degradation using Fenton type processes assisted by UV irradiation: A kinetic study. Journal of Photochemistry and Photobiology A: Chemistry. 2006;181(2–3):195-202.

15. Liu T, Wang Z-L, Zhao L, Yang X. Enhanced chitosan/Fe0-nanoparticles beads for hexavalent chromium removal from wastewater. Chemical Engineering Journal. 2012;189–190:196-202.

16. Wang Y, Zhou D, Wang Y, Zhu X, Jin S. Humic acid and metal ions accelerating the dechlorination of 4-chlorobiphenyl by nanoscale zero-valent iron. Journal of Environmental Sciences. 2011;23(8):1286-92.

17. Izanloo H, Nazari S, Ahmadi Jebelli M, Alizadeh Matboo S, Tashauoei HR, Vakili B, et al. Studying the Polypropylenimine-G2 (PPI-G2) Dendrimer Performance in Removal of Escherichia coli, Proteus Mirabilis, Bacillus Subtilis and Staphylococcus Aureus from Aqueous Solution. Arak Medical University Journal. 2015;18(6):8-16.

18. Poursaberi T, Hassanisadi M, Nourmohammadian F. Application of synthesized nanoscale zero-valent iron in the treatment of dye solution containing basic yellow 28. Prog Color Colorants Coat. 2012;5:35-40.

19. Frost RL, Xi Y, He H. Synthesis, characterization of palygorskite supported zero-valent iron and its application for methylene blue adsorption. Journal of Colloid and Interface Science. 2010;341(1):153-61.

20. Gulnaz O, Sahmurova A, Kama S. Removal of Reactive Red 198 from aqueous solution by Potamogeton crispus. Chemical Engineering Journal. 2011;174(2–3):579-85.

21. Chamarro E, Marco A, Esplugas S. Use of fenton reagent to improve organic chemical biodegradability. Water Research. 2001;35(4):1047-51.

22. Yang GCC, Lee H-L. Chemical reduction of nitrate by nanosized iron: kinetics and pathways. Water Research. 2005;39(5):884-94.

23. Ai L, Zhang C, Liao F, Wang Y, Li M, Meng L, et al. Removal of methylene blue from aqueous solution with magnetite loaded multi-wall carbon nanotube: Kinetic, isotherm and mechanism analysis. Journal of Hazardous Materials. 2011;198:282-90.

24. Li X-q, Elliott DW, Zhang W-x. Zero-Valent Iron Nanoparticles for Abatement of Environmental Pollutants: Materials and Engineering Aspects. Critical Reviews in Solid State and Materials Sciences. 2006;31(4):111-22.

25. Tang WZ, Chen RZ. Decolorization kinetics and mechanisms of commercial dyes by H2O2/iron powder system. Chemosphere. 1996;32(5):947-58.

26. Shu H-Y, Chang M-C, Chang C-C. Integration of nanosized zero-valent iron particles addition with UV/H2O2 process for purification of azo dye Acid Black 24 solution. Journal of Hazardous Materials. 2009;167(1–3):1178-84.

27. Reardon EJ, Fagan R, Vogan JL, Przepiora A. Anaerobic Corrosion Reaction Kinetics of Nanosized Iron. Environmental Science & Technology. 2008;42(7):2420-5.

28. Hsing H-J, Chiang P-C, Chang EE, Chen M-Y. The decolorization and mineralization of Acid Orange 6 azo dye in aqueous solution by advanced oxidation processes: A comparative study. Journal of Hazardous Materials. 2007;141(1):8-16.

29. Mozia S, Tomaszewska M, Morawski AW. Removal of azo-dye Acid Red 18 in two hybrid membrane systems employing a photodegradation process. Desalination. 2006;198(1):183-90.

30. Keshmirizadeh E, Eshaghi S. Removal of Anionic Brown 14 and Cationic Blue 41 dyes via Fenton Process. Journal of Applied Chemical Research. 2015;9(1):83-93.

31. Shojaei S, Shojaei S, Sasani M. The efficiency of eliminating Direct Red 81 by Zero- valent Iron nanoparticles from aqueous solutions using response surface Model (RSM). Modeling Earth Systems and Environment. 2017;3(1):27.

32. Mozia S, Tomaszewska M, Morawski AW. Photocatalytic degradation of azo-dye Acid Red 18. Desalination. 2005;185(1):449-56.

13

Fabrication of Hydrophobic Membrane for the Separation of n-Hexane/Water Mixture Using Novel Oleophilic Nanoparticle and Kevlar Fabric, as a Superior Support

Hanieh Karimnezhad

Polymer Research Center, Department of Chemical Engineering, Razi University, Kermanshah, Iran.

ABSTRACT

The fabrication of functionalized membranes with hydrophobic/oleophilic surfaces for the elimination of n-hexane from water using para-aminobenzoate alumoxane, boehmite-epoxide and a novel nanoparticle, i.e., Stearate Alumoxane by a simple coating technique, is reported here. FTIR was used to characterize nanoparticles. SEM and contact angle measurement analyses were used to identify the nanocomposite membranes. The concentrations of oil in permeate and retentate were measured by UV/vis spectrophotometer. The morphology of Stearate alumoxane nanoparticles was investigated by means of SEM images. The composed film of nanoparticles on the Kevlar fabric was hydrophobic with water contact angle of ~ 145° and oleophilic with oil contact angle of ~ 0º. In addition, the membranes retained stable hydrophobicity and high separation efficiency even after employing for 6 times. Applying these properties, a setup was considered using the functionalized Kevlar fabric to separate oil through down to a collector and leave water drops. Our batch filtration system was exclusively gravity-driven. The achieved separation system can separate the oily water mixture (with the concentration of 20 % (v/v) n-hexane in water), effectively with a separation efficiency of 84%.

Keywords: *Hydrophobic, Nanomaterials Coating, Nanocomposite Membrane, Kevlar Fabric, Oil/water Emulsion.*

INTRODUCTION

Due to the increment in the amount of industrial oily wastewater and contaminated sea water, separation of oil/water mixtures is an essential issue [1-6]. Separation of oil/water mixtures can appear during oil extraction, crude oil production and refinery, petrochemical and metal finishing, textile and leather processing, food processing and lubricant, representing severe harms for the environment [1, 7-9]. Oil products enter the aquatic environment instantly and alter their primary state. Every crude oil type is classified to different compounds. About 50 to 98 % of oil is composed

of hydrocarbons, which are primarily alkanes (paraffin) such as n-hexane [10]. The refineries are one of the places in which hexane and its derivatives are produced. The common application of hexanes is the extraction of oil and grease pollutants from water. Due to the application of hexane in the aquatic environment, the separation of oil pollution from water resources is an essential issue [11]. Conventional methods for this purpose are not satisfactory to solve the problem of oil polluted water due to their disadvantages such as low efficiency, high operation cost, corrosion, and recontamination problems, etc. [12-14].

*Corresponding Author Email: hany_k6589@yahoo.com

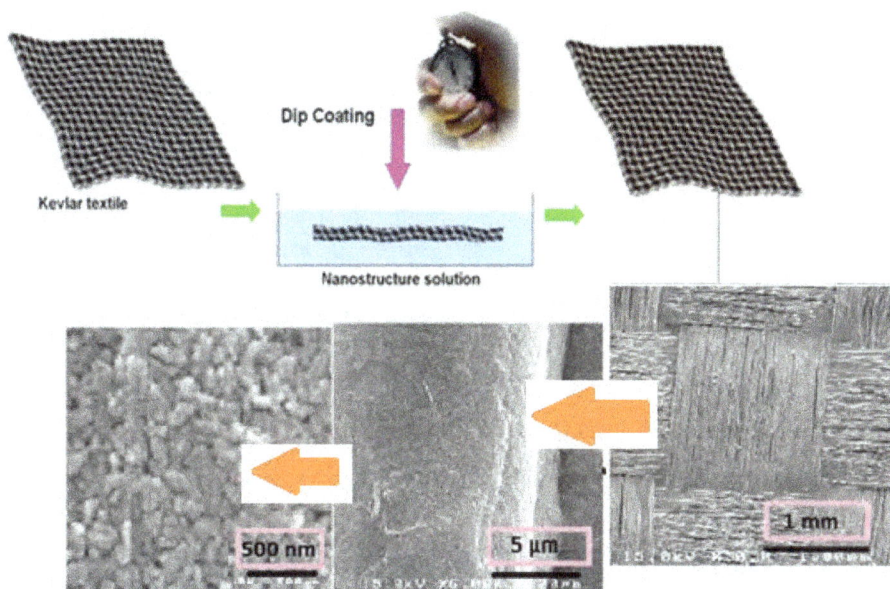

Fig. 1. The preparation process of hydrophobic functionalized Kevlar fabric.

Thus, membrane techniques offer conspicuous advantages in comparison with conventional separation processes [15-16]. Currently, employing functionalized materials and nanoparticles that can efficiently separate oil and water is becoming desire [17-19]. Besides the common techniques of oil separation, the techniques of using "*nano*" are developing, which use carbon nanotubes, aerogels, and nanocomposites, metal, and non-metal nanostructurized oxides, etc. [4-6, 10, 17, 19]. Some of these nanomaterials can be fabricated by safe and low-cost methods and without inducing destruction to the environment [17]. In the recent years, the development of hydrophobic and oleophilic materials for the fabrication of membranes to separate oil from water has attracted much consideration because of their extraordinary separation efficiency and great applicability [1, 10, 20]. The hydrophobic property makes materials reject water entirely, but the oleophilic feature allows oil to permeate easily. Thus, hydrophobic and oleophilic materials can separate oil-water, effectively [1]. Various methods have been developed to fabricate superhydrophobic and superoleophilic materials for oil/water separation consisting of sol–gel process [1, 21-22], electrospinning techniques [1, 23-24], electrochemical deposition [25], etc. In the recent years, a new generation of membrane processes is developing. It means that a filtration medium (support) is selected, afterward it is modified or functionalized with appropriate

materials, and then it is used for separation. Some of these supports are cotton, wools, linens, fibers, porous carbon, sand, diatomite, coal, quartz, and metals including mesh film, filter paper, resin, sawdust, etc. Among them, fiber has attracted much attention. The improvement of fiber properties has made significant benefits [4-5, 26-33]. Only a few kinds of literature have focused on the fabrication of superhydrophobic textiles for oil/water separation [28]. Although these studies have prepared a new opportunity to plan and prepare functionalized fabrics for the separation and filtration requests, superior polymers or complex procedures are needed. Kevlar fabric is one of these special polymers with high mechanical and chemical stability. The hydrophobic and oleophilic textiles with nanocoatings are effectively used for application in the oil/water separation [28].

In this work, I propose a simple technique (dip-coating) [18] for the fabrication of hydrophobic and oleophilic textiles by coating Kevlar fabrics with hydrophobic Stearate Alumoxane. The Kevlar fabrics were pretreated by ethanol, acetone, and distilled water for cleaning of the fiber surface from pollution. Then, Para-aminobenzoate alumoxane (PAB-A) and boehmite – epoxide (Bo-E) nanoparticles were used to provide the Kevlar surface for the addition of Stearate Alumoxane (St-A) nanoparticles as the active top layer. The obtained functionalized textiles revealed superhydrophobic/superoleophilic property, approximately, which can

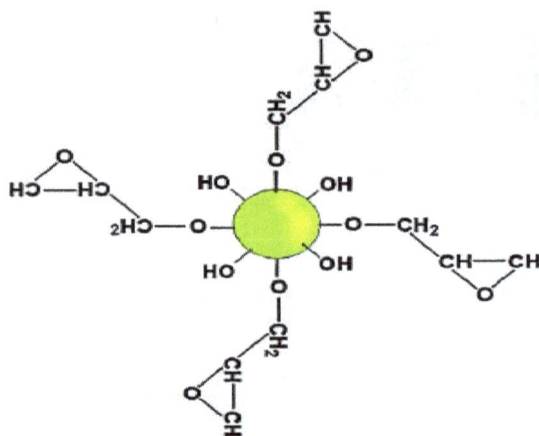

Fig. 2. The chemical structure of Bo-E.

be used for the oil/water separation. The schematic preparation process of the functionalized textiles using the Kevlar fabric as a filter and dip coating as a method of modifying is shown in Fig. 1.

EXPERIMENTAL PROCEDURE
Materials
Aluminum nitrate [Al (NO$_3$)$_3$·9H$_2$O], sodium hydroxide, stearic acid, normal hexane (n-hexane), epichlorohydrin, and para-amino benzoic acid were purchased from Merck. Acetone, ethanol, and distilled water were used as solvents. Kevlar fabric, poly paraphenylene terephthalamide, (Du Pont, USA, with a thickness of 0.28 mm) was used as the membrane support.

Apparatus
Heating was done by electric oven (53 lit. model Digital UNB 400, Memmert, Germany). Digital balance (Model TE212, Made by Sartorious, Germany, Max 210 g, d= 0.01 g) was used to weigh the samples. Sonication of solutions was performed by the ultrasonic bath (240 w–35 kHz, SONREX Digite C Ultrasonic bath, DT52 H model made by the bandelin company, Germany). Emulsions of oil/water were prepared by the ultrasonic processor (UP400S, Hielscher Ultrasound Technology). The specification of the centrifuge is Hettich (Universal 320 R, Germany).

Synthesis of the Nanoparticles
1. Boehmite nanoparticles (γ-AlOOH), as the precursor material, were synthesized based on the sol–gel method published elsewhere [34].
2. PAB-A nanoparticles, utilized as the first

coating layer on the Kevlar fabric, were also synthesized from boehmite and para-aminobenzoic acid according to the procedure described in an earlier study [35].

3. Bo-E was produced pursuant to the protocol explained completely in another investigation [29-30]. The structure of Bo-E is demonstrated in Fig. 2. This Fig. explains the product of the reaction between boehmite and epichlorohydrin in the presence of sodium hydroxide and water. The Cl in the epichlorohydrin and H in the boehmite are separated and form HCl. Thus, the linkage between epichlorohydrin and boehmite is formed.

4. Stearate Alumoxane nanoparticles (St-A): St-A was synthesized through the reaction of boehmite nanoparticles (6 g) and stearic acid (29 g) thoroughly with 165 mL of DMF in a 500 mL round-bottomed flask. The resulting suspension was heated for 5 min then sonicated for 10 min in an ultrasonic and then refluxed for 10 h at its boiling point after which it was allowed to cool to room temperature and then centrifuged (25 °C, 4000 rpm) with ethanol. The resulting solution was kept in the oven at the temperature of 75 °C; finally, the creamy powder was obtained.

Surface Modification of Kevlar Fabrics
The procurement of relative superhydrophobic fabrics was based on 3-step dip coating [18]. At first, roundish segments of Kevlar fabric (5 cm diameter) were washed with ethanol, acetone, and distilled water to remove extra dyes and surface contaminant. Then, the samples were kept in the oven at 50 °C for 10 min to be dried.

Phase 1. 1 g of PAB-A was dissolved in 50 mL of distilled water and sonicated for 10 min in an ultrasonic bath and then, poured into a clean glass petri dish. Kevlar fabrics were immersed in the solution for 1 min. In order to dry the coated fabrics, they were kept in the oven at 100 °C for 14 h.

Phase 2. Here, coated fabrics of the previous step were submerged in the aqueous solution of BO-E nanoparticles with the concentration of 0.1 g/mL for 10 min. It is to be noted that, the Bo-E solution was sonicated for 60 min in an ultrasonic bath. The resulting coated Kevlar fabrics were dried by keeping them in the oven at 100 °C for 3 h.

Phase 3. Surface layer: 1 gr of St-A was dissolved in 30 mL of acetone and sonicated for 10 min in an ultrasonic bath and poured into a Petri dish.

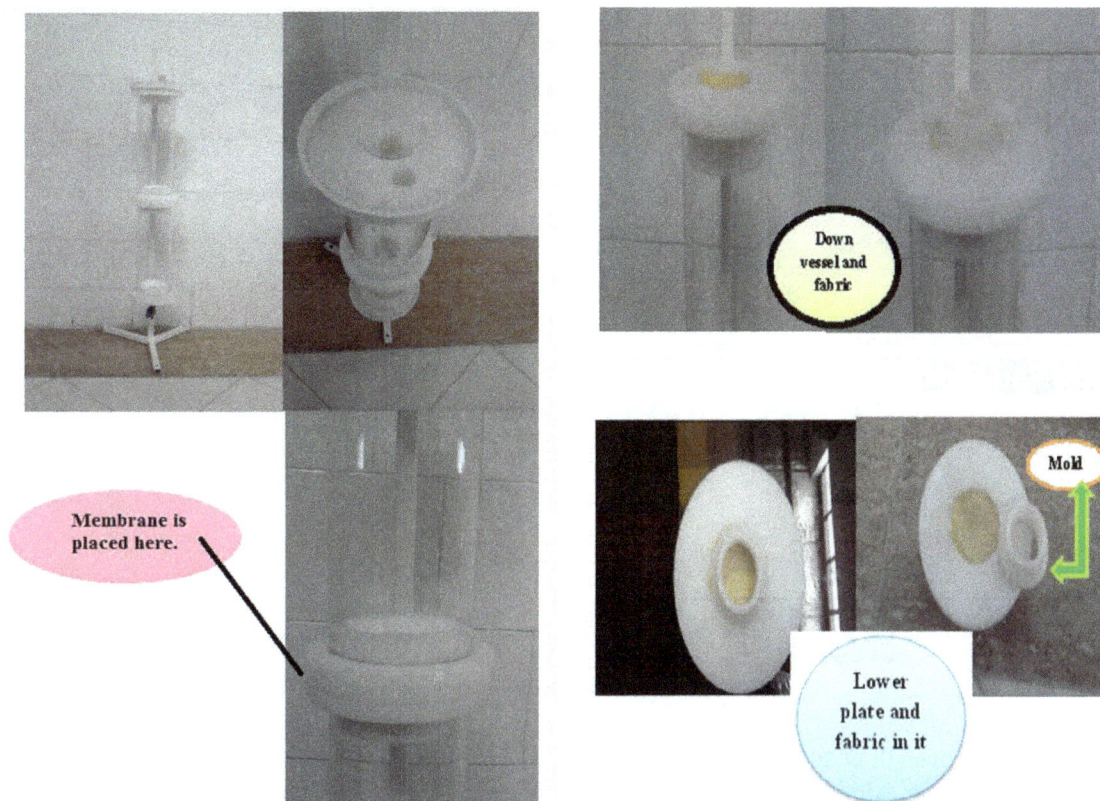

Fig. 3. Experimental setup for separation tests.

Coated fabrics of the previous step (step 2) were immersed in the prepared solution for 10 min. The resulting coated Kevlar fabrics were dried in the oven by keeping them at 100 °C for 24 h.

Both sides of Kevlar fabric (upper and low surfaces) were coated through modification process by dip coating. Thus, if there is a chance for water droplets to pass through the membrane, the beneath surface of modified fabric prevents the passage of water. Accordingly, the separation of oil and water is almost more effective.

Characterization of the St-A Nanoparticles and Membrane

The IR spectrum of the sample (St-A) was obtained through an ABB Bomem Fourier Transform Infrared FTIR (MB-104) spectrophotometer. The morphologies of the synthesized powder and membrane were observed by Scanning Electron Microscopy (SEM, HITACHI, S - 4160). Contact angle measurements were performed using a video–optic contact angle goniometer with a 4 μL water droplet and a 4 μL oil droplet at the ambient temperature to measure the hydrophobicity and oleophilicity of the membranes, respectively.

Experimental Apparatus and Preparation of Oil/Water Emulsions

According to Fig. 3, the gravity-driven experiments for the batch process were conducted by means of a dead-end setup under atmospheric pressure gradient at ambient temperature. As can be seen, the effective surface area of the membrane, i.e., 20.25 cm^2, is placed between two polyethylene plates. The oil removal rate or separation efficiency was calculated using the following equation:

$$\%R = \text{Efficiency (\%)} = [(C_f - C_r)/C_f]*100 \qquad (1)$$

Where C_f and C_r are the oil concentrations in the feed and retentate fluxes, respectively.

The emulsion preparation was carried out at ambient temperature (about 28 °C) without the addition of any surfactant. The oil/water emulsions with the oil concentration of 20 % (v/v) oil in water were prepared by the ultrasonic processor. Ultraviolet-visible spectrophotometer (UV-2100 spectrophotometer) was employed to indicate the absorption or reflectance of oil and accordingly, concentration of oil in mixtures of oil/water to determine the efficiency of the separation. The oil/water emulsion was initially calibrated for different known oil concentrations in terms of absorbance

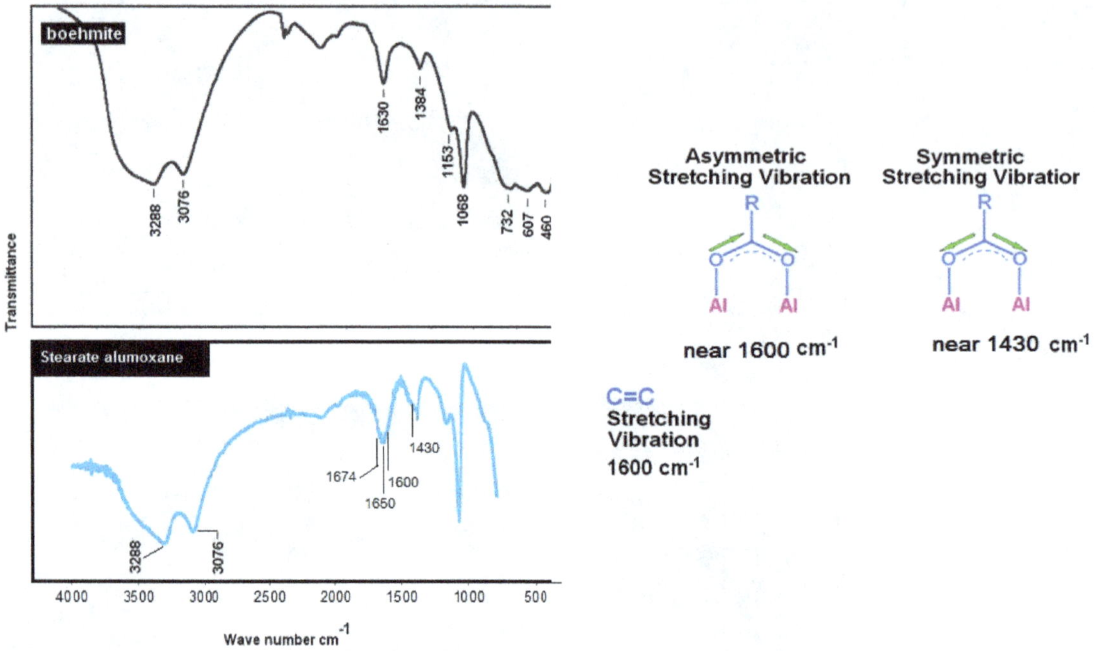

Fig. 4. FT-IR spectra of the boehmite and St-A nanoparticles.

recorded at a wavelength of 200 nm (Which is the wavelength at which maximum absorption occurs).

Reuse

After one step of separation, the membrane was used to separate another emulsion with the same characteristics to understand that if the membrane retains its hydrophobicity property or the linkages disappeared.

RESULTS AND DISCUSSION

Characterization of St-A

FT-IR Analysis of St-A

Fig. 4 illustrates the FT-IR spectra of the boehmite and St-A nanoparticles. As can be seen, two intense bands at 3076 cm^{-1} and 3288 cm^{-1} are related to the stretch frequency of OH groups. The frequencies of 1068 cm^{-1} and 1153 cm^{-1} are related to the symmetrical bending vibrations of hydrogen bands of OH···OH. However, in the St-A spectrum, the boehmite prominent peaks are present which show the amount of unreacted OH groups of boehmite on the surface of St-A. Also, the two broad and prominent peaks around 1600 cm^{-1} and 1430 cm^{-1} indicate that the carboxylate links on nanoparticles surface were formed while reacting with stearic acid. Overspread broad peak at 1650 cm^{-1} is belong to the stretching vibration of C=C of the stearate. Also, peak at 1674 cm^{-1} is attributed to the water in the structure of the particles.

SEM Analysis of St-A

Fig. 5 presents the SEM images of St-A at different magnifications. Stearic acid is a fatty acid with actually amphiphilic property in character. In Fig. 5, at each magnification, two images have been brought that identify two different views of St-A nanoparticles. As Fig. 6 indicates, stearic acid is composed of hydrophilic head and hydrophobic tail. The reaction between the hydroxyl groups (OH) of boehmite with carboxylic groups (COOH) of stearic acid has resulted in the formation of carboxylate linkages on the surface of boehmite. The filamentous modes of nanoparticles in Fig. 6 may show the hydrophobic tail of St-A.

Morphology and Structure of the Modified Alumoxane Membrane

SEM Analysis of the Membrane

Coating of the plain Kevlar fabric with three layers of nanoparticles fills the empty space between the textures of the fabric and makes the fabric uniform and compact, this distinction is shown in Fig. 7. Hydrogen bonding interactions are formed between – NH – CO groups of Kevlar and – NH$_2$ functionalities of PAB-A. Bo-E as the interlayer provides the strength and stability of the surface layer, i.e., St-A layer. The reason for coating of the fabric with PAB-A and Bo-E before applying

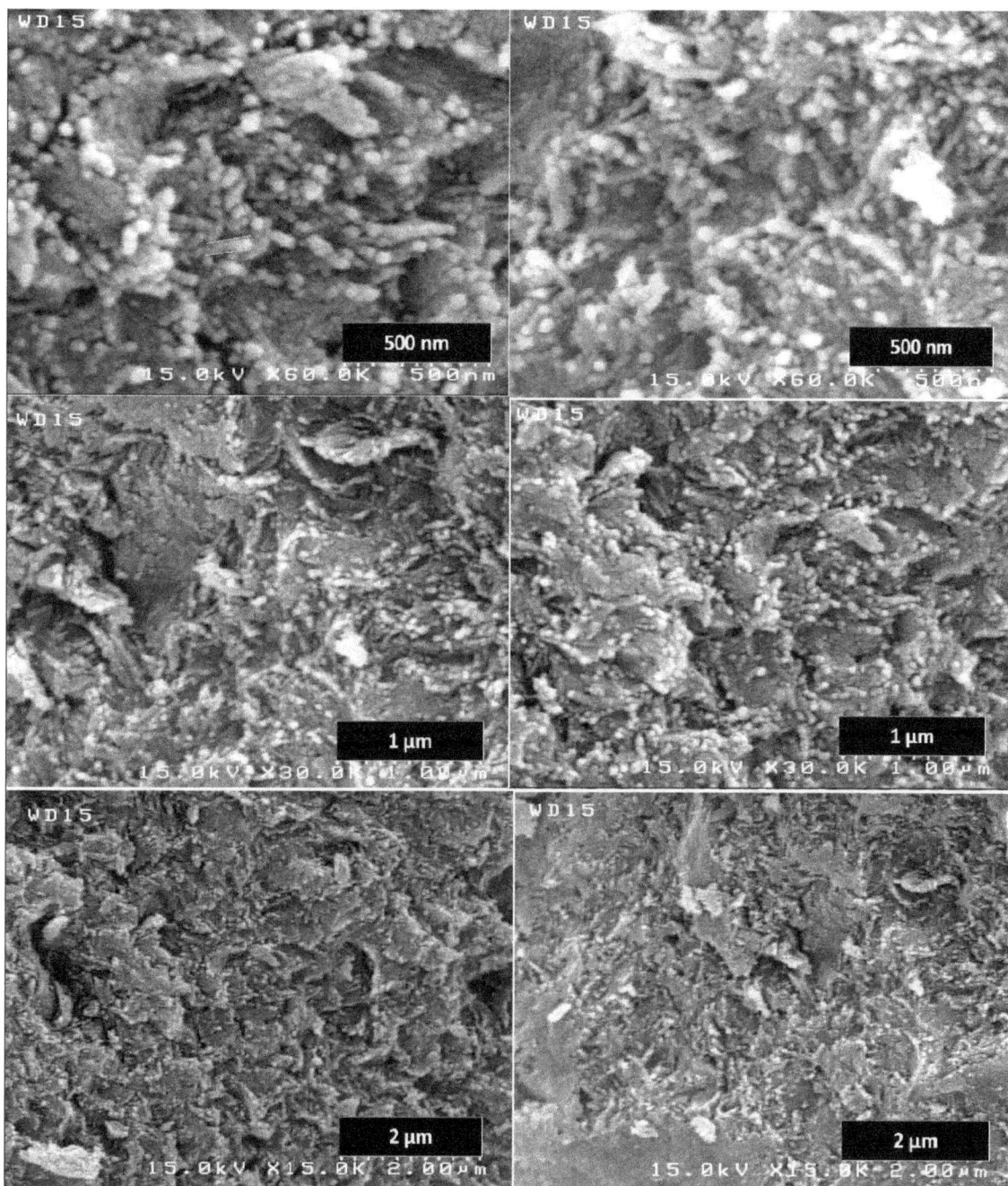

Fig. 5. SEM micrographs of the St-A nanoparticles.

the surface layer is that: if St-A is applied alone, the enough strength between Kevlar and St-A isn't provided and St-A may decompose and separate from fabric. It is remarkable that the linkage between PAB-A and Kevlar fabric is stronger than St-A and fabric. Because the hydrogen bond between $-NH_2-$ and fabric is more considerable. The structures and connections among fabric and nanoparticles are presented in Fig. 8. The

functionalized Kevlar fabric with three layers (Fig. 7) has a more compressed and regular surface in comparison with the un-functionalized (uncoated) one. The gaps between the wraps and woofs of the fabric and the pores on its surface are filled with the coating materials.

Water Contact Angle Measurement

To realize the hydrophobicity/oleophilicity

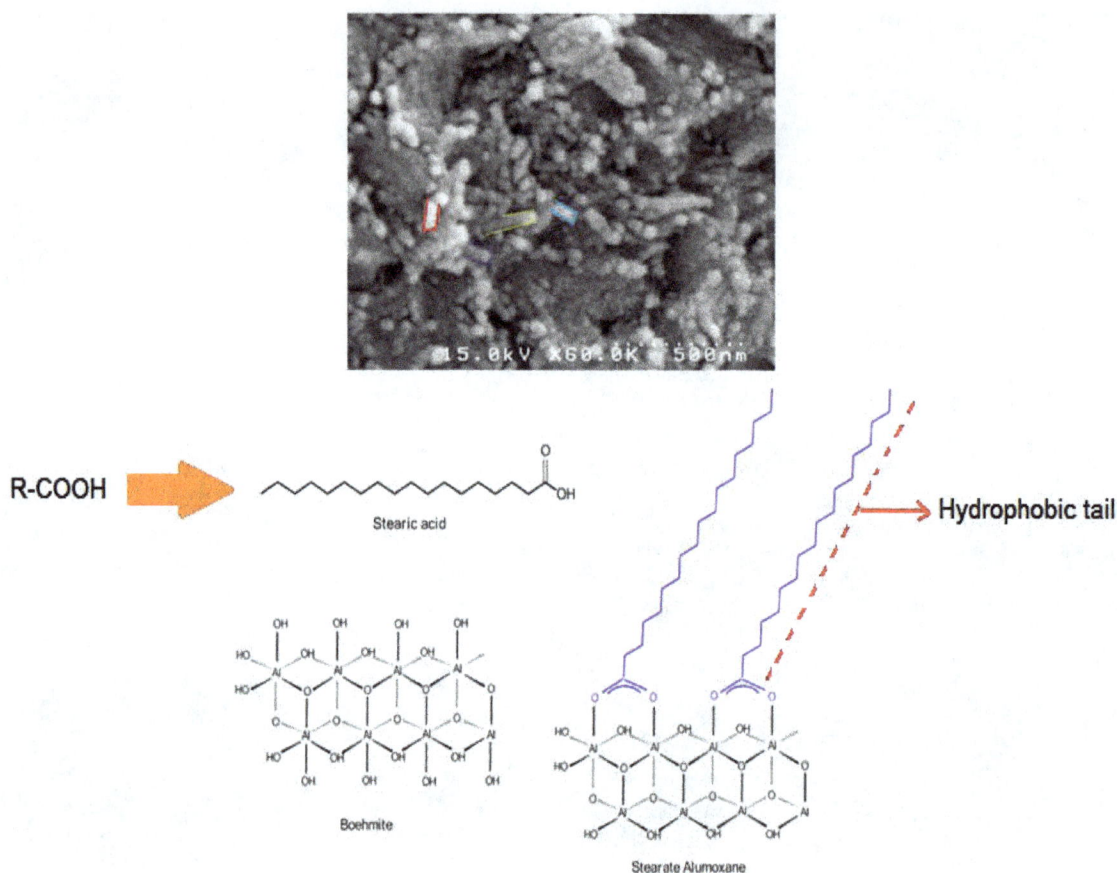

Fig. 6. The chemical structure of St-A.

properties of the prepared membrane, water contact angle test was done. For this purpose, a 4 μL water droplet at ambient temperature was deposited on the sample (A square strip (1.5 cm) of the membrane was positioned face down on a glassy plate.) As can be seen in Fig. 9 (a), WCA is about 145° (Oil contact angle is ~ 0°, oil droplets were immediately absorbed when dripped onto the membrane owning to the hydrophobic/oleophilic character of its surface). According to Fig. 9 (b), WCA demonstrates that the membrane keeps its hydrophobicity even after employing for 6 times. It is noteworthy to mention that plain Kevlar fabric is hydrophilic in nature, due to the amide groups of Kevlar fabric.

The Separation Operation of the Functionalized Fabric

To compare the performance of functionalized (improved) Kevlar fabric by St-A with un-functionalized one, the plain fabric is exposed to the emulsion of oil and water. Then, the efficiency was measured. Investigations indicate that the separation efficiency is modified by the functionalization of Kevlar (Fig. 10). Due to the non-ideality of the membranes and operational conditions, the efficiency isn't 100%, and the possibility of passing water from membrane exists. Thus, the oil rejection is acceptable and almost high. It is concluded that modification of the membrane with PAB-A and Bo-E as a basis for the preparation of condition for adherence of St-A to fabric is effective. Thus, St-A is a material with strong hydrophobicity characteristic.

6 stages of using the same membrane for separation were accomplished. The related data are shown in Fig. 10. The efficiencies are approximately high even after 6 times of using the membrane. The difference between the amounts of the oil rejections is due to:

1. The penetration and entrapment of water molecules on/into the membranes without passing from it. Thus, some of the active sites (oleophilic sites) may be filled and occupied by water molecules and might be lost through the separation process.

2. Moreover, the whole of oil molecules don't

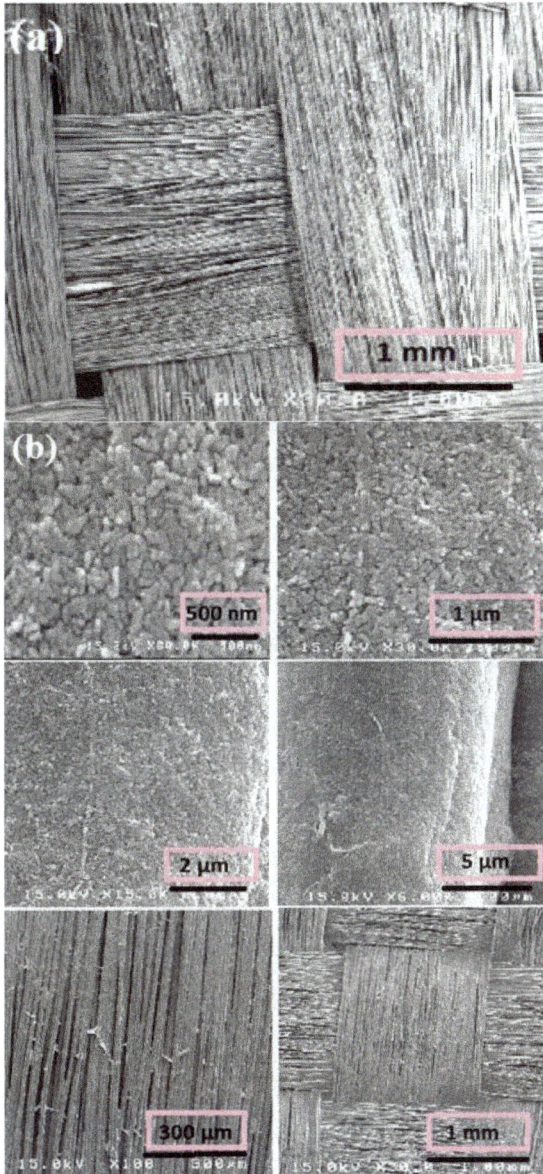

Fig. 7. SEM micrograph of the un-coated Kevlar fabric (a) and the three-layers (PAB-A, Bo-E and St-A) coated Kevlar fabric (b) at different magnifications.

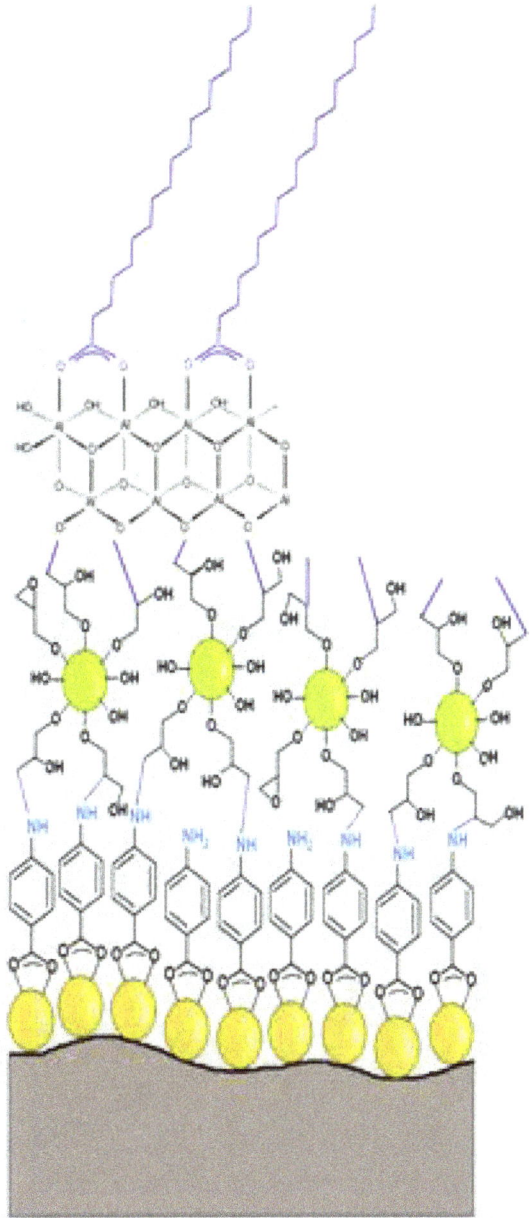

Fig. 8. The three-step coating process of fabric.

pass the membrane and aren't collected in the sample container. It means that according to the oleophilicity of the membrane, some of the oil molecules tend to absorb to membrane without passing.

3. Another reason for nuances in oil rejections (%) may be due to the formation of an oleophilic layer on the membrane which acts as a bridge to absorb oil molecules from the emulsion and allow them to pass through the membrane.

Fig. 11 illustrates the water flux as a function

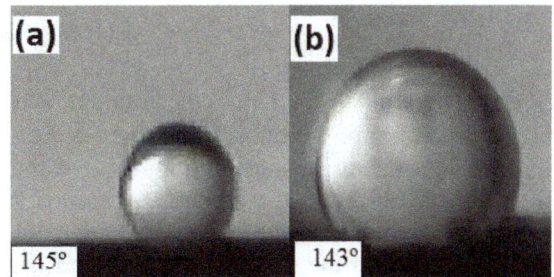

Fig. 9. Contact angle measurements of the functionalized Kevlar fabric (membrane): (a) before using, and (b) after using 6 steps.

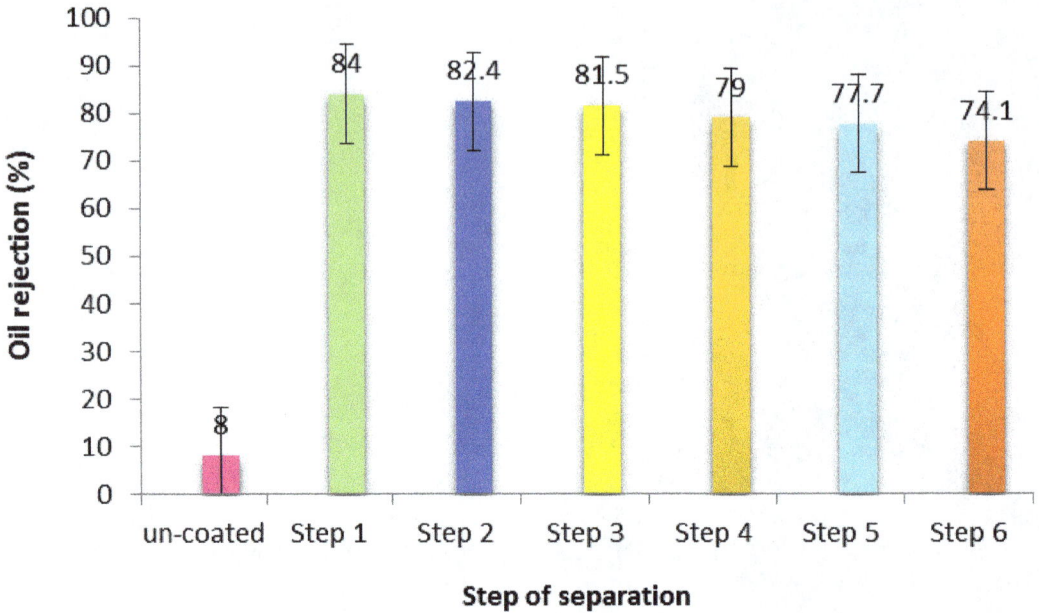

Fig. 10. Oil rejections of the un-functionalized and functionalized Kevlar fabric.

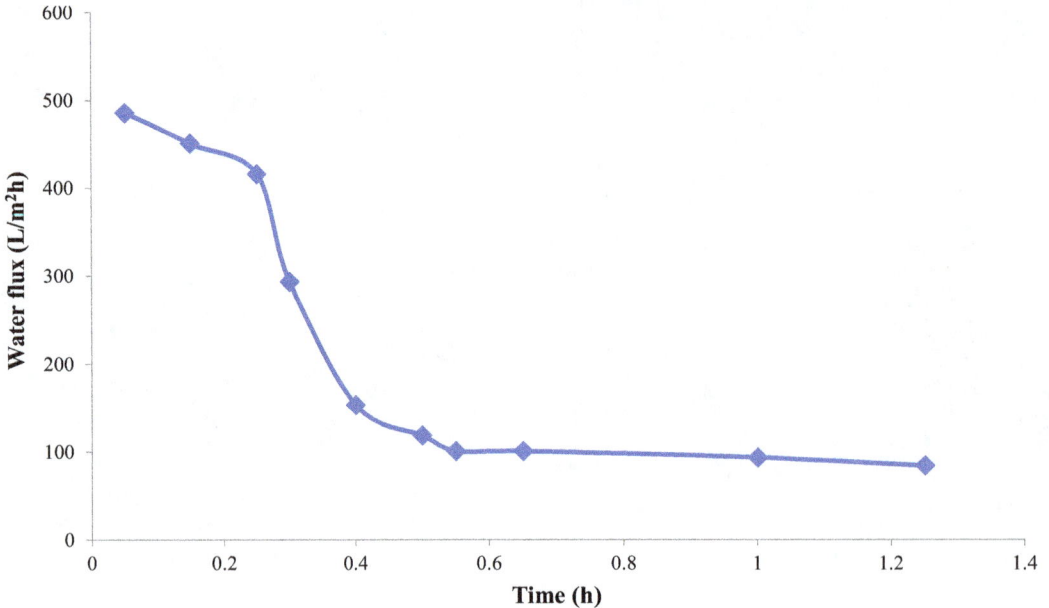

Fig. 11. Time dependence of water flux.

of time. It is understood that the water flux is highly dependent on a number of oil particles pass through the membrane surface according to the superhydrophobicity of membrane and the amount of adsorption depends on the type of interactions occurring between the emulsion droplets and the membrane material such as oleophilic/hydrophilic interactions, hydrogen bonding, Van der Waals interaction and electrostatic effects [36]. According

to Fig. 11, Referring to the initial and final points of the curve, a great reduction in the flux is observed. The flux decreases quickly at the beginning of filtration and then tends to be constant.

A comparison among the oil rejection conducted in this study and other researches has been done in Table 1. Although herein oil rejection is less than the previous one obtained by those researches, it is worthy to mention that this study has some

Table 1. Comparison among nanocomposite membranes performance and those reported by other researchers.

Case	Type of Membrane	Model Oil	Concentration	Oil Rejection (%)	Reference
1	Filter paper prepared by a colloidal deposition method using monodisperse polystyrene NP colloid solution	Hexane	-	~ 100	[1]
2	Filter paper prepared by a colloidal deposition method using monodisperse polystyrene NP colloid solution	Hexadecane	-	~ 100	[1]
3	Filter paper prepared by a colloidal deposition method using monodisperse polystyrene NP colloid solution	Thrichloromethane	-	~ 99	[1]
4	Filter paper prepared by a colloidal deposition method using monodisperse polystyrene NP colloid solution	Silicone oil	-	~ 100	[1]
5	Filter paper prepared by a colloidal deposition method using monodisperse polystyrene NP colloid solution	Gasoline	-	~ 99.5	[1]
6	Polytetrafluoroethylene polyphenylene sulfide composite coating mesh film	Diesel oil and Kerosene	0.05 wt-%	85	[19]
7	Polytetrafluoroethylene polyphenylene sulfide composite coating mesh film	Diesel oil and Kerosene	0. 8 wt-%	99	[19]
8	Polytetrafluoroethylene polyphenylene sulfide composite coating mesh film	Diesel oil and Kerosene	6 wt- %	99	[19]
9	Novel fluorosilanes grafted ceramic membranes C8	Water washing cars colle from a station of gas oil	110 (mg/l)	100	[20]
10	Modification of the symmetric plane Al_2O_3 membranes by ethyltrimethoxysilane (C_2)	Kerosene	570 ppm of water into	77.8	[37]
11	Modification of the symmetric plane Al_2O_3 membranes by hexyltrimethoxysilane (C_6)	Kerosene	570 ppm of water into	85.2	[37]
12	Modification of the symmetric plane Al_2O_3 membranes by octyltrimethoxysilane (C_8)	Kerosene	570 ppm of water into	84.2	[37]
13	Modification of the symmetric plane Al_2O_3 membranes by hexade-cyltrimethoxysilane (C_{16})	Kerosene	570 ppm of water into	84.3	[37]
14	Acid-etched then POTS-modified copper mesh	rapeseed oil	100 wt-%	96.2	[38]
15	Functionalized Kevlar fabric	n-hexane	20% (v/v) oil in water	84	This study

advantages such as:

1) Having simple set up,

2) The presence of strong and superior support (Kevlar fabric), which can be utilized in different industries. Even though, it is expensive,

3) Since the hydrophobic chain of the surface layer is long, this membrane may be used to separate other types of oil and oleophilic compounds such as kerosene, oleic acid, toluene, other hydrocarbons (like octane), etc.,

4) The preparation methods of this membrane and the precursor materials aren't complex,

5) Even after 6 steps of using one membrane, the efficiency didn't change significantly,

6) Compared with other approaches for fabricating materials for oil/water separation, this method is able to fabricate feasible materials for oil/water separation easily and economically. Consequently, the prepared membrane is a promising candidate for oil/water separation.

CONCLUSIONS

Consequential achievements from this research are summarized as below:

A membrane was prepared with two properties of hydrophobicity and oleophilicity.

Oleophilicity was provided using a novel nanoparticle as a top layer, called *Stearate alumoxane.*

In order to prepare a membrane with high adherence of St-A, two other nanoparticles, PAB-A and Bo-E, were selected as the first and middle coating layers, respectively.

Dip-coating was used as the method to coat the support.

Modified fabric can easily absorb the oil, which then forms a film, due to the strong adhesion properties of the oil molecules.

The oil removal of this functionalized membrane reached about 84%.

The prepared membrane was reused 6 times to investigate the capability of the membrane and especially the nanomaterials in the separation process. The oil rejection didn't decrease significantly.

This technique is adequate (from the aspects of economical, method of production, chemical, and mechanical stability, etc.) for the preparation of notable nanomaterials, consequently, application of them in the modification of membrane for oil/water mixtures separation.

CONFLICT OF INTEREST

The authors declare that there are no conflicts of interest regarding the publication of this manuscript.

REFERENCES

1. Du C, Wang J, Chen Z, Chen D. Durable superhydrophobic and superoleophilic filter paper for oil–water separation prepared by a colloidal deposition method. Applied Surface Science. 2014;313:304-10.

2. Chen W, Su Y, Zheng L, Wang L, Jiang Z. The improved oil/water separation performance of cellulose acetate-graft-polyacrylonitrile membranes. Journal of Membrane Science. 2009;337(1–2):98-105.

3. Xue C-H, Ji P-T, Zhang P, Li Y-R, Jia S-T. Fabrication of superhydrophobic and superoleophilic textiles for oil–water separation. Applied Surface Science. 2013;284:464-71.

4. Jiang Y, Hou J, Xu J, Shan B. Switchable oil/water separation with efficient and robust Janus nanofiber membranes. Carbon. 2017;115:477-85.

5. Ma W, Guo Z, Zhao J, Yu Q, Wang F, Han J, et al. Polyimide/cellulose acetate core/shell electrospun fibrous membranes for oil-water separation. Separation and Purification Technology. 2017;177:71-85.

6. Cheng Q, Ye D, Chang C, Zhang L. Facile fabrication of superhydrophilic membranes consisted of fibrous tunicate cellulose nanocrystals for highly efficient oil/water separation. Journal of Membrane Science. 2017;525:1-8.

7. Abbasi M, Salahi A, Mirfendereski M, Mohammadi T, Pak A. Dimensional analysis of permeation flux for microfiltration of oily wastewaters using mullite ceramic membranes. Desalination. 2010;252(1–3):113-9.

8. Yi XS, Yu SL, Shi WX, Sun N, Jin LM, Wang S, et al. The influence of important factors on ultrafiltration of oil/water emulsion using PVDF membrane modified by nano-sized TiO2/Al2O3. Desalination. 2011;281:179-84.

9. Su C, Xu Y, Zhang W, Liu Y, Li J. Porous ceramic membrane with superhydrophobic and superoleophilic surface for reclaiming oil from oily water. Applied Surface Science. 2012;258(7):2319-23.

10. Kharisov BI, Dias HVR, Kharissova OV. Nanotechnology-based remediation of petroleum impurities from water. Journal of Petroleum Science and Engineering. 2014;122:705-18.

11. Mohammed RR, Ibrahim IAR, Taha AH, McKay G. Waste lubricating oil treatment by extraction and adsorption. Chemical Engineering Journal. 2013;220:343-51.

12. Daiminger U, Nitsch W, Plucinski P, Hoffmann S. Novel techniques for iol/water separation. Journal of Membrane Science. 1995;99(2):197-203.

13. Hong A, Fane AG, Burford R. Factors affecting membrane coalescence of stable oil-in-water emulsions. Journal of Membrane Science. 2003;222(1–2):19-39.

14. Chen W, Peng J, Su Y, Zheng L, Wang L, Jiang Z. Separation of oil/water emulsion using Pluronic F127 modified polyethersulfone ultrafiltration membranes. Separation and Purification Technology. 2009;66(3):591-7.

15. Kubota N. Polymer membranes. Filtration membranes for

water treatment. Purasuchikkusu Eji. 2010;56:72-5.

16. Fakhru'l-Razi A, Pendashteh A, Abdullah LC, Biak DRA, Madaeni SS, Abidin ZZ. Review of technologies for oil and gas produced water treatment. Journal of Hazardous Materials. 2009;170(2–3):530-51.

17. Qu X, Alvarez PJJ, Li Q. Applications of nanotechnology in water and wastewater treatment. Water Research. 2013;47(12):3931-46.

18. Maguire-Boyle SJ, Barron AR. A new functionalization strategy for oil/water separation membranes. Journal of Membrane Science. 2011;382(1–2):107-15.

19. Qin F, Yu Z, Fang X, Liu X, Sun X. A novel composite coating mesh film for oil-water separation. Frontiers of Chemical Engineering in China. 2009;3(1):112-8.

20. Khemakhem M, Khemakhem S, Ben Amar R. Emulsion separation using hydrophobic grafted ceramic membranes by. Colloids and Surfaces A: Physicochemical and Engineering Aspects. 2013;436:402-7.

21. Yang H, Pi P, Cai Z-Q, Wen X, Wang X, Cheng J, et al. Facile preparation of super-hydrophobic and super-oleophilic silica film on stainless steel mesh via sol–gel process. Applied Surface Science. 2010;256(13):4095-102.

22. Su C, Xu Y, Zhang W, Liu Y, Li J. Porous ceramic membrane with superhydrophobic and superoleophilic surface for reclaiming oil from oily water. Applied Surface Science. 2012;258(7):2319-23.

23. Wang L, Yang S, Wang J, Wang C, Chen L. Fabrication of superhydrophobic TPU film for oil–water separation based on electrospinning route. Materials Letters. 2011;65(5):869-72.

24. Lin J, Shang Y, Ding B, Yang J, Yu J, Al-Deyab SS. Nanoporous polystyrene fibers for oil spill cleanup. Marine Pollution Bulletin. 2012;64(2):347-52.

25. Shutao W, Yanlin S, Lei J. Microscale and nanoscale hierarchical structured mesh films with superhydrophobic and superoleophilic properties induced by long-chain fatty acids. Nanotechnology. 2007;18(1):015103.

26. Briscoe BJ, Luckham PF, Jayarajah JN, Akeju T. Separation of emulsions using fibrous fabric. Colloids and Surfaces A: Physicochemical and Engineering Aspects. 2000;163(2–3):151-64.

27. Lim T-T, Huang X. In situ oil/water separation using hydrophobic–oleophilic fibrous wall: A lab-scale feasibility study for groundwater cleanup. Journal of Hazardous Materials. 2006;137(2):820-6.

28. Xue C-H, Ji P-T, Zhang P, Li Y-R, Jia S-T. Fabrication of superhydrophobic and superoleophilic textiles for oil–water separation. Applied Surface Science. 2013;284:464-71.

29. Karimnezhad H, Salehi E, Rajabi L, Azimi S, Derakhshan AA, Ansari M. Dynamic removal of n-hexane from water using nanocomposite membranes: Serial coating of para-aminobenzoate alumoxane, boehmite-epoxide and chitosan on Kevlar fabrics. Journal of Industrial and Engineering Chemistry. 2014;20(6):4491-8.

30. Karimnezhad H, Rajabi L, Salehi E, Derakhshan AA, Azimi S. Novel nanocomposite Kevlar fabric membranes: Fabrication characterization, and performance in oil/water separation. Applied Surface Science. 2014;293:275-86.

31. Zhang L, Zhang Z, Wang P. Smart surfaces with switchable superoleophilicity and superoleophobicity in aqueous media: toward controllable oil/water separation. NPG Asia Mater. 2012;4:e8.

32. Maiti S, Mishra IM, Bhattacharya SD, Joshi JK. Removal of oil from oil-in-water emulsion using a packed bed of commercial resin. Colloids and Surfaces A: Physicochemical and Engineering Aspects. 2011;389(1–3):291-8.

33. Cambiella Á, Ortea E, Ríos G, Benito JM, Pazos C, Coca J. Treatment of oil-in-water emulsions: Performance of a sawdust bed filter. Journal of Hazardous Materials. 2006;131(1–3):195-9.

34. Rajabi L, Derakhshan AA. Room Temperature Synthesis of Boehmite and Crystallization of Nanoparticles: Effect of Concentration and Ultrasound. Science of Advanced Materials. 2010;2(2):163-72.

35. Derakhshan AA, Rajabi L, Karimnezhad H. Morphology and production mechanism of the functionalized carboxylate alumoxane micro and nanostructures. Powder Technology. 2012;225:156-66.

36. Chakrabarty B, Ghoshal AK, Purkait MK. Ultrafiltration of stable oil-in-water emulsion by polysulfone membrane. Journal of Membrane Science. 2008;325(1):427-37.

37. Gao N, Ke W, Fan Y, Xu N. Evaluation of the oleophilicity of different alkoxysilane modified ceramic membranes through wetting dynamic measurements. Applied Surface Science. 2013;283:863-70.

38. Guo W, Zhang Q, Xiao H, Xu J, Li Q, Pan X, et al. Cu mesh's super-hydrophobic and oleophobic properties with variations in gravitational pressure and surface components for oil/water separation applications. Applied Surface Science. 2014;314:408-14.

14

Arsenic (III) Adsorption Using Palladium Nanoparticles from Aqueous Solution

Farzane Arsiya[1], Mohammad Hossein Sayadi[1,], Sara Sobhani[2]*

[1]Department of Environmental Sciences, School of Natural Resources and Environment, University of Birjand, Birjand, Iran.

[2]Department of Chemistry, College of Science, University of Birjand, Birjand, Iran.

ABSTRACT

The presence of Arsenic in drinking water is the greatest threat to health effects especially in water. The purpose of this study is application of green palladium nanoparticles for removal of trivalent Arsenic from aqueous solutions and also the impact of some factors such as retention time, pH, concentration of palladium nanoparticles and Arsenic concentrations was studied. The values for Arsenic removal from aqueous solutions were measured by furnace atomic adsorption spectrometry (Conter AA700). In the study, Langmuir and Freundlich isotherm models and pseudo-second order kinetic model were studied. The results of optimization is shown that 0.5 g of nanoparticles can removed %99.8 of Arsenic with initial concentration of 0.5 g/l, in 5 minutes at pH=4. Langmuir model, Freundlich model (R^2=0.94) and pseudo-second order kinetic model (R^2=0.99) shown high correlation for removing of Arsenic from aqueous solutions. It was found, palladium nanoparticles can be used as an efficient method to remove Arsenic from aqueous solutions in a short time.

Keywords: *Alga, Drinking Water, Heavy Metals, Initial Concentration, Synthesis*

INTRODUCTION

Heavy metals are accounted to cause universal and serious issues that have very high toxicity for the health of human being as well as the environment. The heavy metals whose specific weight is higher than 5 g/cm³ have been extensively used in the industries which unfortunately can accumulate in the environment and body tissues of living beings[1, 2]. Though heavy metals are classified into two groups of essential and non-essential elements, nickel, iron, copper, zinc, and manganese are namely the essential elements

required for growth and development of plants, but Arsenic, cadmium, mercury, chrome, and lead are not essential for the living beings [3, 4].

Arsenic is a metalloid found naturally in water, air, earth's crust (rocks and soils) and living systems. Probably presence of Arsenic in drinking water, especially in the underground waters, is the biggest threat for the therapeutics which causes issues such as pulmonary issues, skin lesions and bladder cancer [5-7]. Nevertheless, the emergence of this element in waters is due to human and natural activities[8] and is dependent on temperature, pH, oxidation, and soluble compound[9] The natural

*Corresponding Author Email: mh_sayadi@ birjand.ac.ir

Table 1. Experimental design

specification of experiments	Contact time effect	pH	Arsenic concentration effect	PdNPs concentration effect
Contact time effect	2, 5, 10,15, 20 and 30 min	7	0.5 g/l	1 g/l
pH	5 min	3, 4, 5, 6, 7, 8	0.5 g/l	1 g/l
Arsenic concentration effect	5 min	4	0.5, 0.7, 0.8, 1, 1.5 and 2 g/l	1 g/l
PdNPs concentration effect	5 min	4	0.5 g/l	0.1, 0.25, 0.5, 1, 2 and 4 g/l

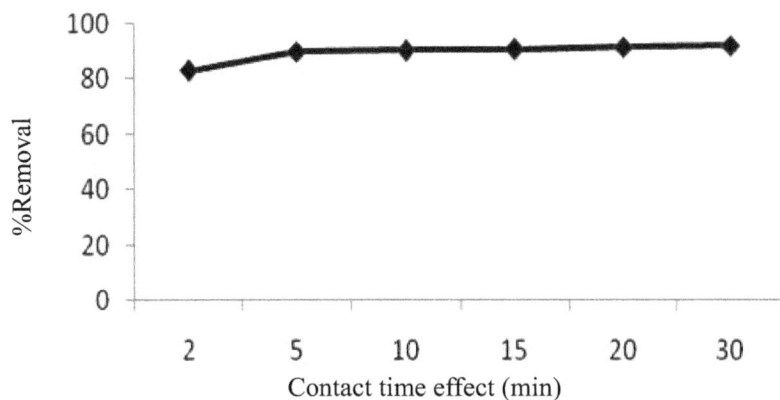

Fig. 1. Effect of contact time on adsorption.

concentration of this element lower than 1-2 µg/l are natural in waters [10]. Generally, Arsenic is found in non-organic forms as a five valence arsenate oxyanions [As (V) or Arsenate) or three valence Arsenite [As (III) or Arsenite) in natural waters wherein Arsenite has higher impairments and toxicity (25-60 times more) for human beings and environment [11]. Some countries have a lowest standard Arsenic level (10 g/l or 10 ppb), although most of them *viz.* China, Bangladesh, Egypt, India, Indonesia, Philippines, and others have a highest standard Arsenic level (50 g/l or 50 ppb) in their regulatory systems[12].

Therefore, some of the general methods that are used for Arsenic adsorption include chemical oxidation, iron sheathed sand and gravel, adsorption, ion exchange resin, activated alumina, membrane processes, nanofiltration and etc. Although some of these advanced methods are frequently expensive [13, 14], but adsorption is a cost-effective simple and efficient method for the adsorption of heavy metals ions from the waste [12, 15]. Till date for the Adsorption of Arsenic from aqueous solutions the nanoparticles such as zero valence iron nanoparticles[16, 17], dioxide titanium nanocrystals [18], titanium oxide nanoparticles [19], Europium [20], copper oxide [21], gold [22], and others have been used at the global level and in even in Iran some studies have been carried out

in this field [23-25]. Palladium nanoparticles are used as catalysts to reduce vehicle pollution and have many potential applications for organic and inorganic regeneration reactions. Besides that, palladium nanoparticles have high reactivity and acceptable adsorption capability due to the high specific surface area. As well, these nanoparticles are used to correct and adsorption environmental pollution[26, 27].

The main aim of present study was Adsorption of Arsenic from aqueous solutions *via* palladium bio-nanoparticles and determination of optimal conditions of adsorption. In this study, the effect of retention time, pH, different concentrations of palladium nanoparticles and Arsenic has been investigated.

Material and methods

The palladium bio-nanoparticles were prepared using *Chlorella Vulgaris* alga extract procured from the Natural Resources and Environment Division of Birjand University. Pd NPs which is used were spherical with good monodispersity and crystalline in nature. The average particle size was 15nm[28].

In this study, for the preparation of sample water containing Arsenic (III), NaAsO$_2$ was used. For the preparation of Arsenic stock solution (1000 ppm) using the atomic mass of the considered salt (129.9 g/mol), 1.73 g of Sodium Arsenite salt was dissolved

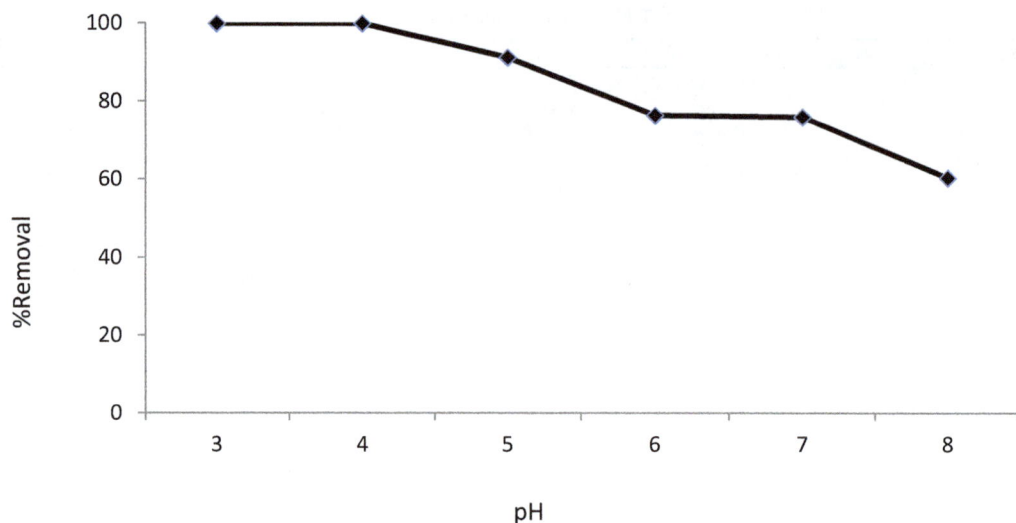

Fig. 2. Effect of pH on adsorption.

in 1-liter deionized water. The entire experiment was carried out at a temperature of 25±1 °C in 150 mm Erlenmeyer and was shaken at 200 rpm [16]. In these experiments, the influence of factors such as retention time (2, 5, 10, 15, 20, 30 minutes), pH (3, 4, 5, 6, 7, 8), concentration of Arsenic pollutant (0.5, 0.7, 0.8, 1, 1.5 and 2 g/l) and concentration of palladium nanoparticles (0.1, 0.25, 0.5, 1, 2 and 4 g/l) was assessed. The specifications of the conducted experiments in each stage are shown in Table 1. In order to adjust the considered pH, NaOH and 0.1 molar HCL were used. In each stage of the experiment, a variable parameter and other constant parameters were considered so that an optimal condition of each parameter is obtained. Even for the preparation of samples initially, 5 ml of sample was taken and kept immovable for a period of 5 minutes so that the particles are settled. Then its supernatant was filtered three times *via* filter paper (0.45 microns pore diameter) with the help of a vacuum pump. Finally, the concentration assessment of this solution was carried out *via* furnace atomic absorption spectrometer (Conter AA700).

Results and Discussion
Results of Contact Time Effect and pH
Arsenic (III) adsorption percent *via* palladium nanoparticles has been demonstrated in Fig. 1. In this experiment the sampling was carried out in definite periods from the solution containing 0.5 g/l of Arsenic and 1 g/l palladium nanoparticles. Even, Fig. (2) shows the effect of different pH in

definite period (5 minutes) on the adsorption of Arsenic (III).

As observed in Fig. 1, initially, with time lapse the Arsenic level reduces with a higher rate and after 5 minutes this rate becomes approximately constant. This is in a manner that in a period of 5 minutes 89.8% and after the lapse of 30 minutes 92.1% of Arsenic is eliminated. The reason for this reduction rate can be considered as very high oxidation state of palladium nanoparticles and an as well reduction amount of contamination in the aqueous solution. Therefore, an optimal adsorption of Arsenic was obtained after the lapse of 5 minutes and it was used for conduction of next experiments. Even Martinson and Reddy in their study showed that adsorption rate of three and five valence Arsenic from the aqueous solution with the help of copper oxide nanoparticles was very high in the initial 30 minutes and after that reduced [21].

Even as per Fig. 2 it can be stated that Arsenic adsorption percent from the aqueous solution with pH increase, reduces in a manner that this rate in pH 3 and 4 is 99.9% and at pH 5, 6, 7 and 8 is 91.2, 76.4, 75.9 and 60.3%, respectively. It may be due to low pH, H^+ ions that exist in the environment compete with the metals to bind to the active site of adsorbent and so the metals with higher positively charge adsorbed simply [29]. Thus, for the conduction of experiments to adsorption Arsenic, a suitable acidic pH accelerates the reaction. For this reason, in this study for conduction of more experiments pH=4 was used. Even Akin and associates in their studies demonstrated that

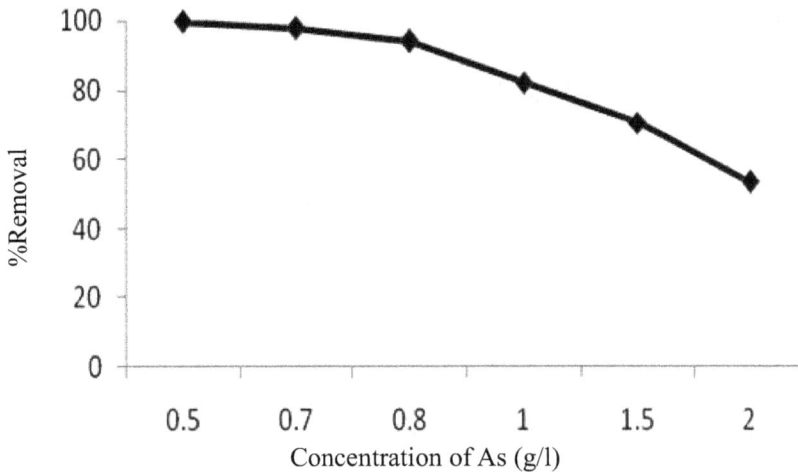

Fig. 3. Effect of initial Arsenic concentration on adsorption.

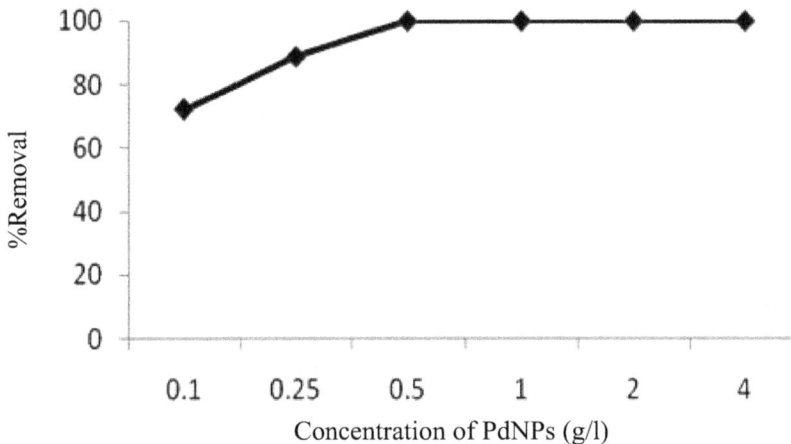

Fig. 4. Effect of adsorbent dose on adsorption.

with pH increase, the Arsenic adsorption percent from aqueous solutions reduces with the help of magnetic nanoparticles such that the maximum Arsenic adsorption was obtained at pH=2.5 [15].

Results of Arsenic Contamination Effect

To assess the preliminary concentration of Arsenic contamination, the experiments were carried out in different Arsenic (III) concentrations and an amount of 1 g/l adsorbent and time and optimal pH determined in the two previous stages. The results of this stage of the experiment are depicted in Fig. 3. As demonstrated in Fig. 3 in the preliminary concentration 0.5 g/l, adsorption percent is 99.8% and later with an increase of concentration amount the adsorption percent also reduces. As with increase of three valence preliminary Arsenic concentration, the adsorption

percent reduces from 99.8 to 53.4% wherein it could be due to effective reduction level of nanoparticles for the Arsenic adsorption. Rahmani and associates in their study to eliminate Arsenic *via* zero valence iron nanoparticles showed that with an increase of Arsenic preliminary concentration from 1 to 30 mg/l, the adsorption percent reduces from 100 to 88.3 percent, wherein these results are in concurrence with the present results[29].

Results of Adsorption Dose Effect

Adsorbent dose (palladium nanoparticles) is one of the important parameters in Arsenic (III) adsorption from the aqueous solution. For this reason, the experiments were carried out in values 0.1, 0.25, 0.5, 1, 2 and 4 g/l of nanoparticle and in the solution containing 0.5 g/l Arsenic (III). Fig. 4 also shows that with concentration increase of

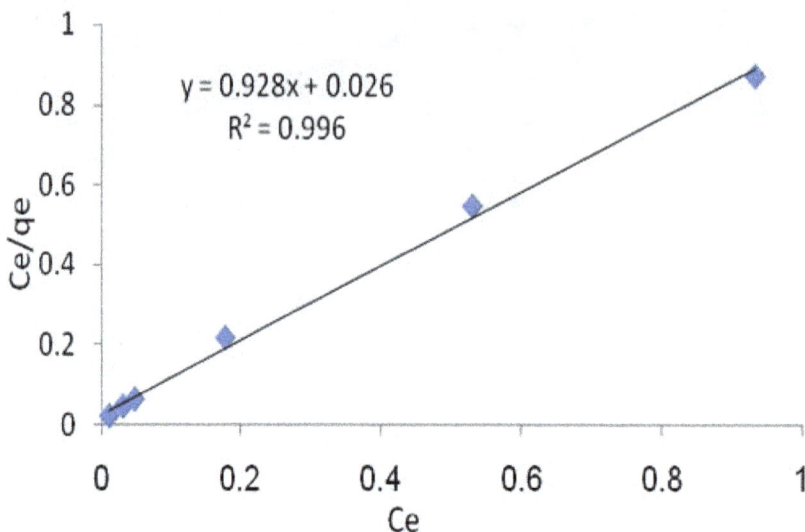

Fig. 5. Linear curve of Langmuir adsorption isotherm for As adsorption using the prepared adsorbent..

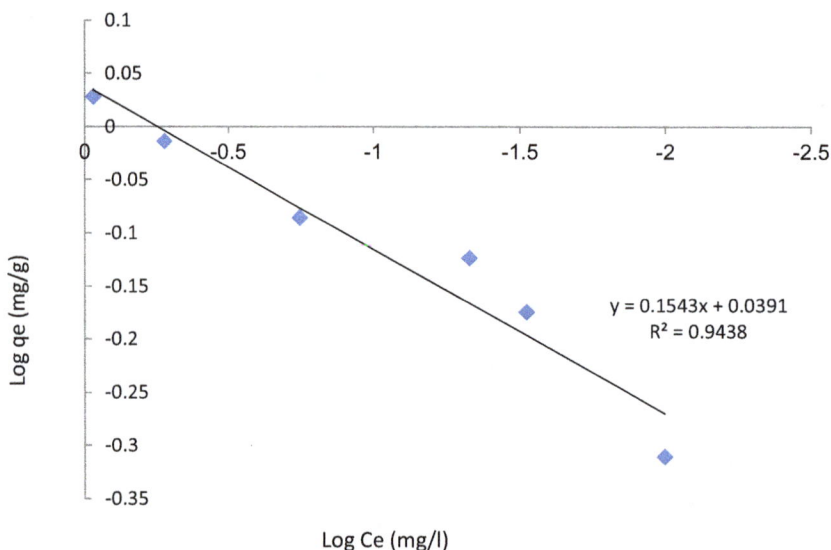

Fig. 6. Linear curve of Freundlich adsorption isotherm for As adsorption using the prepared adsorbent.

nanoparticles, the adsorption percent also increases till it reaches a saturation point. As in concentrations of 0.5 g and above, the existing Arsenic level in the aqueous solution was almost eliminated (99.9%) 100%. Therefore, it can be safely mentioned that initially with an increase of adsorbent dose the Arsenic adsorption percent also increased and later reaches a stable level. Goswami and associates in their study, for Adsorption of Arsenic in different concentrations *via* copper (II) oxide nanoparticles, showed that Arsenic adsorption percent increases

with the increase of adsorbent amount and then reaches a constant quantum [30].

Isotherm and Kinetic Studies

In order to analyze absorbed isotherms, the obtained data were analyzed with the help of Langmuir and Freundlich adsorption isotherms. The linear equation of Langmuir and Freundlich adsorption isotherm was equivalent to the equations 1 and 2, respectively [15, 31]. Equation 1:

Table 2. Langmuir and Freundlich isotherm parameters for the sorption of As(III)

| | Isotherm parameters Sorption | | | | |
| Freundlich model | | | Langmuir model | | |
K_f	n	R^2	K_L	q_m	R^2
1.09	6.48	0.94	35.04	1.09	0.99

Table 3. Pseudo first and second-order reaction of adsorption

| Pseudo first-order reaction of adsorption | | | Pseudo-second-order reaction of adsorption | | |
k_1	q_e	R^2	k_2	q_e	R^2
1.19	2.11	0.75	4.71	0.50	0.99

$$\frac{C_e}{q_e} = \frac{1}{q_m k_L} + \frac{C_e}{q_m}$$

Equation 2:

$$Log q_e = Log K_F + \frac{1}{n} Log C_e$$

Wherein K_L is Langmuir constant, K_F and n are Freundlich constants, C_e is a concentration of adsorbing substance in liquid phase after reaching an equilibrium state in mg/l and q_e is the quantity of adsorbed substance in a unit mass of adsorbent in mg/g.

Even the Pseudo first and second order kinetic equations were based on equations 3 and 4, respectively.
Equation 3:

$$Log (q_e - q_t) = Log (q_e) - \frac{K_2}{2.303} t$$

Equation 4:

$$\frac{t}{q_t} = \frac{1}{K_2 q_e^2} + \frac{1}{q_e} t \qquad (4)$$

Wherein in these equations, q_e is equivalent to the amount of adsorbed salts in an equilibrium state (mg/g), q_t is the amount of adsorbed salts at time t (mg/g), K_1 and K_2 are Pseudo first and second order kinetic equilibrium rate constants.

In this study, the isothermal Langmuir and Freundlich models, as well as the Pseudo first and second order kinetic adsorption models, were used to determine the control mechanism of adsorbent processes. The constant coefficient and correlation coefficient of Langmuir and Freundlich adsorption isotherms and Pseudo second order kinetic is depicted in Tables 2 and 3. The Figs. 5 and 6 show the Langmuir and Freundlich adsorption isotherms and Fig. 7 shows the Pseudo second order kinetic adsorption. With the comparison of the correlation coefficient of both the isotherms, it can be concluded that Arsenic follows both the isotherm

adsorption models. The correlation coefficient in Langmuir and Freundlich models were equivalent to 0.99 and 0.94, respectively. Even Koohpayezadeh and associates in their study for the Adsorption of Arsenic took the assistance of iron nanoparticles from the Langmuir and Freundlich model which shows that Arsenic abides both the models [24, 32].

Even the adsorption data better follows the Pseudo order two kinetic model in relation to the Pseudo order one kinetic (Fig. 7). Akin and associates in their study to eliminate arsenate with the help of iron nanoparticles showed that Arsenic adsorption follows the Pseudo order two equation and its R^2 amount was equivalent to 0.99 [15].

CONCLUSION

The results of present research clearly showed that palladium nanoparticles synthesized *via* green alga *Chlorella Vulgaris* in a short time period have a significant ability to eliminate Arsenic from the aqueous solution. This is in a manner that it introduces these particles as a suitable option for the Adsorption of Arsenic. In this study, an optimal condition for Adsorption of Arsenic from aqueous solution with the help of palladium nanoparticles was determined. The results demonstrated that in a short period of 5 minutes and at pH=4, 0.5 g of nanoparticle has the capability to eliminate 99.8% of Arsenic from the aqueous solution. Arsenic follows the two models Langmuir and Freundlich adsorption isotherm as well as Pseudo order two kinetic equation, in a manner that adsorption isotherms R^2 were equivalent to 0.94 and for Pseudo second order equation equivalent to 0.999, respectively.

ACKNOWLEDGEMENT

This research paper is part of MSc thesis of Environmental Sciences department of the University of Birjand. The valuable collaboration

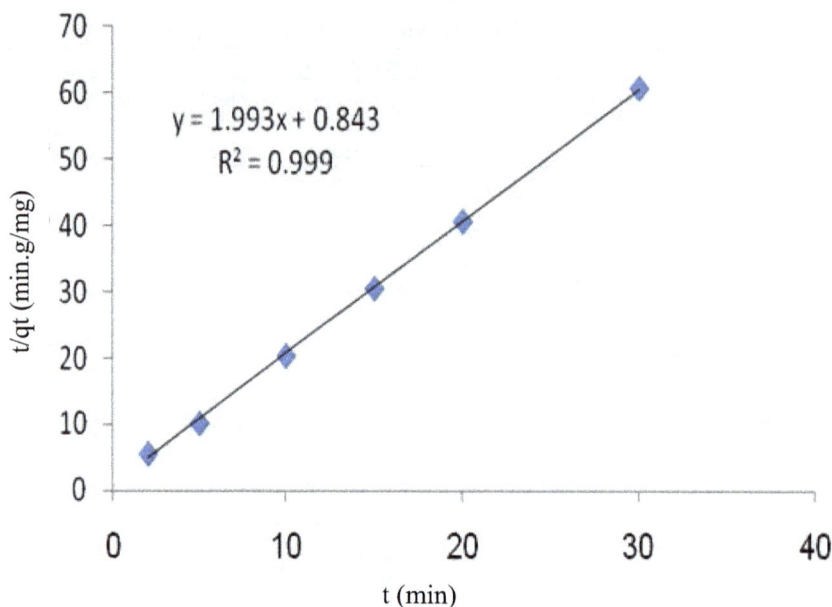

$$y = 1.993x + 0.843$$
$$R^2 = 0.999$$

Fig. 7. Pseudo-second-order kinetics for lead removal using the prepared adsorbent.

of faculty authorities of Natural Resources and Environmental Sciences, University of Birjand to provide necessary facilities for conduction of this study is highly appreciated.

CONFLICT OF INTEREST

The authors declare that there are no conflicts of interest regarding the publication of this manuscript.

REFERENCES

1. Sayadi MH, Rezaei MR. Impact of land use on the distribution of toxic metals in surface soils in Birjand city, Iran. Proceedings of the International Academy of Ecology and Environmental Sciences. 2014;4(1):18-29.

2. Sayadi MH, Rezaei MR, Rezaei A. Fraction distribution and bioavailability of sediment heavy metals in the environment surrounding MSW landfill: a case study. Environmental Monitoring and Assessment. 2014;187(1):4110.

3. Sayadi MH, Rezaei MR, Rezaei A. Sediment Toxicity and Ecological Risk of Trace Metals from Streams Surrounding a Municipal Solid Waste Landfill. Bulletin of Environmental Contamination and Toxicology. 2015;94(5):559-63.

4. Sayadi MH, Shabani M, Ahmadpour N. Pollution Index and Ecological Risk of Heavy Metals in the Surface Soils of Amir-Abad Area in Birjand City, Iran. Health Scope. 2015;4(1):ee21137.

5. Ghaleno OR, Sayadi M, Rezaei M, Kumar CP, Somashekar R, Nagaraja B. Potential ecological risk assessment of heavy metals in sediments of water reservoir case study: Chah Nimeh of Sistan. Proc Int Acad Ecol Environ Sci. 2015;5(4):89-96.

6. Mohan D, Pittman Jr CU. Arsenic removal from water/ wastewater using adsorbents—A critical review. Journal of

Hazardous Materials. 2007;142(1–2):1-53.

7. Sayadi M, Torabi S. Geochemistry of soil and human health: A review. Pollution Research. 2009;28(2):257-62.

8. Smedley PL, Kinniburgh DG. A review of the source, behaviour and distribution of arsenic in natural waters. Applied Geochemistry. 2002;17(5):517-68.

9. Nordstrom DK. Worldwide Occurrences of Arsenic in Ground Water. Science. 2002;296(5576):2143-5.

10. Organization WH. Guidelines for drinking-water quality. Vol. 1, Recommendations. 3rd ed: World Health Organization; 2004.

11. Wang P, Sun G, Jia Y, Meharg AA, Zhu Y. A review on completing arsenic biogeochemical cycle: Microbial volatilization of arsines in environment. Journal of Environmental Sciences. 2014;26(2):371-81.

12. Nguyen TV, Vigneswaran S, Ngo HH, Pokhrel D, Viraraghavan T. Specific Treatment Technologies for Removing Arsenic from Water. Engineering in Life Sciences. 2006;6(1):86-90.

13. Bhargavi RJ, Maheshwari U, Gupta S. Synthesis and use of alumina nanoparticles as an adsorbent for the removal of Zn(II) and CBG dye from wastewater. International Journal of Industrial Chemistry. 2015;6(1):31-41.

14. Bissen M, Frimmel FH. Arsenic — a Review. Part II: Oxidation of Arsenic and its Removal in Water Treatment. Acta hydrochimica et hydrobiologica. 2003;31(2):97-107.

15. Akin I, Arslan G, Tor A, Ersoz M, Cengeloglu Y. Arsenic(V) removal from underground water by magnetic nanoparticles synthesized from waste red mud. Journal of Hazardous Materials. 2012;235–236:62-8.

16. Kanel SR, Manning B, Charlet L, Choi H. Removal of Arsenic(III) from Groundwater by Nanoscale Zero-Valent Iron. Environmental Science & Technology. 2005;39(5):1291-8.

17. Rahmani A, Ghaffari H, Samadi M. A comparative

study on arsenic (III) removal from aqueous solution using nano and micro sized zero-valent iron. Iranian Journal of Environmental Health Science & Engineering. 2011;8(2):157-66.

18. Pena M, Meng X, Korfiatis GP, Jing C. Adsorption Mechanism of Arsenic on Nanocrystalline Titanium Dioxide. Environmental Science & Technology. 2006;40(4):1257-62.

19. Li Q, Easter NJ, Shang JK. As(III) Removal by Palladium-Modified Nitrogen-Doped Titanium Oxide Nanoparticle Photocatalyst. Environmental Science & Technology. 2009;43(5):1534-9.

20. Ghosh D, Luwang MN. Arsenic detection in water: YPO4:Eu3+ nanoparticles. Journal of Solid State Chemistry. 2015;232:83-90.

21. Martinson CA, Reddy KJ. Adsorption of arsenic(III) and arsenic(V) by cupric oxide nanoparticles. Journal of Colloid and Interface Science. 2009;336(2):406-11.

22. Shrivas K, Shankar R, Dewangan K. Gold nanoparticles as a localized surface plasmon resonance based chemical sensor for on-site colorimetric detection of arsenic in water samples. Sensors and Actuators B: Chemical. 2015;220:1376-83.

23. JANBAZ FM, KHOLGHI M, HORFAR A, HAGHSHENAS D. EXPERIMENTAL INVESTIGATION OF ARSENIC REMOVAL BY USING FE NANO PARTICLES IN BATCH EXPERIMENT. journal of Environmental Studies. 2014;39(4):149-56.

24. Koohpayehzadeh H, Torabian A, Nabi Bidhendi G, Habashi N. Nanoparticle Zere-valent Iron Affect on As (V) Removal from Drinking Water. Journal of Water and Wastewater(parallel title); Ab va Fazilab (in persian). 2012;23(3):60-7.

25. Olyaie E, Banejad H, Rahmani AR, Afkhami A, Khodaveisi J.

Feasibility study of using Calcium Peroxide Nanoparticles in Arsenic Removal from Polluted Water in Agriculture and It's Effect on the Irrigation Quality Parameters Iranian Journal of Health and Environment. 2012;5(3):319-30.

26. Adlim M, Abu Bakar M, Liew KY, Ismail J. Synthesis of chitosan-stabilized platinum and palladium nanoparticles and their hydrogenation activity. Journal of Molecular Catalysis A: Chemical. 2004;212(1–2):141-9.

27. De Corte S, Hennebel T, Fitts JP, Sabbe T, Bliznuk V, Verschuere S, et al. Biosupported Bimetallic Pd–Au Nanocatalysts for Dechlorination of Environmental Contaminants. Environmental Science & Technology. 2011;45(19):8506-13.

28. Arsiya F, Sayadi MH, Sobhani S. Green synthesis of palladium nanoparticles using Chlorella vulgaris. Materials Letters. 2017;186:113-5.

29. Rahmani A, Ghaffari H, Samadi M. Removal of Arsenic (III) from Contaminated Waterby Synthetic Nano Size Zerovalent Iron. World Academy of Science, Engineering and Technology, International Journal of Environmental, Chemical, Ecological, Geological and Geophysical Engineering. 2010;4(2):96-9.

30. Goswami A, Raul PK, Purkait MK. Arsenic adsorption using copper (II) oxide nanoparticles. Chemical Engineering Research and Design. 2012;90(9):1387-96.

31. Sartape AS, Mandhare AM, Jadhav VV, Raut PD, Anuse MA, Kolekar SS. Removal of malachite green dye from aqueous solution with adsorption technique using Limonia acidissima (wood apple) shell as low cost adsorbent. Arabian Journal of Chemistry.

32. Sajadi F, Sayadi MH, Hajiani M. Study of optimizing the process of Cadmium adsorption by synthesized silver nanoparticles using Chlorella vulgaris. Journal of Birjand University of Medical Sciences. 2016;23(2):119-29.

15

Study of the Adsorption of Amido Black 10B Dye from Aqueous Solution Using Polyaniline Nano-adsorbent: Kinetic and Isotherm Studies

Marjan Tanzifi*, Mohsen Mansouri, Maryam Heidarzadeh, Kobra Gheibi

Department of Chemical Engineering, Faculty of Engineering, University of Ilam, Ilam, Iran

ABSTRACT

In the present study, adsorptive properties of Polyaniline (PAn) were investigated for Amido Black 10B dye in aqueous solution. Different variables, including adsorption time, adsorbent dosage, solution pH and initial dye concentration were changed, and their effects on dye adsorption onto PAn nano-adsorbent were investigated. The study yielded the result that an increase in pH decreases the adsorption efficiency of nano-adsorbent. Also, Dye adsorption capacity increased with increase in the initial dye concentration. Optimum adsorption time and nano-adsorbent dosage were obtained 30 min and 0.1 gr, respectively. Kinetic studies illustrated that the Amido Black 10B dye adsorption process onto PAn nano-adsorbent followed the pseudo-second-order model, which indicates that the adsorption process is chemisorption-controlled. Also, adsorption equilibrium data were fitted to Freundlich isotherm. The maximum dye adsorption capacity, predicted by the Langmuir isotherm, was 142.85 mg/g. Moreover, Dubinin-Radushkevich isotherm showed that the adsorption of dye onto PAn nano-adsorbent is a chemisorption process.

KEYWORDS: *Amido Black 10B; Isotherm; Kinetic; Nano-adsorbent; Polyaniline*

INTRODUCTION

Dyes have been widely used for many years for various applications such as: textile, pigment, paint and etc. About 1.6 million tons of dyes are produced per year and 10–15% of this amount enters the wastewater stream [1,2]. Dyes are organic compounds containing one or more benzene rings. Such toxic materials cause irreparable damage to the environment and humans, such as cancer, mutagenesis, etc. Therefore, removal of dyes from water is essential and necessary [3]. Typically, dye is removed from water and wastewater using several

methods such as electrochemical treatment [4], membrane filtration [5], photocatalytic degradation [6], as well as adsorption [7-9]. Adsorption technique is an appropriate and economic way to produce water with high quality. In adsorption process, elimination of pollutant from wastewater is conducted via binding it to an organic or inorganic adsorbent. The binding can be done by ion exchange, electrostatic, Vander Waals, etc. Dye adsorption process through adsorbent is dependent on various conditions such as adsorption time, pH of solution, particle size of adsorbent, temperature

---

* Corresponding Author Email: m.tanzifi@ilam.ac.ir

and presence of surfactants [10]. Application of nano-adsorbents in adsorption of contaminants from wastewater, due to their high specific surface area, high adsorption and desorption capacity and high reactivity, has received extensive consideration in recent years [11-13]. Different nano-adsorbents, including TiO$_2$/chitosan nanocomposite [14], flower-shaped Zinc oxide nanoparticles (ZON) [15], copper oxide nanoparticle loaded on activated carbon (Cu$_2$O-NP-AC) [16], magnetic oxidized multiwalled carbon nanotube-k-carrageenan-Fe$_3$O$_4$ nanocomposite [17], graphene/Fe$_3$O$_4$/chitosan nanocomposite [18], have been used for removal of dye from wastewater.

Among the variety of industrial dyes, Amido Black 10B is a high toxicity dye which applied to both of natural and synthetics fibers namely, wool, cotton, silk, polyesters, rayon and acrylics. This diazo dye causes respiratory diseases and irritation of skin and eye [19]. Generally, a Little research has been done regarding the elimination of Amido Black 10B dye from wastewater by adsorption process. In the study conducted by Garg et al (2015), zeolite synthesized from fly ash was used as an adsorbent for the uptake of Amido Black 10B dye. It was found that optimum zeolite dosage and contact time were 10g/L and 6 hr, respectively. Also, maximum dye adsorption was obtained at low pH in the range 2-5 [19]. The results of the research carried out by Zhang et al (2016) showed that Zr (IV) surface-immobilized cross-linked chitosan/bentonite composite are highly efficient in removal of Amido Black 10B dye from aqueous solution. It was found that adsorption data fitted the Langmuir isotherm and the maximum adsorption capacity was reported to be 418.4 mg/g at natural solution pH (pH=6) and 298 K [20].

Polyaniline is one of the most important conductive polymers which has ion exchange properties and is capable of removing various contaminants such as heavy metals, nitrate, organic materials as well as dyes from water and wastewater. Various factors, including polymer synthesis conditions, the presence of surfactant, size of polymer and size and type of the dopant affect on ion exchange properties of polyaniline [21-23]. In the research carried out by Sharma et al (2016), polyaniline used for the adsorptive removal of cationic (crystal violet) and anionic (methyl orange) dyes from aqueous solutions. The adsorption capacity was obtained up to 245 and 220 mg/g for crystal violet and methyl orange, respectively. Furthermore, both dyes followed pseudo second order kinetic and Langmuir isotherm models [24]. Bhaumik etal (2013) showed that polypyrrole–polyaniline nanofibre is an effective adsorbent for removal of Congo red dye from aqueous solution. The maximum adsorption capacity was found to be 222.22 mg/g at 25 °C [25].

In the present work, the adsorption of Amido Black 10B dye from water using polyaniline nano-adsorbent is experimentally studied. The experiments were conducted to scrutinize the effect of different experimental parameters including pH of solution, nano-adsorbent dosage, initial concentration and adsorption time on adsorption efficiency of dye. Furthermore, kinetic and isotherm studies of Amido Black 10B adsorption on polyaniline nano-adsorbent were carried out.

EXPERIMENTAL

Materials and Instruments

Aniline monomer, ammonium peroxydisulfate, sulfuric acid, sodium hydroxide, sodium carboxymethyl cellulose and Amido Black 10B pigment were purchased from Merck (Germany). Aniline monomer was distilled once before polymerization to remove the impurities. The following instruments were also utilized in the process of the study: a magnetic stirrer (model HMS 8805, Iran), digital scale (model Traveler TA30), scanning electron microscope (SEM) (model KYKY-EM3200, China) and Fourier-transform infrared (FTIR) spectrometer (model VERTEX 70; Bruker, Germany). Moreover, UV–visible spectroscopy (UV-VIS) (model Perkin Elmer, lambda 25) was used to determine the concentration of dye in the solution. The spectroscopy was calibrated using standard Amido Black 10B solutions (0.25-15 ppm).

Synthesis of Nano-adsorbent

In order to prepare PAn nano-adsorbent, 2.5 g of ammonium peroxydisulfate as oxidizing agent, was added to 100 ml of 1M sulfuric acid containing sodium carboxymethyl cellulose (0.1 g) as a surfactant. The solution was stirred by a magnetic stirrer for 30 minutes. Afterwards, 1 ml of aniline monomer was added drop-wise to the solution. The solution was filtered after 5 hr. In order to remove

Fig. 1. Scanning electron microscopy of Polyaniline nano-adsorbent

oligomers, impurities and acid, the polymer was washed several times with distilled water. The final product was dried for 48 hours in an oven at 45 °C and converted into a fine powder.

Dye Adsorption Experiments

Different parameters including adsorbent dosage, pH, adsorption time and initial dye concentration were changed, and their impacts on the efficiency of PAn nano-adsorbent in removing Amido Black 10B dye from solution was investigated. Also, kinetic and isotherm studies of dye adsorption were carried out. In all experiments, solution volume and stirring speed were 50 ml and 500 rpm. Sulfuric acid and sodium hydroxide were used to change the pH of solution. A specified amount of nano-adsorbent was added to the initial solution with certain dye concentration. The solution was next stirred using a magnetic stirrer for a certain time. Afterwards, it was filtered, and the concentration of dye in solution was determined using a UV-Vis spectrophotometer at 618 nm maximum wavelength.

Removal efficiency of dye was determined using the following equation:

$$(\%) = \frac{c_i - c_f}{c_i} \times 100 \qquad (1)$$

Where, C_i (mg/l) and C_f (mg/l) are the initial and final dye concentrations, respectively. q_t (mg/g) is

dye adsorption capacity at time (t), and q_e (mg/g) is the amount of adsorption at equilibrium, which were calculated as follows:

$$q_t = (C_i - C_t) \times \frac{V}{m} \qquad (2)$$

$$q_e = (C_i - C_e) \times \frac{V}{m} \qquad (3)$$

Where, C_t is the concentration of dye at time (t); C_e is the equilibrium concentration of dye; V is the solution volume (ml) and m is the PAn nano-adsorbent mass (gr).

RESULTS AND DISCUSSION

Characterization of Nano-adsorbent

The morphology of the synthesized nano-adsorbent was investigated using scanning electron microscopy (SEM). Fig. 1 shows the SEM image of PAn nano-adsorbent. As shown in figure, the synthesized adsorbent particles feature nano-scaled size (The average size is about 59 nm), uniform distribution, and spherical shape. Presence of surfactant in polymer synthesis environment, lead to a decrease in particle size. surfactant can either form a chemical bond with polymer or be physically adsorbed, consequently preventing the excessive growth of the polymer chain and accumulated mass of particles during polymerization. Nano-adsorbent has high specific surface area and consequently

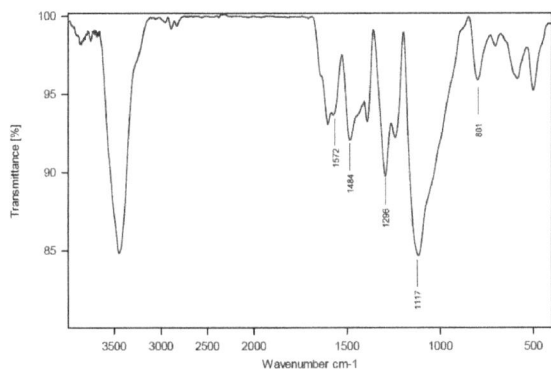

Fig. 2. Fourier Transform Infrared Spectroscopy of Polyaniline nano-adsorbent

high adsorption capacity.

Fig. 2 shows the infrared spectrum of PAn nano-adsorbent within the range of 450-4000 Cm^{-1}. As can be seen, the FTIR spectrum of nano-adsorbent features peaks at wavelength 1572, 1484, 1296, 1117, 801 Cm^{-1} ascribed to (C=C stretching vibration of the quinoid ring), (C=C stretching vibration of the benzenoid ring), (C-N stretching vibration), (C-H in-plane deformation), and (C-H out-of-plane deformation), respectively [26].

Effect of pH

The pH of solution is one of the main factors affecting the adsorption of adsorbate since adsorbent surface charge and the degree of ionization of adsorbate are influenced by pH. For investigation the effect of PH of solution on adsorption efficiency, the range of pH was considered to be 2-10. Adsorbent dosage, initial concentration and volume of dye solution were 0.1 gr, 30mg/l and 50 ml, respectively. The result was illustrated in Fig. 3. As it is shown in the figure, adsorption efficiency decreases with increase in pH value of solution. An increase in the pH of solution leads to an increase in the negative charge density of the adsorbent surface area. The electrostatic repulsion between the negatively charged pigment and negatively charged surface of the adsorbent reduces dye adsorption. However, decreasing the pH of solution increases the adsorption efficiency of nano-adsorbent. The reason might lie in the fact that at low pH, active sites in the structure of nano-adsorbent can be protonated. As a result, the positive charge density of adsorbent surface area increases, and consequently, dye adsorption efficiency increases due to electrostatic attraction.

More dye adsorption happened at pH values lower than 6. Hence, the optimum pH of Amido Black 10B dye solution is in the range 2–6. Therefore, the pH of 6 (natural solution pH) was used for all other experiments.

Effect of Nano-adsorbent Dosage

Adsorbent dosage is one of the effective parameters in determining dye adsorption efficiency. Gaining higher adsorption efficiency with less adsorbent dosage reduces the adsorption cost. In order to evaluate the effect of nano-adsorbent dosage on dye adsorption, different amounts of PAn (0.02-0.2 gr) was added to 50 ml of 30 mg/l dye solution. The effect of nano-adsorbent dosage on dye adsorption efficiency is shown in Fig. 4. As can be seen, an increase in nano-adsorbent dosage leads to a rise in adsorption efficiency. Dye adsorption efficiency per 0.1 g of nano-adsorbent dosage was measured 95%, and then remains constant in the range of 0.1-0.2 gr. Therefore, the optimum nano-adsorbent dosage was considered to be   0.1 g.

Effect of Adsorption Time and Dye Adsorption Kinetics

Fig. 5 illustrates the effect of adsorption time on dye adsorption efficiency. The effect of adsorption time on dye adsorption efficiency, was done with the change in time from 1 to 60 min. As can be seen, increasing the adsorption time from 1 min to 30 min causes the adsorption efficiency to increase and remain constant afterwards. As a result, the optimum adsorption time of PAn nano-adsorbent was determined to be 30 min.

The results of the change of adsorption time were analyzed to obtain information about the kinetics of adsorption onto PAn nano-adsorbent. The adsorption kinetics of Amido Black 10B dye onto PAn nano-adsorbent was investigated using the three equations: pseudo-first-order, pseudo-second-order, and Weber–Morris.

Pseudo-first-order kinetic model is based on the assumption that the process of adsorption can be controlled by weak interaction between adsorbate and adsorbent surface (physical adsorption). Linear form of pseudo-first-order equation [27] is expressed as follows:

$$log(q_e - q_t) = logq_e - \left(\frac{K_1}{2.303}\right)t \qquad (4)$$

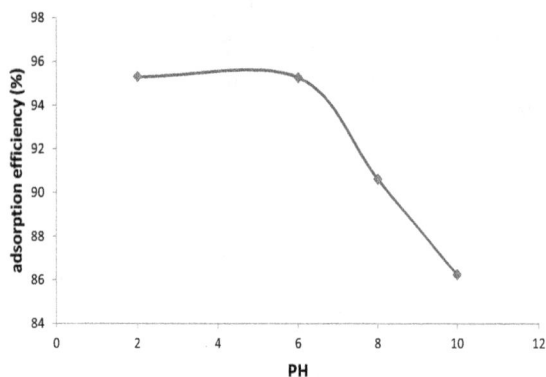

Fig. 3. Effect of pH on adsorption efficiency of Amido Black 10B dye onto Polyaniline nano-adsorbent

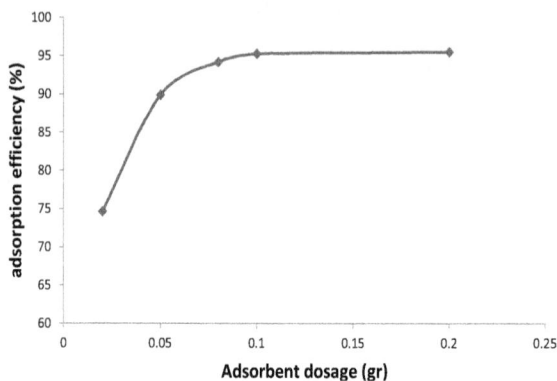

Fig. 4. Effect of adsorbent dosage on adsorption efficiency of Amido Black 10B dye onto Polyaniline nano-adsorbent

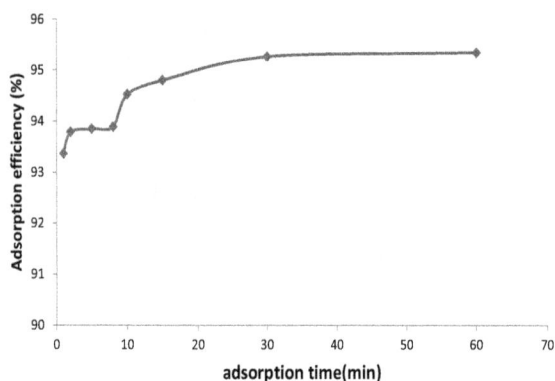

Fig. 5. Effect of adsorption time on adsorption efficiency of Amido Black 10B dye onto Polyaniline nano-adsorbent

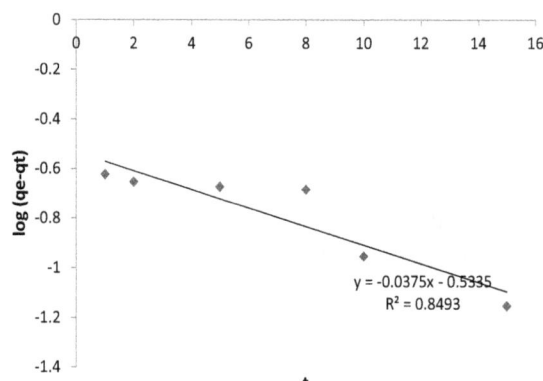

Fig. 6. Pseudo-first-order plot of Amido Black 10B adsorption onto Polyaniline nano-adsorbent

Where q_e (mg/g) and q_t (mg/g) are the amount of dye adsorbed at equilibrium and at time (t); and K_1 is the rate constant of pseudo-first-order equation. Fig. 6 shows the plot of the pseudo-first-order model for dye adsorption on PAn nano-adsorbent. The correlation coefficient and rate constant of the pseudo-first-order model were presented in Table 1.

The adsorption kinetic of Amido Black 10B dye was also investigated using pseudo-second-order model. This kinetic model assumes that the process of dye adsorption is chemisorption. This model is expressed as follows [28]:

$$q_t = \frac{K_2 q_e^2 t}{1 + K_2 q_e t} \qquad (5)$$

Where, K_2 is the rate constant of pseudo-second-order model. This equation is presented in four linear forms, i.e. type 1, type 2, type 3, and type 4 which were shown in Table 1. Moreover, the plots of pseudo-second-order model for dye adsorption through PAn nano-adsorbent were

presented in Figs. 7a-7d. As the figures show, the pseudo-second-order model, type 1 results in a better correlation coefficient in comparison to the other types of pseudo-second-order model.

In order to determine whether intraparticle diffusion is a rate-controlling step in dye adsorption, Weber-Morris model was used for the analysis of the kinetic data. The Weber-Morris model can be written as follows [29]:

$$q_t = K_{id}(t)^{0.5} + C \qquad (6)$$

Where, K_{id} is rate constant of Weber-Morris model; and C is a constant that gives an idea about the thickness of the boundary layer. If the plot q_t versus $t^{0.5}$ is linear, the process of dye adsorption is controlled by diffusion resistance. When the plot passes through the origin, it indicates that intraparticle diffusion is the only rate-controlling step. The slope and intercept of the plot (Fig. 8) can be used to calculate values for the constants K_{id} and C, respectively. The values were presented in Table 1.

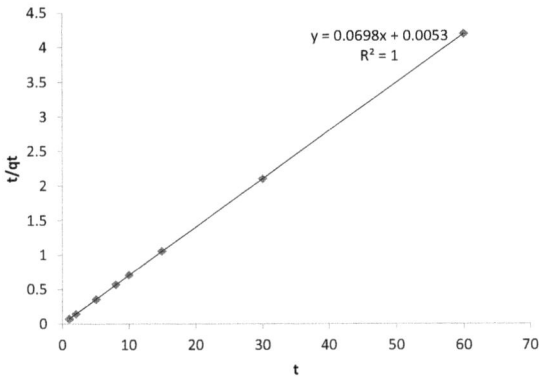

Fig. 7(a). Pseudo-second-order plot of Amido Black 10B adsorption onto Polyaniline nano-adsorbent (type 1)

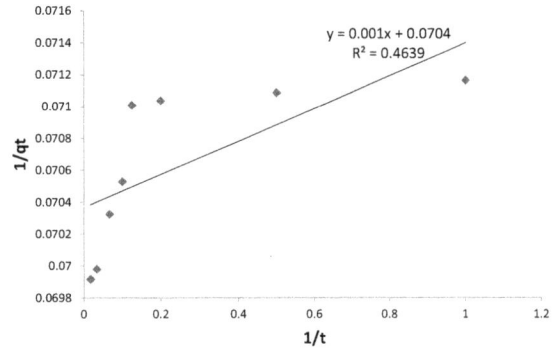

Fig. 7(b). Pseudo-second-order plot of Amido Black 10B adsorption onto Polyaniline nano-adsorbent (type 2)

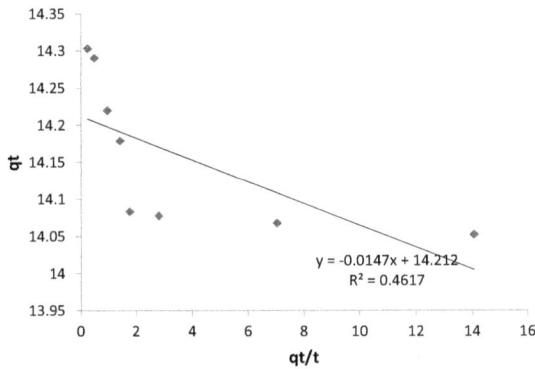

Fig. 7(c). Pseudo-second-order plot of Amido Black 10B adsorption onto Polyaniline nano-adsorbent (type 3)

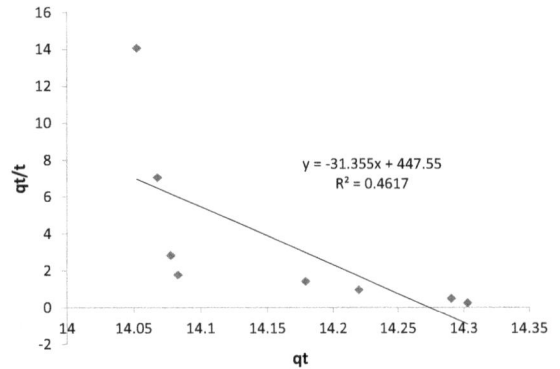

Fig. 7(d). Pseudo-second-order plot of Amido Black 10B adsorption onto Polyaniline nano-adsorbent (type 4)

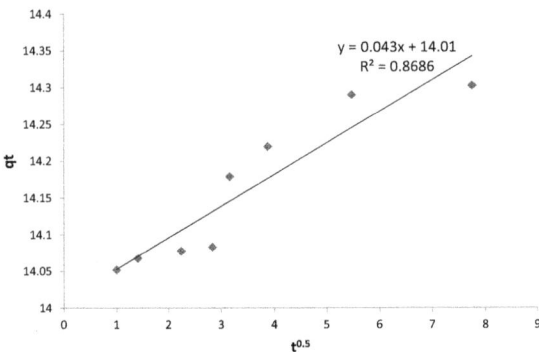

Fig. 8. Weber-Morris plot of Amido Black 10B adsorption onto Polyaniline nano-adsorbent

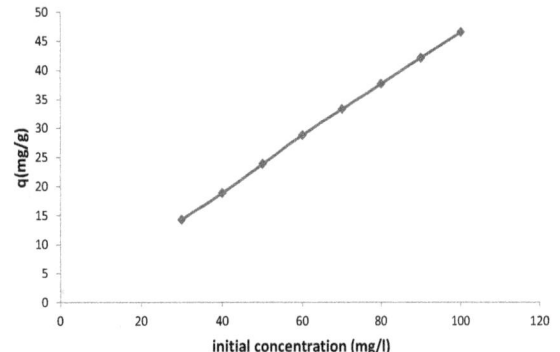

Fig. 9. Effect of initial concentration on adsorption capacity of Amido Black 10B dye onto Polyaniline nano-adsorbent

The correlation coefficient (R^2) of the three kinetic models, i.e. pseudo-first-order, pseudo-second-order (type 1), and Weber–Morris equations for dye adsorption on PAn nano-adsorbent was obtained as 0.8493, 1, and 0.8686, respectively (Table 1). So, the experimental data were well described by the pseudo-second-order kinetic model, which indicates that the process of Amido Black 10B dye adsorption onto PAn nano-adsorbent is chemisorption-controlled. Also, dye adsorption capacity obtained through pseudo-second-order kinetic model was very close to its experimental value.

Table 1. Kinetic constants for Amido Black 10B dye adsorption

Kinetic model	Parameters	R^2
Pseudo-first-order	$q_e(Experimental) = 14.29$ $q_e = 0.2927, K_1 = 0.00863$	0.8493
Pseudo-second-order	$q_e(Experimental) = 14.29$	
Type 1: $\frac{t}{qt} = \frac{1}{K_2 qe^2} + \frac{1}{qe}t$	$q_e = 14.32, K_2 = 0.9192$	1
Type 2: $\frac{1}{qt} = \frac{1}{qe} + \frac{1}{K_2 qe^2}\left(\frac{1}{t}\right)$	$q_e = 14.20, K_2 = 4.9561$	0.4639
Type 3: $qt = qe - \frac{1}{K_2 qe}\left(\frac{qt}{t}\right)$	$q_e = 14.21, K_2 = 4.7866$	0.4617
Type 4: $\frac{qt}{t} = K_2 qe^2 - K_2(qe)qt$	$q_e = 14.27, K_2 = 2.1969$	0.4617
Weber-Morris	$C = 14.01, K_{id} = 0.043$	0.8686

Table 2. Isotherm constants for Amido Black 10B dye adsorption

Isotherm model	Parameters	R^2
Freundlich	$n = 0.79157, K_F = 0.04794$	0.9579
Langmuir		
Type 1: $\frac{C_e}{q_e} = \frac{1}{K_L q_m} + \frac{1}{q_m}C_e$	$q_m = 102.04, K_L = 0.1259, R_L = 0.07358 - 0.20933$	0.7478
Type 2: $\frac{1}{q_e} = \frac{1}{q_m K_L C_e} + \frac{1}{q_m}$	$q_m = 142.85, K_L = 0.0810, R_L = 0.10989 - 0.29154$	0.9243
Type 3: $q_e = q_m - \frac{q_e}{K_L C_e}$	$q_m = 66.837, K_L = 0.2525, R_L = 0.03809 - 0.11661$	0.3731
Type 4: $\frac{q_e}{C_e} = K_L q_m - K_L q_e$	$q_m = 127.61, K_L = 0.0942, R_L = 0.09597 - 0.26137$	0.3731
Temkin	$B = 20.483, K_T = 0.33786$	0.9574
Dubinin–Radushkevick	$\beta = 6 \times 10^{-9}, X_m = 103.87, E = 9.12$	0.8474

Effect of Initial Concentration and Dye Adsorption Isotherm

In order to investigate the effect of initial concentration on adsorption capacity, Amido Black 10B dye solutions with the initial concentration of 30-100 mg/l were prepared. Fig. 9 shows the plot of the adsorption capacity based on the initial concentration of dye. As can be seen, the adsorption capacity of nano-adsorbent increased with increasing the initial concentration. When the initial concentration of the solution changed from 30 mg/l to 100 mg/l, the adsorption capacity of PAn nano-adsorbent increased from 14.29 to 46.55 mg/g.

The relationship between dye concentration in solution and the amount of dye adsorbed on the solid phase at equilibrium was described by isotherm models. In this study, four adsorption isotherm models namely Langmuir, Freundlich, Temkin, and Dubinin–Radushkevich were applied to explain adsorption isotherm of Amido Black 10B dye onto PAn nano-adsorbent.

Langmuir isotherm model assumes adsorption energy is independent of surface coverage and adsorption is limited to a monolayer. Langmuir isotherm model is expressed as follows [30]:

$$q_e = \frac{q_m K_L C_e}{1 + K_L C_e} \tag{7}$$

Here, q_m is the maximum adsorption capacity (mg/g); and K_L is adsorption constant of Langmuir

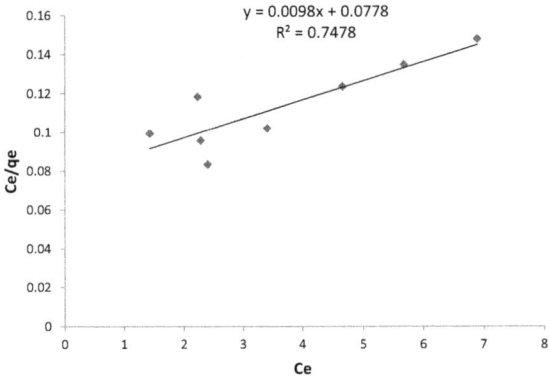

Fig. 10 (a). Langmuir adsorption isotherm of Amido Black 10B onto Polyaniline nano-adsorbent (type 1)

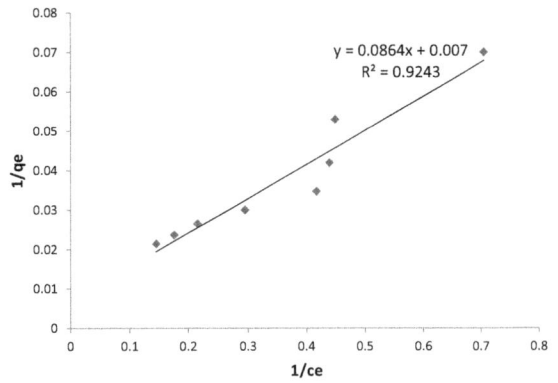

Fig. 10 (b). Langmuir adsorption isotherm of Amido Black 10B onto Polyaniline nano-adsorbent (type 2)

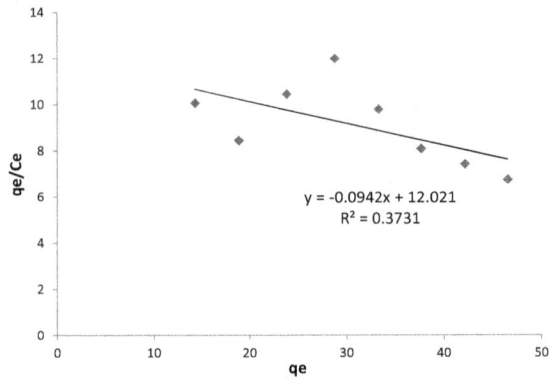

Fig. 10 (c). Langmuir adsorption isotherm of Amido Black 10B onto Polyaniline nano-adsorbent (type 3)

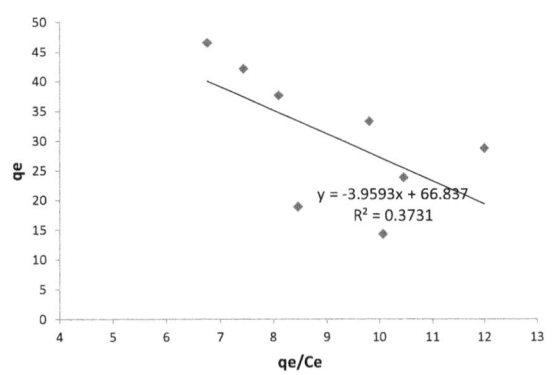

Fig. 10 (d). Langmuir adsorption isotherm of Amido Black 10B onto Polyaniline nano-adsorbent (type 4)

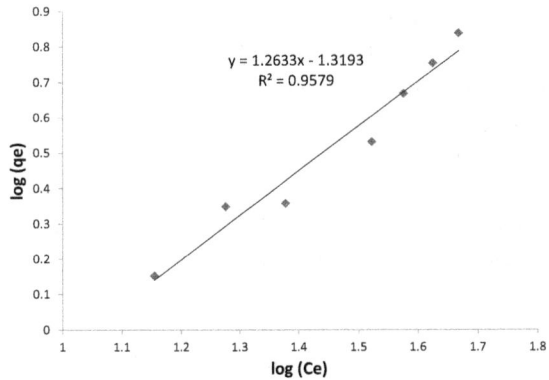

Fig. 11. Freundlich adsorption isotherm of Amido Black 10B onto Polyaniline nano-adsorbent

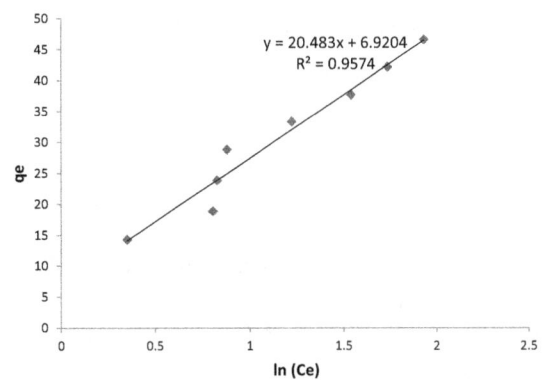

Fig. 12. Temkin adsorption isotherm of Amido Black 10B onto Polyaniline nano-adsorbent

isotherm. Langmuir isotherm can be linearized into four different types, as displayed in Table 2. The value of maximum adsorption capacity and constant of Langmuir isotherm were calculated from intercept and slope of Linear Langmuir plots (Figs. 10a-10d). As the figures show, Linear Langmuir equations, type

2 has a higher correlation coefficient in comparison to the other linear equations. Separation factor (R_L) is a dimensionless constant which expresses the essential features of the Langmuir isotherm [31]:

$$R_L = \frac{1}{1+K_L c_i} \tag{8}$$

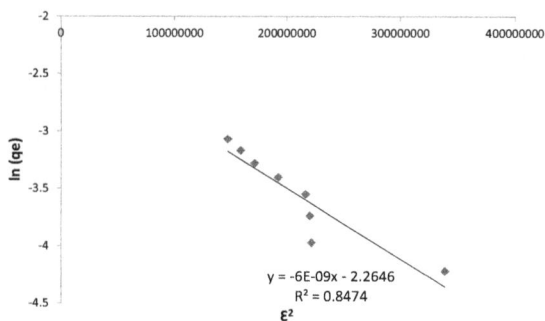

Fig. 13. D-R adsorption isotherm of Amido Black 10B onto Polyaniline nano-adsorbent

In this equation, C_i is the initial concentration of dye; and K_L is constant of Langmuir isotherm. Dye adsorption process onto PAn nano-adsorbent is favorable when the value of R_L obtained in the range 0-1. The separation factors from Linear Langmuir equation (type 2) obtained in the range of 0.10989-0.29154.

The Freundlich isotherm is an empirical model based on the adsorption on heterogeneous surface. This model does not predict the maximum adsorption. Also it is suitable for multilayer adsorption. The Freundlich equation is described as follows [32]:

$$q_e = K_F(C_e)^{1/n} \qquad (9)$$

Where, K_F and $1/n$ are Freundlich constant and adsorption intensity, respectively. Freundlich equation can be linearized as follows:

$$log(q_e) = log(K_F) + \frac{1}{n}log(c_e) \qquad (10)$$

Freundlich constants (K_F) and n were calculated from the intercept and the slope of the linearized plots of $log(q_e)$ versus $log(c_e)$ The plot of Freundlich model was presented in Fig. 11.

Temkin Isotherm contains a factor that represents interaction between nano-adsorbent and dye. The linear form of Temkin isotherm is expressed as follows [33]:

$$q_e = B \ln(K_T) + Bln\,(C_e) \quad , \quad B = \frac{RT}{b_T} \qquad (11)$$

Where, K_T is equilibrium binding constant of Temkin isotherm(L/g); b_T is constant of Temkin isotherm; R is the universal gas constant (8.314 J/mol.K); T is temperature in Kelvin(293.15 K); and B is constant related to the adsorption heat(J/mol). Values of B and K_T can be obtained via plot of q_e versus $ln(C_e)$ (Fig. 12), which were represented in

Table 2.

Dubinin–Radushkevich isotherm [34] is used to investigate the type of Amido Black 10B dye adsorption on PAn nano-adsorbent (physical or chemical). The linear form of D-R model is presented in the following equation:

$$ln(q_e) = ln(X_m) - \beta\varepsilon^2 \quad where \quad \varepsilon = RT \ln(1 + \frac{1}{C_e}) \quad (12)$$

In these equations, C_e, e, X_m and b are the dye equilibrium concentration, polanyi potential, maximum adsorption capacity and constant related to the adsorption energy, respectively. The value of b calculated from the slope of $Ln(q_e)$ versus ε^2 plot (Fig 13). The average adsorption energy, E (kJ/mol), obtained by b is calculated by the following equation:

$$E = (2\beta)^{-0.5} \qquad (13)$$

If the value of E is between 8 and 16 kJ/mol, the adsorption process is chemisorption whereas if E<8 kJ/mol, the adsorption process is of physical nature. The results of the study indicated that the process of Amido Black 10B dye adsorption onto PAn nano-adsorbent is chemisorption.

As shown in Table 2, Freundlich adsorption isotherm shows the best fit to Amido Black 10B dye adsorption data on PAn nano-adsorbent because this model has a higher correlation coefficient (R^2= 0.9579).

CONCLUSION

PAn nano-adsorbent was prepared through chemical polymerization and used as an adsorbent of Amido Black 10B dye from aqueous solution. SEM and FTIR were used to examine the morphology and chemical structure of the synthesized nano-adsorbent. The results of SEM indicated that the synthesized adsorbent particles feature nano-scaled size, uniform distribution, and spherical shape. Such characteristics have been attributed to high efficiency adsorbents. Moreover, FTIR spectrum has proved the formation of PAn nano-adsorbent. The adsorption experiments indicated that PAn nano-adsorbent has high efficiency in Amido Black 10B dye adsorption. The research also yielded the result that optimum adsorption time and adsorbent dosage of PAn nano-adsorbent was 30 min and 0.1 gr, respectively. Dye adsorption was influenced by pH of solution, that is, adsorption efficiency decreases by increasing of pH. Also, dye adsorption capacity increased with increase in the initial dye concentration. The kinetic data illustrated that the adsorption process was controlled by pseudo-second-order model, which indicates that the dye adsorption is chemisorption in nature. Also, the adsorption capacity which calculated by this model (type 1) was 14.32 mg/g, which is close to the experimental value (14.29 mg/g). The results of isotherm studies revealed that the experimental data were best represented by Freundlich isotherm model. The correlation coefficient of this model was obtained 0.9579. The maximum adsorption capacity obtained from Langmuir model, was 142.85 mg/g. The separation factors were in the range of 0-1, indicating that the adsorption of Amido Black 10B dye onto PAn nano-adsorbent was favorable. Moreover, the results obtained from D-R Isotherm showed that the process of dye adsorption onto PAn nano-adsorbent was chemical adsorption. The average adsorption energy was obtained 9.12 KJ/mol.

CONFLICT OF INTEREST

The authors declare that there are no conflicts of interest regarding the publication of this manuscript.

REFERENCES

1. Liu, J., D. Guo, Y. Zhou, Z. Wu, W. Li, F. Zhao and X. Zheng, 2011. Identification of ancient textiles from Yingpan, Xinjiang, by multiple analytical techniques. Journal of Archaeological Science. 38 (7): 1763–1770.
2. Hunger, K., 2003. Industrial Dyes-Chemistry, Properties, Application, Wiley.
3. Daneshvar, N., D. Salari and A.R., Khataee, 2003. Photocatalytic degradation of azo dye acid red 14 in water: investigation of the effect of operational parameters. Journal of Photochemistry and Photobiology A: Chemistry, 157 (1): 111-116.
4. Elaissaoui, I., H. Akrout, S. Grassini, D. Fulginiti and L. Bousselmi, 2016. Role of SiO$_x$ interlayer in the electrochemical degradation of Amaranth dye using SS/PbO$_2$ anodes. Materials and Design, 110: 633–643.
5. Alventosa-deLara, E., S. Barredo-Damas, M.I. Alcaina-Miranda and M.I. Iborra-Clar, 2012. Ultrafiltration technology with a ceramic membrane for reactive dye removal: optimization of membrane performance, Journal of Hazardous Materials. 209–210: 492–500.
6. Sabbaghi. S. and F. Doraghi, 2016. Photo-Catalytic degradation of Methylene blue by ZnO/SnO$_2$ nanocomposite. Journal of water and environmental nanotechnology, 1 (1): 27-34.
7. Lin, Y.F. and F.L. Liang, 2016. ZrO$_2$/carbon aerogel composites: A study on the effect of the crystal ZrO$_2$ structure on cationic dye adsorption. Journal of the Taiwan Institute of Chemical Engineers, 65: 78–82.
8. Liu, K., L. Chen, L. Huang and Y. Lai, 2016. Evaluation of ethylenediamine-modified nanofibrillated cellulose/chitosan

composites on adsorption of cationic and anionic dyes from aqueous solution. Carbohydrate Polymers, 151: 1115–1119.
9. Liu, F., Z. Guo, H. Ling, Z.H. Huang and D. Tang, 2016. Effect of pore structure on the adsorption of aqueous dyes to ordered mesoporous carbons. Microporous and Mesoporous Materials, 227: 104–111.
10. Ramalho, P.A., M.C. Roberto and A.P. Cavaco, 2005. Degradation of dyes with mocroorganisms: studies with ascomycete yeasts. Universidade Do Minho.
11. Tian, J., J. Xu, F. Zhu, T. Lu, C. Su and G. Ouyang, 2013. Application of nanomaterials in sample preparation. Journal of Chromatography A., 1300: 2-16.
12. Rahmanzadeh, L. M. Ghorbani and M. Jahanshahi, 2016. Effective removal of hexavalent mercury from aqueous solution by modified polymeric nanoadsorbent. Journal of water and environmental nanotechnology, 1 (1): 1-8.
13. Samadi, S., R. Motallebi and M. Nasiri Nasrabadi, 2016. Synthesis, characterization and application of Lanthanide metal-ion-doped TiO$_2$/bentonite nanocomposite for removal of Lead (II) and Cadmium (II) from aquatic media. Journal of water and environmental nanotechnology, 1 (1): 35-44.
14. Kamal, T., Y. Anwar, S.H. Bahadar Khan, M.T. Saeed Chani and A.M. Asiri, 2016. Dye adsorption and bactericidal properties of TiO$_2$/chitosan coating layer. Carbohydrate Polymers, 148: 153–160
15. Kataria, N., V.K. Garg, M. Jain and K. Kadirvelu, 2016. Preparation, characterization and potential use of flower shaped Zinc oxide nanoparticles (ZON) for the adsorption of Victoria Blue B dye from aqueous solution. Advanced Powder Technology, 27 (4): 1180–1188.
16. Agarwal, S., I. Tyagi, V.K. Gupta, A.R. Bagheri, M. Ghaedi, A. Asfaram, S. Hajati and A.A. Bazrafshan, 2016. Rapid adsorption of ternary dye pollutants onto copper (I) oxide nanoparticle loaded on activated carbon: Experimental optimization via response surface methodology. Journal of Environmental Chemical Engineering, 4 (2): 1769–1779.
17. Duman, O., S. Tunç, T. Gürkan Polat, B. Kancı Bozoğlan, 2016. Synthesis of magnetic oxidized multiwalled carbon nanotube-κ-carrageenan-Fe$_3$O$_4$ nanocomposite adsorbent and its application in cationic Methylene Blue dye adsorption. Carbohydrate Polymers, 147: 79–88.
18. Van Hoa, N., T.H. Trung Khong, T.T. Hoang Quyen and T.S. Trung, 2016. One-step facile synthesis of mesoporous graphene/Fe$_3$O$_4$/chitosan nanocomposite and its adsorption capacity for a textile dye. Journal of Water Process Engineering, 9: 170–178.
19. Garg, A., M. mainrai, V. kumar bulasara and S. barman, 2015. Experimental Investigation on Adsorption of Amido Black 10B Dye onto Zeolite Synthesized from Fly Ash. Chemical Engineering Communications, 202 (1):123–130.
20. Zhang, L., P. Hu, J. Wang and R. Huang, 2016. Adsorption of Amido Black 10B from aqueous solutions onto Zr (IV) surface-immobilized cross-linked chitosan/bentonite composite. Applied Surface Science, 369: 558–566.
21. Weidlich, C., K.M. Mangold and K. Juttner, 2001. Conducting polymers as ion-exchangers for water purification. Electrochimica Acta, 47 (5): 741–745.
22. Zhou, Q., Y. Wang, J. Xiao, H. Fan, 2016. Adsorption and removal of bisphenol A, α-naphthol and β-naphthol from aqueous solution by Fe$_3$O$_4$@polyaniline core–shell nanomaterials. Synthetic Metals, 212: 113–122.
23. Hojjat Ansari, M. and J. Basiri Parsa, 2016. Removal of nitrate from water by conducting polyaniline via electrically

switching ion exchange method in a dual cell reactor: Optimizing and modeling. Separation and Purification Technology, 169: 158–170.

24. Sharma, V., P. Rekha and P. Mohanty, 2016. Nanoporous hypercrosslinked polyaniline: An efficient adsorbent for the adsorptive removal of cationic and anionic dyes. Journal of Molecular Liquids, 222: 1091–1100.

25. Bhaumik, M., R. McCrindle and A. Maity. 2013. Efficient removal of Congo red from aqueous solutions by adsorption onto interconnected polypyrrole–polyaniline nanofibres. Chemical Engineering Journal, 228: 506–515.

26. Ghorbani, M., M. Soleimani Lashkenari and H. Eisazadeh, 2011. Application of polyaniline nanocomposite coated on rice husk ash for removal of Hg(II) from aqueous media. Synthetic Metals, 161: 1430– 1433.

27. Lagergren, S., 1898. About the theory of so-called adsorption of soluble substances. handlingar, 24 (4): 1-39.

28. Ho, Y.S. and G. McKay, 1999. Pseudo-second order model for sorption processes. Process Biochemistry, 34 (5): 451–465.

29. Weber, W.J. and J.C. Morris, 1963. Kinetics of Adsorption on Carbon from Solution. Journal of the Sanitary Engineering Division, 89 (2): 31-60.

30. Langmuir, I., 1916. The Constitution and Fundamental Properties of Solids and Liquids. Part1. Solids. Journal of the American chemical Society, 38 (11): 2221-2295.

31. Weber, T.W. and R.K. Chakraborti, 1974. Pore and solid diffusion models for fixed bed adsorbents. AIChE Journal, 20 (2): 228–238.

32. Freundlich, H.M.F., 1906. Uber die adsorption in losungen. Zeitschrift fur Physikalische Chemie-Leipzig., 57 (A): 385-470.

33. Tempkin, M.J. and V. Pyzhev, 1940. Kinetics of Ammonia Synthesis on Promoted Iron Catalysts. Acta Physicochimica URSS, 12: 217-222.

34. Dubinin M.M. and L.V. Radushkevich, 1947. The equation of the characteristic curve of activated charcoal. Physical Chemistry Section, U.S.S.R., 55: 331-337.

16

Treatment of Petrochemical wastewater by Modified electro-Fenton Method with Nano Porous Aluminum Electrode

Maryam Adimi [1]*, Maziyar Mohammad Pour [1], Hassan Fathinejad Jirandehi [2]

[1] *Department of Chemical Engineering, Farahan Branch, Islamic Azad University, Farahan, Iran.*
[2] *Young researchers and Elite Club, Farahan Branch, Islamic Azad University, Farahan, Iran.*

ABSTRACT

This research reported a study on COD removal from petrochemical wastewater (ml/l) by the electro-Fenton process via the effects of different parameters such as reaction time, current density, pH, H_2O_2/Fe^{2+} molar ratio, and volume fraction of H_2O_2. For this purpose, first, the Nanopores on the aluminum electrode surface were prepared as the AAO films were fabricated using the two-step anodization of 6063 aluminum alloy sheets at ambient temperature in sulfuric acid and phosphoric acid electrolyte solutions respectively.The nanostructures created on electrode confirmed by Scanning Electron Microscopy (SEM). Then, Efficiency of electrochemical oxidation process was tested by COD determination via electrolyte cell contain waste water, Fe^{2+}, H_2O_2 and AAO electrode based on experimental design. The optimum COD removal (65.03%) was obtained at pH of 2.96, the reaction time of 89.51 min, the current density of 69.57 mA, the H_2O_2/Fe^{2+} molar ratio of 3.42 and volume fraction of H_2O_2 to petrochemical wastewater of 1.93 (ml/l).

Keywords: *COD, Electro-Fenton, Nanoporous Aluminum Electrode.*

INTRODUCTION

Petrochemical plants, including polluting industries environment through emissions of carbon monoxide, hydrogen and synthesis gas etc. The petrochemical industry, including oil refining, petrochemical processing, and natural gas production, generates large amounts of wastewater [1]. Such wastewater is usually characterized by significant concentrations of suspended solids, chemical oxygen demand (COD), oil and grease, sulfide, ammonia, phenols, hydrocarbons, benzene, toluene, ethylbenzene, xylene and polycyclic aromatic hydrocarbons (PAHs) [2,3]. Large quantities of wastewater are generated from petrochemical industries. The discharge of petrochemical wastewater (PCW) could cause

*Corresponding Author Email: maryam_ad1354@yahoo.com

serious environmental pollution and human health concerns [4].The treatment method used for petrochemical wastewater is generally contained pretreatment for improving the biodegradation and reduction of toxicity, via different methods of biological treatment such as activated sludge process, anoxic–oxic (A/O) process, fluidized bed reactor, membrane bioreactor and biofilm process [5,6,7]. Ultrasonic, flocculation, Fenton and ozone oxidation, as well as anaerobic hydrolysis–acidification, are methods that used for pretreatment processes [8].

Several methods are used for removal of COD from petrochemical wastewater, low sludge generation, the high removal efficiency of pollutants and on-site generation of hydrogen peroxide (H_2O_2)

are advantages that advanced oxidation processes (AOPs) among them are considered as promising technologies with [9,10]. These methods are based on the strong oxidant formation, mostly the hydroxyl radical (OH·) a powerful oxidizing agent of organic contaminant materials via processes of photochemical, chemical, and electrochemical [11]. In the AOP methods, the electro-Fenton (EF) way is able to catalysis total mineralization of organic compounds for short time; this way for produce the hydroxyl radical in solution requirement Fe^{2+} ions and oxygen. The raw materials are suitable to treat wastewater. Since they generally include specified amount of Fe^{2+} ions, therefore, the oxygen required for the preparation of air bubbles from compressed air [12, 13]. The E-Fenton method has intricate in the reaction mechanism. However, the main reactions of production of hydroxyl radical can be explained in the following equations [14]:

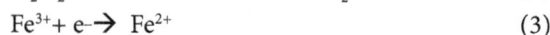

$$O_2+2H^++2e- \rightarrow H_2O_2 \qquad (1)$$
$$H_2O_2+Fe^{2+}+H^+ \rightarrow Fe^{3+}+OH· +H_2O \qquad (2)$$
$$Fe^{3+}+ e- \rightarrow Fe^{2+} \qquad (3)$$

Hydroxyl radicals, with a standard potential of 2.8 V and a half lifetime of about 9-10 seconds, enable fast and non-selective oxidation of a wide range of molecules of environmental concern. Usually cathode material for electro-Fenton processes evaluated base on reducing ability oxygen to hydrogen peroxide and ferric ions reducing to ferrous ions and in next step, promoting the reaction leading to OH radical production. Electro-generation of hydrogen peroxide has different applications such as for electro-Fenton application, direct oxidation, disinfection, and nanomaterial based cathodes [15, 16, 17, 18]. Torrades and coworkers reported using of Fenton and photo-Fenton for the treatment of textile wastewater [19]. Cruz-González and coworkers used the electro-Fenton method for Acid Yellow 36 decolorization [20] and Li reported electro-Fenton used for COD reduction of tannery industrials waste water [21]. removal of synthetic dyes waste water by electro-Fenton and photo-electro-Fenton method [22].also in 2014 reported using of Al and Fe electrode in petrochemical wastewater treatment by electro-Fenton processes [23]. Flores and coworkers reported wastewater removal of olive oil industrials by electro-Fenton and pho electro-Fenton via diamond electrode [24].

Anodic Aluminum Oxide (AAO) was produced via an electrochemical process by anodization of aluminum (Al). Ordered AAO has resistant because of chemical and mechanical properties [25]. Nanopore structures demonstrate a dramatic increase in a surface-to-volume ratio that enhances the signals corresponding to an interaction between analytic and surfaces [25].All these properties make AAO an excellent candidate with exciting opportunities for development of advanced, smart, simple, cost-effective electrodes for Electrochemical wastewater treating applications.

In this work, the removal of chemical oxygen demand (COD) from petrochemical wastewater (PCWW) was experimentally done by using electro-Fenton (EF). It seems using of nano-porosity in the electrode surface can help to improve the act of purification. Due to the very small size of proposition AAO electrode as well as their high active surface it was suggested that also these electrodes use as an auxiliary agent in order to accelerate the production of hydroxyl ions in the reaction medium.

EXPERIMENTAL
Chemicals and materials
All chemicals used here were of reagent grade and used without further purification. Sodium sulfate anhydrous, ferric chloride, sodium borohydride, sulfuric acid, reduced iron powder and other chemical reagents were all of the analytical grades and purchased from Shanghai Chemical Reagents Company. All the solutions were prepared with ultrapure water obtained from a Millipore Milli-Q system with resistivity > 18 MX cm at 25 °C. Their initial pH was adjusted with analytical grade sulfuric acid supplied by Merck. Heptahydrate ferrous sulfate used as a catalyst was of analytical grade purchased from Merck.

Preparation of the Nanoporous anodic aluminum oxide (AAO)
The AAO films were fabricated using the two-step anodization of 6063 aluminum alloy sheets (1mm thickness). The aluminum sheet was initially cut into 1cm×5cm pieces and degreased in acetone, without further thermal treatment or chemical polishing. The first anodization step was then carried out on the aluminum specimen, suspended in the electrolyte as an anode, under the constant current density of 5mA/cm^2 for 10 h. Another aluminum specimen was used as the cathode. Sulfuric acid solution (0.4M concentration)

was used as the electrolyte, and the electrolyte temperature was ambient. The formed AAO film was chemically removed by immersing the specimen in 0.4M phosphoric acid solution for 1 h. The second anodization step was subsequently conducted under the same condition mentioned before for the first step, to produce the final AAO film with a regular Nanopore array. Some final samples were immersed in 0.2M phosphoric acid to widen the pores. Finally, the specimens were rinsed several times with deionized water and then dried in the air [26].

SEM of Nanoporous Anodic Aluminum Oxide

Fig. 1 shows that nanopores are formed on aluminum oxide. From the end of anodic alumina pores with hexagonal condensed arrays of nanopores, after unfolding of pores, nanostructures were arrayed under anodizing and produced in the acidic electrolyte at ambient temperature. The size of nanopores between 223.81 nm to 448.76 nm.

Electro-Fenton experiments

The electrolysis was performed in a small, open, cylindrical, one-compartment electrochemical reactor of 6-cm diameter and 400 mL capacity, stirred by a magnetic during the treatment to enhance the mass transport towards electrodes. The solution pH was fixed to 3 since much higher pH values hamper the development of Fenton-based systems due to the Fe $(OH)_3$ precipitation, which leads to both the decrease of dissolved iron ion [27]. The solution pH was measured with a Cyber Scan pH 1500 pH-meter from Eutech Instruments. In each run, 250 ml of wastewater was placed in an electrolytic cell and desired amounts of iron (Fe2+) and hydrogen peroxide (H_2O_2) were added before the electrical current was turned on. Then, AAO electrodes were placed in the reactor and solutions were mixed at 350 rpm [28].

RESULTS AND DISCUSSION
Statistical analysis

For experiments design, the Design Expert software (RSM Method, version 7) was applied to minimize the number of experiments. The operating conditions ranges were pH of 2-5, the current density of 25-80 mA/cm², the reaction time of 10-90 min, the H_2O_2/Fe^{2+} molar ratio of 0.5-5 and volume fraction of H_2O_2 to petrochemical wastewater of 0.25-2.14 ml/l. 46 experiments were designed by the software and carried out as illustrated in Table 1.

The reduced models for describing COD

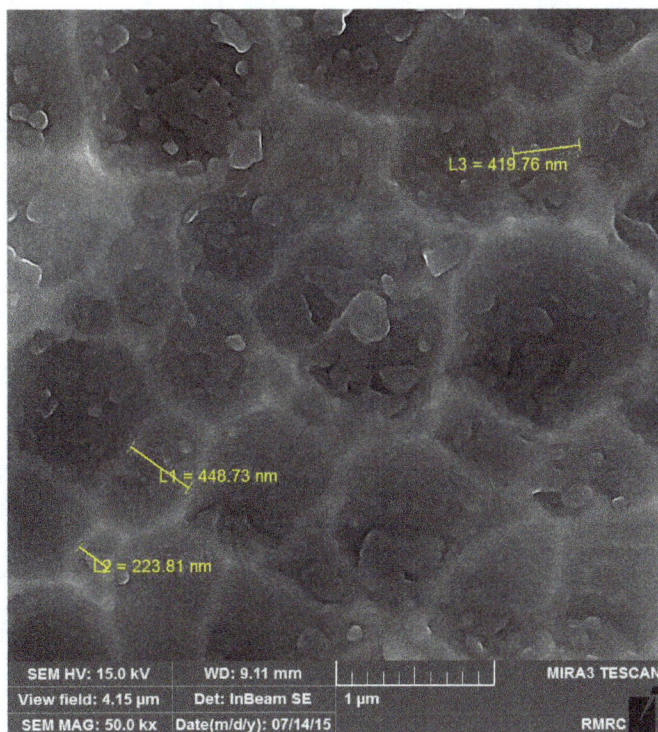

Fig. 1. SEM image of Nano porous anodic aluminum oxide

Table1. Experiments design and obtained results

Run	pH	molH2O2 / molFe 2 +	mlH2O2 / Litpw	Reaction time	Current density	COD removal (%)	
						Exp.	Pred.
1	3.5	2.75	2.14	50	80	45.21	46.16
2	3.5	2.75	1.22	10	80	36.65	35.86
3	2	2.75	1.22	50	80	46.01	43.2
4	3.5	5	0.3	50	52.5	31.26	33.83
5	2	2.75	0.3	50	52.5	38.33	38.61
6	3.5	5	1.22	50	25	26.50	27.05
7	5	2.75	0.3	50	52.5	30.74	31.05
8	3.5	0.5	1.22	50	25	26.54	28.68
9	2	2.75	1.22	10	52.5	37.85	35.23
10	3.5	2.75	1.22	90	25	40.06	40.53
11	2	5	1.22	50	52.5	42.56	44.44
12	3.5	2.75	1.22	50	52.5	47.83	46.62
13	5	2.75	1.22	90	52.5	40.98	42.89
14	3.5	5	1.22	90	52.5	59.06	56.61
15	3.5	2.75	0.3	90	52.5	42.54	42.3
16	5	2.75	2.14	50	52.5	37.41	38.01
17	3.5	0.5	1.22	50	80	34.12	35.58
18	3.5	5	2.14	50	52.5	44.25	44.43
19	3.5	2.75	0.3	50	80	37.51	35.97
20	3.5	2.75	1.22	50	52.5	47.86	46.62
21	3.5	2.75	1.22	90	80	52.64	54.07
22	5	2.75	1.22	50	80	36.48	37.24
23	3.5	2.75	1.22	50	52.5	45.63	46.62
24	3.5	2.75	1.22	10	25	26.59	27.31
25	3.5	0.5	1.22	90	52.5	43.81	41.63
26	3.5	0.5	1.22	10	52.5	29.37	31.37
27	3.5	2.75	2.14	90	52.5	62.95	63.26
28	2	2.75	1.22	90	52.5	52.24	56.73
29	2	2.75	2.14	50	52.5	51.73	49.57
30	3.5	2.75	1.22	50	52.5	43.13	46.62
31	3.5	2.75	0.3	10	52.5	31.26	32.08
32	3.5	5	1.22	10	52.5	28.18	29.43
33	5	0.5	1.22	50	52.5	30.96	32.86
34	3.5	2.75	0.3	50	25	23.49	21.66
35	3.5	2.75	1.22	50	52.5	43.14	46.62
36	2	2.75	1.22	50	25	36.56	37.02
37	3.5	2.75	1.22	50	52.5	48.75	46.62
38	3.5	0.5	1.22	50	52.5	39.62	37.77
39	5	5	1.22	50	52.5	42.56	41.88
40	5	2.75	1.22	10	52.5	31.09	32.95
41	3.5	2.75	2.14	50	25	33.59	36.4
42	3.5	5	1.22	50	80	43.58	43.24
43	3.5	0.5	0.3	50	52.5	33.06	32.45
44	3.5	2.75	2.14	10	52.5	27.17	29.36
45	2	0.5	1.22	50	52.5	40.58	39.42
46	5	2.75	1.22	50	25	24.98	23.1

removal percentage can be presented:

COD Removal (%)= +46.62 +9.03A +5.60B -4.01C +4.81D +2.57E +1.25AB -2.89AC +5.92AD+4.23A E +1.99BC -1.14BD +2.32BE -1.00CD+1.00C+1.32DE -0.91A^2 -7.45 B^2 -3.19 C^2 -4.13D^2 -5.15E^2   (4)

Where A, B, C, D and E are pH, H_2O_2/Fe^{2+} molar ratio, volume fraction of H_2O_2 to petrochemical wastewater (ml/l), reaction time (min) and current density (mA/cm^2), respectively. R^2 and adjusted R^2 are close to one (R^2=0.9747 and R^2 adj.=0.9544). According to the results, 97.47% of the variations for COD removal were explained by the independent variables in the model. R value is more close to 100% then it proves the high significance of model [29].

The adequacy of developed mathematical models to the experiment was examined by the diagnostic plots such as predicted plot versus actual one (Fig. 2) and normal percent probability graphs (Fig. 3). The predicted plot versus actual one showed that actual values were distributed near the straight line Fig. 2. It indicates that actual values are very close to the predicted ones. These plots show very good agreement between the observed data and the correlated ones obtained from the models [30].

Effective parameters on the electro-Fenton process
The impact of time

Fig. 3 shows that reaction time (min) has a positive impact on the progress of the electro-Fenton process. But its effectiveness decreases with increasing time, so after the optimum time, process efficiency does not change with time considerably. That's why optimal reaction time (min) is obtained at almost 2/3 of the total time [31]. Over time the amount of reactive material in the reaction medium decreases and reduces the pollutant removal rate. To achieve perfect efficiency, the system should be given enough time to be able to produce enough amount of OH ions and take place the act of purification of organic pollutants in the environment. By increasing the time it seems that the system achieves to a chemical equilibrium and the maximum removal occurs at the beginning of this balance and the passage of time do not have much impact on improving the treatment process [32].

The effect of pH on COD

Fig. 5 shows the effect of pH on COD removal. As you can see the effect of this parameter within the acidic range is more than basic and to increase

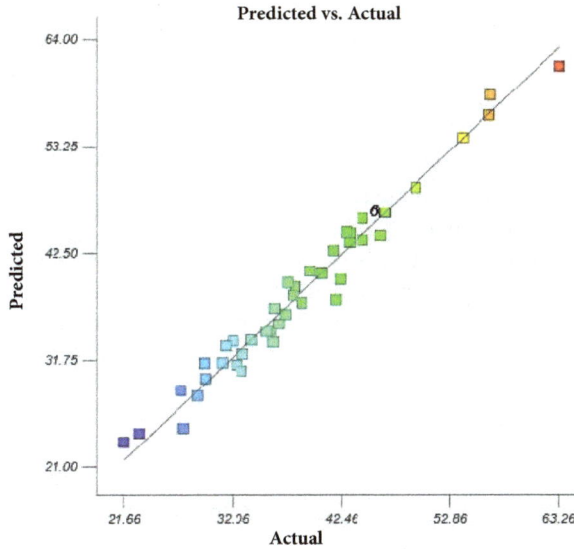

Fig. 2. Predicted vs. actual values plot for COD removal

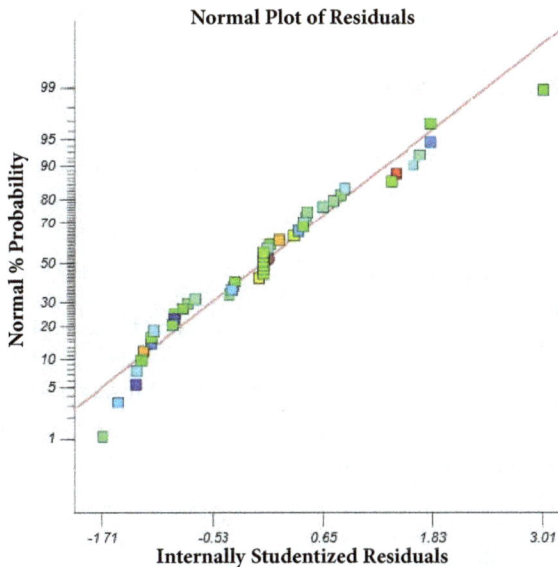

Fig. 3. Normal probability vs. internally studentized residuals values for COD removal

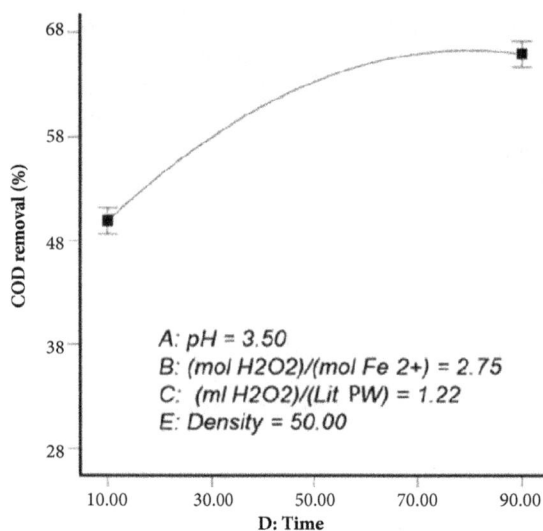

Fig. 4. The effect of time on the reduction of COD parameter

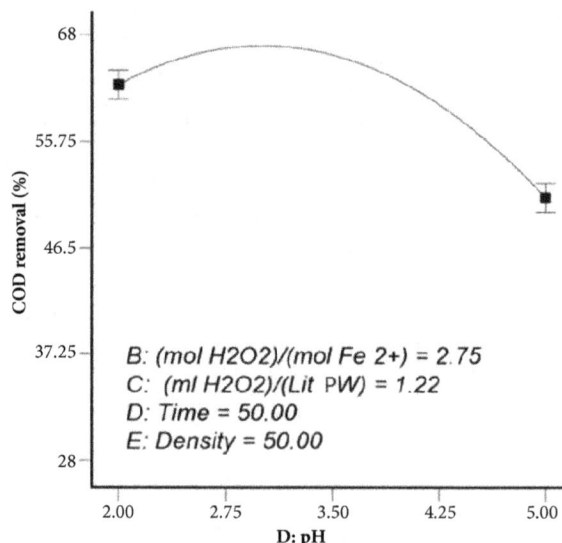

Fig. 5. The effect of pH on the removal of COD parameter

to about 2.75, the effect of this factor increases and then began to decline. With increasing pH, the rate of treatment drops sharply. In fact, this is the optimal operating point that away from it, its impact decreases. Obviously, the Fenton reaction occurs at low pH. At Fenton reactions, the highest efficiency obtains when the pH is around 2–4 [33]. Low pH is suitable for the production of hydrogen peroxide because to convert dissolved oxygen into hydrogen peroxide absorb its required proton of the acidic solution [34]. For pH >4 the oxidation rate significantly decreased meanly because ferrous ions are unstable and they easily form ferric ions and precipitate [35]. On the other hand, at very acidic pH, below pH 3, the electrogenerated hydrogen peroxide turns into an oxonium $H_3O_2^+$ by capturing one proton. $H_3O_2^+$ is electrophilic, leading to the decreasing rate of reaction between H_2O_2 and Fe^{2+} and consequently fewer hydroxyl radicals are produced [36, 37].

Effect of volume fraction on COD removal

The two factors are related to each other so in examining their effects on the COD removal, they are analyzed together. Increasing of $(\frac{mlH2O2}{LitPW})$ ratio with a fixed volume of wastewater increase the amount of H_2O_2. The increasing leads to an increase in the ratio of $\frac{molH2O2}{molFe2+}$ and thus increase of the reactive material in the reaction medium. As the figures indicate, increasing of reactive material to the optimal point leads to increasing the amount of COD removal and when passes the optimal point, causing wasting reactant materials. Using these

two ratios, the required amount of Fe^{2+} and H_2O_2 is obtained. The Fe^{3+} concentration is a significant parameter in the electro-Fenton process. Fig. 6 represents the influence of molar ratio on the act of purification. Since Fenton's reaction is of first grade, the growth of the reactants concentration will develop the reaction rate therefore, and concentration of hydrogen peroxide is dependent on the rise of the Fe^{2+} amount. As can be seen, at low doses, the efficiency ratio is low and this is because the amount of available material is not sufficient for the production of OH and after enhancement, the filtration rate is added till it comes to the optimal point and after passing through this point we see the opposite effect. This negative influence might be due to the increase of the reactions that scavenged hydroxyl radicals, just like Eq. (5)[38]. In addition, hydroperoxyl radicals ($HO_2^\bullet$), which is a less oxidizing agent than $^\bullet OH$, would be formed by reaction (6) and(7) as the Fe^{3+} concentration increased [39].Another reason for the lower removal efficiencies at high Fe^{3+} concentration can be the formation of a yellow precipitate of $Fe(OH)_3$, which is deposited on the electrode surface. This deposited Fe^{3+} cannot be reconverted to Fe^{2+}, so the result is less Fe^{2+} available to react with the H_2O_2 to produce hydroxyl radicals [39].

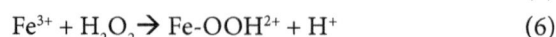

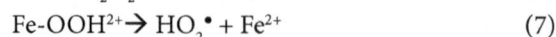

$$Fe^{2+} + {}^\bullet OH \rightarrow Fe^{3+} + OH^- \qquad (5)$$

$$Fe^{3+} + H_2O_2 \rightarrow Fe\text{-}OOH^{2+} + H^+ \qquad (6)$$

$$Fe\text{-}OOH^{2+} \rightarrow HO_2^\bullet + Fe^{2+} \qquad (7)$$

Fig. 7 shows the effect of the volume ratio on the treatment parameter. As can be seen, it has an

A: pH = 3.50
C: (ml H2O2)/(Lit PW) = 1.22
D: Time = 50.00
E: Density = 50.00

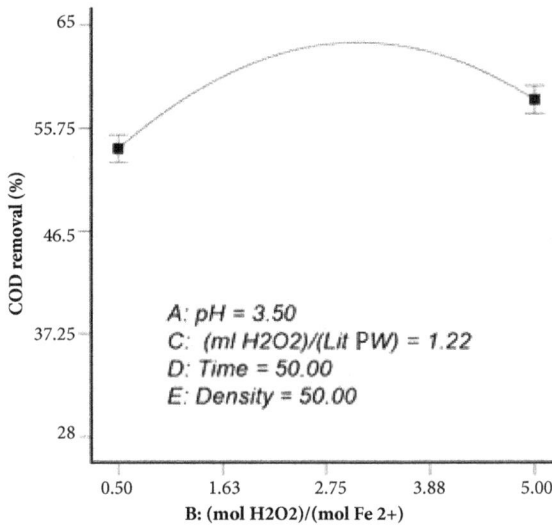

Fig. 6. Effect of $\frac{molH2O2}{molFe2+}$ molar ratio on COD removal

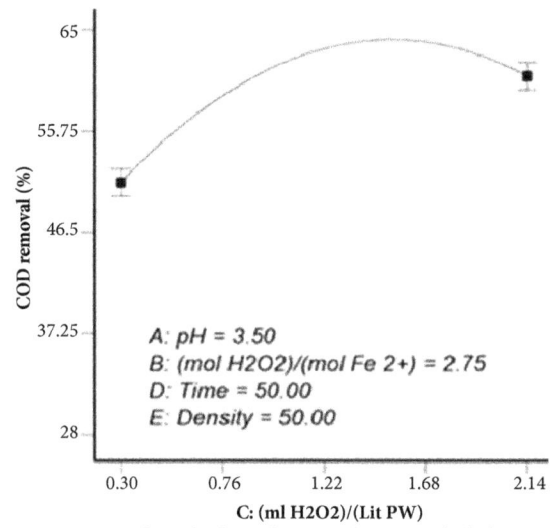

A: pH = 3.50
B: (mol H2O2)/(mol Fe 2+) = 2.75
D: Time = 50.00
E: Density = 50.00

Fig. 7. Effect of volume fraction of H_2O_2 to alcoholic wastewater ($\frac{mlH2O2}{LitPW}$) on COD removal

Table 2. Optimum data for both electro-Fenton processes {with the modified electrodes (current research) and traditional electro-Fenton [23]}

Reaction time (min)	Current density (mA/cm²)	pH	Volume fraction of H₂O₂ to Petrochemical wastewater (ml/l)	H₂O₂/Fe²⁺ molar ratio	COD removal (%)		EF type
					Statistical data	Experimental data	
89.51	69.57	2.96	1.93	3.42	65.03	64.15	Current research
78.97	68.65	3.06	2.14	4.99	53.94	51.23	[23]

optimum point. At the small amounts of this ratio, there is no much impact on removal due to the low quantity of hydrogen peroxide in the environment. COD removal can be enhanced by increasing the H_2O_2 concentration. A further increase from optimum point decreases the removal efficiency. At a high level of H_2O_2, the decrease in removal efficiency was due to the hydroxyl radical scavenging effect of H_2O_2 (Eqs. (8) and (9)) and hydroxyl radical recombination (Eq. (10)) [40, 41].

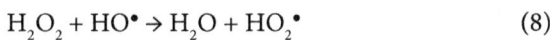

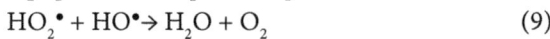

$$H_2O_2 + HO^\bullet \rightarrow H_2O + HO_2^\bullet \qquad (8)$$
$$HO_2^\bullet + HO^\bullet \rightarrow H_2O + O_2 \qquad (9)$$
$$HO^\bullet + HO^\bullet \rightarrow H_2O_2 \qquad (10)$$

The slope of this graph is greater than the molar ratio graph and by developing a small amount of this ratio we see a significant increase in the amount of treatment. So the hydrogen peroxide has higher efficiency comparing to the iron salts in the experiment environment.

Optimization and validation

Numerical optimization was used to determine the optimum parameters for maximum COD removal. The software automatically gives the optimum conditions. In this software, numerical optimization is used to achieve the maximum amounts of COD removal. In fact, all variables are targeted to be in a range and COD removal are a goal to be maximized. The found data were shown in Table 2. There are very good agreement between experimental optimum data and statistical ones. Table 2 compares our results with the traditional electro-Fenton data (with plain aluminum electrodes) at the optimum conditions, as well [28]. According to this table, modified AAO electrodes increased 11% COD removal. It follows that the creation of Nano surface in a test environment without changes in operating conditions the system can apply appropriate efficiency. Therefore, special attention given to this issue is justified. By comparing the optimal experimental conditions in the presence of Nanoporosity and lack of it, it is observed that time needed to achieve optimal conditions, decreases. In fact, it seems that the presence of Nanoporous in the environment, accelerate the rate of production of hydroxide ions and purification process. But it is evident that in this case the current density increases and the increase may be due to the fact that reactions related to the production of iron ions and disinfectant radicals requires high energy consumption and thus increase the current density and it is because of the reaction time is reduced and treatment is fastened. In the case of pH, it is clear that the operation was carried out in the more acidic range. In the case of the reactive substances which present themselves in the parameters of molar ratio and volume ratio, it is observed that there has been a marked decline and this can be one of the major advantages of using Nanoporous materials, because by creating Nano-porosity can save a significant amount of raw materials is used in the process. This impact is likely to be this way, Nanoporous by creating a higher level, helps to decompose hydrogen peroxide with higher efficiency and thereby reducing the consumption of hydrogen peroxide andiron catalyst.

CONCLUSIONS

In this study, the electro-Fenton method was proposed to treat effluent from industrial activities conducted in Arak Petrochemical Company. Finally, the operating conditions were optimized and presented to achieve the best results with the highest amount of COD reduction. The optimum COD removal (65.03%) was obtained at pH of 2.96, reaction time of 89.51 min, current density of 69.57 mA, H_2O_2/Fe^{2+} molar ratio of 3.42 and volume fraction of H_2O_2 to Petrochemical wastewater of 1.93 (ml/l) and it can be seen that the modified electro-Fenton process could optimally remove COD from petrochemical wastewater up to 65.03% while it was around 53.94% in the traditional electro-Fenton, as it is observed, treatment using the second method has been improved by 11%[28]. It seems that the Nano-porosity can be used as an auxiliary agent in order to accelerate the decomposition of hydrogen peroxide and ultimately lead to increase in production of hydroxyl ions. Due to the effectiveness of electro-Fenton operation, this method can be proposed as an easy way to petrochemical wastewater treatment.

CONFLICT OF INTEREST

The authors declare that there are no conflicts of interest regarding the publication of this manuscript.

REFERENCES

1. Asatekin A, Mayes AM. Oil Industry Wastewater Treatment with Fouling Resistant Membranes Containing Amphiphilic Comb Copolymers. Environmental Science & Technology. 2009;43(12):4487-92.

2. Tobiszewski M, Tsakovski S, Simeonov V, Namieśnik J. Chlorinated solvents in a petrochemical wastewater treatment plant: An assessment of their removal using self-organising maps. Chemosphere. 2012;87(8):962-8.

3. Pérez G, Fernández-Alba AR, Urtiaga AM, Ortiz I. Electro-oxidation of reverse osmosis concentrates generated in tertiary water treatment. Water Research. 2010;44(9):2763-72.

4. Chen C, Yu J, Yoza BA, Li QX, Wang G. A novel "wastes-treat-wastes" technology: Role and potential of spent fluid catalytic cracking catalyst assisted ozonation of petrochemical wastewater. Journal of Environmental Management. 2015;152:58-65.

5. Jianlong W, Hanchang S, Yi Q. Wastewater treatment in a hybrid biological reactor (HBR): effect of organic loading rates. Process Biochemistry. 2000;36(4):297-303.

6. Guo J, Ma F, Chang C-C, Cui D, Wang L, Yang J, et al. Start-up of a two-stage bioaugmented anoxic–oxic (A/O) biofilm process treating petrochemical wastewater under different DO concentrations. Bioresource Technology. 2009;100(14):3483-8.

7. Yang Q, Xiong P, Ding P, Chu L, Wang J. Treatment of petrochemical wastewater by microaerobic hydrolysis and anoxic/oxic processes and analysis of bacterial diversity. Bioresource Technology. 2015;196:169-75.

8. Lin C-K, Tsai T-Y, Liu J-C, Chen M-C. Enhanced biodegradation of petrochemical wastewater using ozonation and bac advanced treatment system. Water Research. 2001;35(3):699-704.

9. Labiadh L, Oturan MA, Panizza M, Hamadi NB, Ammar S. Complete removal of AHPS synthetic dye from water using new electro-fenton oxidation catalyzed by natural pyrite as heterogeneous catalyst. Journal of Hazardous Materials. 2015;297:34-41.

10. Paramo-Vargas J, Camargo AME, Gutierrez-Granados S, Godinez LA, Peralta-Hernandez JM. Applying electro-Fenton process as an alternative to a slaughterhouse effluent treatment. Journal of Electroanalytical Chemistry. 2015;754:80-6.

11. Gençten M, Özcan A. A detailed investigation on electro-Fenton treatment of propachlor: Mineralization kinetic and degradation intermediates. Chemosphere. 2015;136:167-73.

12. Brillas E, Sirés I, Oturan MA. Electro-Fenton Process and Related Electrochemical Technologies Based on Fenton's Reaction Chemistry. Chemical Reviews. 2009;109(12):6570-631.

13. Vasconcelos VM, Ponce-de-León C, Nava JL, Lanza MRV. Electrochemical degradation of RB-5 dye by anodic oxidation, electro-Fenton and by combining anodic oxidation–electro-Fenton in a filter-press flow cell. Journal of Electroanalytical Chemistry. 2016;765:179-87.

14. Song S, Wu M, Liu Y, Zhu Q, Tsiakaras P, Wang Y. Efficient and Stable Carbon-coated Nickel Foam Cathodes for the Electro-Fenton Process. Electrochimica Acta. 2015;176:811-8.

15. Petrucci E, Da Pozzo A, Di Palma L. On the ability to electrogenerate hydrogen peroxide and to regenerate ferrous ions of three selected carbon-based cathodes for electro-Fenton processes. Chemical Engineering Journal. 2016;283:750-8.

16. Qiang Z, Chang J-H, Huang C-P. Electrochemical generation

of hydrogen peroxide from dissolved oxygen in acidic solutions. Water Research. 2002;36(1):85-94.

17. Pozzo AD, Palma LD, Merli C, Petrucci E. An experimental comparison of a graphite electrode and a gas diffusion electrode for the cathodic production of hydrogen peroxide. Journal of Applied Electrochemistry. 2005;35(4):413-9.

18. Davarnejad R, Azizi J. Alcoholic wastewater treatment using electro-Fenton technique modified by Fe2O3 nanoparticles. Journal of Environmental Chemical Engineering. 2016;4(2):2342-9.

19. Torrades F, García-Montaño J. Using central composite experimental design to optimize the degradation of real dye wastewater by Fenton and photo-Fenton reactions. Dyes and Pigments. 2014;100:184-9.

20. Cruz-González K, Torres-López O, García-León A, Guzmán-Mar JL, Reyes LH, Hernández-Ramírez A, et al. Determination of optimum operating parameters for Acid Yellow 36 decolorization by electro-Fenton process using BDD cathode. Chemical Engineering Journal. 2010;160(1):199-206.

21. Kurt U, Apaydin O, Gonullu MT. Reduction of COD in wastewater from an organized tannery industrial region by Electro-Fenton process. Journal of Hazardous Materials. 2007;143(1–2):33-40.

22. J.M. Peralta-Hernández CAM-H, Guzmán-Mar JL, Hernández-Ramírez aA. Recent advances in the application of electro-Fenton and photo electro- Fenton process for removal of synthetic dyes in wastewater treatment. Journal of Environmental Engineering and Management. 2009;19(5):257-65.

23. Davarnejad R, Mohammadi M, Ismail AF. Petrochemical wastewater treatment by electro-Fenton process using aluminum and iron electrodes: Statistical comparison. Journal of Water Process Engineering. 2014;3:18-25.

24. Flores N, Sirés I, Garrido JA, Centellas F, Rodríguez RM, Cabot PL, et al. Degradation of trans-ferulic acid in acidic aqueous medium by anodic oxidation, electro-Fenton and photoelectro-Fenton. Journal of Hazardous Materials. 2016;319:3-12.

25. Md Jani AM, Losic D, Voelcker NH. Nanoporous anodic aluminium oxide: Advances in surface engineering and emerging applications. Progress in Materials Science. 2013;58(5):636-704.

26. Santos A, Kumeria T, Losic D. Nanoporous anodic aluminum oxide for chemical sensing and biosensors. TrAC Trends in Analytical Chemistry. 2013;44:25-38.

27. Moghadam H, Samimi A, Behzadmehr A. Effect of Nanoporous Anodic Aluminum Oxide (AAO) Characteristics On Solar Absorptivity. Transport Phenomena in Nano and Micro Scales. 2013;1(2):110-6.

28. Tsantaki E, Velegraki T, Katsaounis A, Mantzavinos D. Anodic oxidation of textile dyehouse effluents on boron-doped diamond electrode. Journal of Hazardous Materials.

2012;207–208:91-6.

29. Davarnejad R, Nikseresht M. Dairy wastewater treatment using an electrochemical method: Experimental and statistical study. Journal of Electroanalytical Chemistry. 2016;775:364-73.

30. Oturan N, Zhou M, Oturan MA. Metomyl Degradation by Electro-Fenton and Electro-Fenton-Like Processes: A Kinetics Study of the Effect of the Nature and Concentration of Some Transition Metal Ions As Catalyst. The Journal of Physical Chemistry A. 2010;114(39):10605-11.

31. Ballesteros Martín MM, Sánchez Pérez JA, Casas López JL, Oller I, Malato Rodríguez S. Degradation of a four-pesticide mixture by combined photo-Fenton and biological oxidation. Water Research. 2009;43(3):653-60.

32. Pera-Titus M, García-Molina V, Baños MA, Giménez J, Esplugas S. Degradation of chlorophenols by means of advanced oxidation processes: a general review. Applied Catalysis B: Environmental. 2004;47(4):219-56.

33. Masomboon N, Ratanatamskul C, Lu M-C. Chemical oxidation of 2,6-dimethylaniline by electrochemically generated Fenton's reagent. Journal of Hazardous Materials. 2010;176(1–3):92-8.

34. Rosales E, Pazos M, Longo MA, Sanromán MA. Electro-Fenton decoloration of dyes in a continuous reactor: A promising technology in colored wastewater treatment. Chemical Engineering Journal. 2009;155(1–2):62-7.

35. Jiang C-c, Zhang J-f. Progress and prospect in electro-Fenton process for wastewater treatment. Journal of Zhejiang University-SCIENCE A. 2007;8(7):1118-25.

36. Panizza M, Barbucci A, Delucchi M, Carpanese MP, Giuliano A, Cataldo-Hernández M, et al. Electro-Fenton degradation of anionic surfactants. Separation and Purification Technology. 2013;118:394-8.

37. Umar M, Aziz HA, Yusoff MS. Trends in the use of Fenton, electro-Fenton and photo-Fenton for the treatment of landfill leachate. Waste Management. 2010;30(11):2113-21.

38. Lin H, Zhang H, Wang X, Wang L, Wu J. Electro-Fenton removal of Orange II in a divided cell: Reaction mechanism, degradation pathway and toxicity evolution. Separation and Purification Technology. 2014;122:533-40.

39. Pajootan E, Arami M, Rahimdokht M. Discoloration of wastewater in a continuous electro-Fenton process using modified graphite electrode with multi-walled carbon nanotubes/surfactant. Separation and Purification Technology. 2014;130:34-44.

40. Rosales E, Pazos M, Sanromán MA. Advances in the Electro-Fenton Process for Remediation of Recalcitrant Organic Compounds. Chemical Engineering & Technology. 2012;35(4):609-17.

41. Muruganandham M, Swaminathan M. Decolourisation of Reactive Orange 4 by Fenton and photo-Fenton oxidation technology. Dyes and Pigments. 2004;63(3):315-21.

17

Synthesis of Highly Effective Novel Graphene Oxide-Polyethylene Glycol-Polyvinyl Alcohol Nanocomposite Hydrogel For Copper Removal

Eman Serag[1], Ahmed El-Nemr[1,]*, Azza El-Maghraby[2]

[1]*Marine Pollution Department, Environmental Division, National Institute of Oceanography and Fisheries, Kayet Bey, El-Anfoushy, Alexandria, Egypt*
[2] *Fabrication Technology Department, Advanced Technology and New Materials Institute, City for Scientific Research and Technology Application, Alexandria, Egypt*

ABSTRACT

A novel Graphene oxide-polyethylene glycol and polyvinyl alcohol (GO-PEG-PVA) triple network hydrogel were prepared to remove Copper(II) ion from its aqueous solution. The structures, morphologies, and properties of graphene oxide (GO), the composite GO-PEG-PVA and PEG-PVA were characterized using FTIR, X-ray diffraction, Scanning Electronic Microscope and Thermal Gravimetric analysis. A series of systematic batch adsorption experiments were conducted to study the adsorption property of GO, GO-PEG-PVA hydrogel and PEG-PVA hydrogel under different conditions (e.g. pH, contact time and Cu^{2+} ions concentration). The high adsorption capacity, easy regeneration, and effective adsorption–desorption results proved that the prepared GO-PEG-PVA composite hydrogel could be an effective adsorbent in removing Cu^{2+} ion from its aqueous solution. The maximum adsorption capacities were found to be 917, 900 and 423 mg g^{-1} for GO-PEG-PVA hydrogel, GO and PEG-PVA hydrogel, respectively at pH 5, 25 °C and Cu^{2+} ions' concentration 500 mg l^{-1}. The removal efficiency of the recycled GO-PEG-PVA hydrogel were 83, 81, 80 and 79% for the first four times, which proved efficient reusability.

Keywords: *Copper; Graphene Oxide; Hydrogel; Polyethylene Glycol; Polyvinyl Alcohol*

INTRODUCTION

Heavy metals are considered of great concern due to their extreme toxicity even at low concentrations because of their tendency to accumulate in organisms and the food chain [1]. The disposal of wastewater indiscriminately is an environmental problem worldwide due to the negative impacts of polluted water on human health [2]. Effluents of industries such as chemical, metallurgical, mining, and dying, as well as battery manufacturing that contain different types of toxic heavy metal ions such as As, Cd, Co, Cr, Ni, Hg, and Pb [1-4]. Therefore, it is necessary to treat industrial effluents contaminated by heavy metals prior to their discharge into the environment [5, 6]. Different sorbent materials have been prepared and studied extensively to remove heavy metal ions [7-11].

Although many techniques were employed for the treatment of wastewater containing heavy metals, it is important to report that the selection of the most suitable treatment method for metal removal from wastewater depends on different parameters such

* Corresponding Author Email: *ahmed.m.elnemr@gmail.com*

as pH, metal concentration, environmental impact and economics parameter [1-3].

Graphene oxide (GO) is of great interest as a result of its low price, easy access, and graphene conversion ability. Graphene can be prepared by oxidation of graphite to graphene oxide to intersperse the carbon layers with oxygen molecules and then reduce GO to separate the carbon layers completely into individual or a few layers to give graphene. Therefore, graphene oxide is a by-product of this process, because when the oxidizing agents react with graphite, the inter-planar spacing between the layers of graphite increases. The completely oxidized graphite can be dispersed in a base solution such as water to produce graphene oxide [12-14]. Also, different studies have been made for photodegradation and waste treatment [15-18].

The primary synthetic method of GO is the conventionally-modified Hummers method, where proportional amounts of oxidants, such as $KMnO_4$, $NaNO_3$ and concentrated H_2SO_4, are mixed in order with the graphite [19]. Considering the new functional groups that have been introduced to the surface of GO and its high surface area, the GO nano-sheets should have a high-sorption capacity of heavy metal ions from their aqueous solutions [19].

The GO chitosan composite powders [20], chitin GO hybrid composite [21] and GO chitin composite foams [22] have been prepared for metal ions and dye removals. However, it is somewhat tough to separate the GO powder adsorbents from wastewater due to their high solubility and dispersibility, which may lead to secondary environmental problems. Therefore, it is very urgent to synthesize and design stable composite materials that are easy to separate expediently and efficiently from the treated wastewater, as well as easy to regenerate without much complication.

Poly (vinyl alcohol) (PVA) is a water-soluble polymer with non-toxicity, biocompatibility and biodegradability characteristics. Hence, PVA has been manufactured into a variety of adsorbent types, such as ion-exchange film and hydrogel [23]. PVA is a semi-crystal hydrophilic polymer consisting of one hydroxyl group in each repeat unit and that representing desirable adsorbent structure. Like other hydrogels, the PVA hydrogel shows faster adsorption kinetics for removing heavy metals from aqueous solution [24]. PEG, as a biodegradable synthetic material, shows good biocompatibility and low immunogenicity and it is nontoxic. PEG is soluble in both water and many organic solvents [25].

In this work, we aim to synthesize graphene oxide and a novel nano-composite graphene oxide-polyethylene glycol and polyvinyl alcohol (GO-PEG-PVA) hydrogel to use in the safe removal of water pollutants without leaving any toxic residual. GO and synthesized composite hydrogel were separately investigated as adsorbents for the removal of Cu(II) ions from water through batch adsorption.

MATERIAL AND METHOD
Chemicals and reagents
Graphite Flakes (acid treated 99%, Asbury Carbons), sodium nitrate (98%, Nice Chemicals), potassium permanganate (99%, RFCL), hydrogen peroxide (30% wt, Emplura), sulphuric acid (98%, ACS), PVA (molecular weight, 72,000; degree of hydrolysis, 98.0–98.8 mol %) and polyethylene glycol were purchased from Sigma–Aldrich Company. Copper sulphate pentahydrate and glutaraldehyde were obtained from a local company and used without further purifications.

Synthesis of graphene oxide (GO)
Modified Hummer's method was used in the synthesis of graphene oxide from graphite powder [13]. In brief, 1.0 g of graphite and 0.5 g of $NaNO_3$ were well mixed together and then 23 ml of concentrated H_2SO_4 was added slowly under constant stirring. After stirring for 1 hour, 3 g of $KMnO_4$ was added in portions to the solution while keeping the reaction temperature below 20 °C to prevent overheating and explosion. Then, the obtained reaction mixture was stirred at about 35 °C for 12 hours followed by adding 500 ml of distilled water in the presence of vigorous stirring. The reaction mixture was left to react at 98 °C for 40 minutes further. That was then followed by adding 50 ml of 30% H_2O_2 drop by drop. The reaction mixture was cooled, centrifuged and washed with 1N HCl aqueous solution to remove the metal ions and further washed with double distilled water until the solution pH reached 6.5. The resulting yellow-brown graphene oxide solid was dried under reduced pressure for 24 hours (Fig. 1) [26, 27].

Preparation of Hydrogel
GO-PEG-PVA hydrogel was synthesized by fully dissolving 10.0 g of PVA in 100 ml of double distilled

water at a temperature of 90 °C under magnetic stirring. Then, it was left to cool down to room temperature. That was later followed by adding 100 mg of GO under vigorous magnetic stirring. 5 ml of PEG was added to the reaction mixture, followed by a cross-linking using 10 ml of 2.5% glutaraldehyde solution in 1% HCl, which acted as the cross-linking agent and catalyst, respectively. Stirring continued for 15 minutes. The reaction mixture was kept in an oven at 80 °C to produce the GO-PEG-PVA hydrogel (Fig. 2a) [28]. PEG-PVA hydrogel without GO was also prepared following the same method for comparison (Fig. 2b).

Characterization techniques

The samples were characterized by FTIR spectra using Bruker VERTEX 70 spectrophotometer with ATR platinum unit, and X-ray diffraction analysis using XRD-7000, Schimadzu Corp., Columbia, MD that operated with Cu Ka radiation (λ D 0.154060 nm) that was generated at 30 kV and 30 mA. Scans were done at 2° min⁻¹ for 2θ values between 10 and 100 degrees. Thermogravimetric analysis was obtained using TGA50 Schimadzu, Japan. This measurement was carried out at a heating rate of 10 °C min⁻¹ and under nitrogen gas. The morphologies of the samples were observed under Scanning Electron Microscope (JEOL, Model JSM 6360LA, Japan).

Adsorption Experiments
Preparation of stock solution

1.978 g of copper sulfate pentahydrate ($CuSO_4 \cdot 5H_2O$) was dissolved in 1,000 ml of double distilled water to prepare the stock solution of 500 mg l⁻¹. The other solutions with different

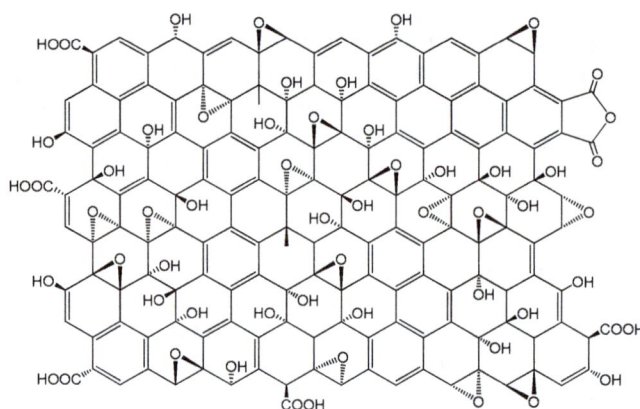

(a)   (b)

Fig. 1: (a) Photographic image of as-prepared GO solution (b) Proposed model of GO structure.

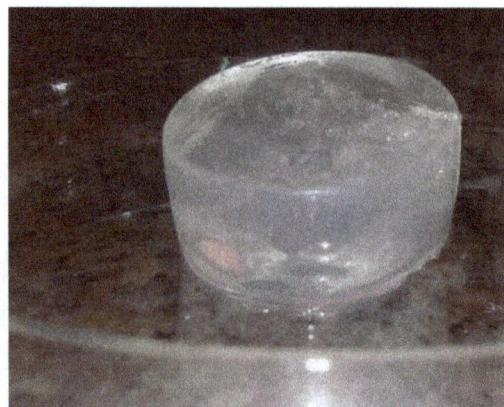

(a)   (b)

Fig. 2: (a) Photographic image of PVA-PEG-GO hydrogel (b) photographic image of PVA-PEG hydrogel

concentrations were obtained by successive dilution. The pH of the solutions was adjusted using 1.0M HCl or 1.0M NaOH.

Batch adsorption experiment

In typical batch experiments with GO (100 mg), the hydrogel corresponds to 100 mg GO while the hydrogel without GO (the control which had been carried out to correct any adsorption of metal on PVA or PEG surface) swelled in 100 ml of standard solution. A concentration of 500 mg l^{-1} of Cu^{2+} ions on a rotary shaker at 150 rpm and 250 ml capped conical flasks were all employed under the required contact time at room temperature ($25^\circ C \pm 1^\circ C$). The collected samples were filtered through 0.45 micron filters and the concentration of Cu^{2+} ions was immediately determined through a UV Spectrophotometer (Analyticjena Spekol 1300 UV-VIS Spectrophotometer, Model No 45600-02, Cole Parmer Instrument Co., USA) using 1-(2-pyridylazo)-2-naphthal (PAN) solution at the maximum wavelength λ_{max} for the colored complex solution; the result was determined at 550 nm [29].

The effect of pH was studied at different pH values (1, 3, 5 and 7) at initial Cu^{2+} ion concentration of 500 mg l^{-1} and 100 mg dosage of GO, the hydrogel corresponded to 100 mg GO and to that without GO. The effect of contact time on Cu^{2+} ions removal was conducted by studying the removal percentage at different times from 10 to 100 minutes using the dosage of 100 mg GO and hydrogel with and without GO at the optimum pH 5.

To investigate the effect of Cu^{2+} ions' initial concentration, the experimental work was conducted at Cu^{2+} ion initial concentration between 50 to 500 mg l^{-1} dosages at pH 5 for 60 minutes. A number of Cu^{2+} ions adsorbed at equilibrium q_e (mg g^{-1}) were calculated using the following equation 1 [28].

$$q_e = \frac{(C_0 - C_e) \times V}{m} \tag{1}$$

Where C_0, represents the initial ion concentration (mg g^{-1}), C_e is the equilibrium Cu^{2+} ion concentration (mg g^{-1}), m is the mass (g) of the adsorbent and V is the solution volume in L.

For the recyclability experiment, the adsorbed Cu^{2+} ions on the adsorbent were eluted with 0.1 M HCl solution. The adsorbent was then treated with 0.1 M NaOH solution to neutralize the hydronium ion (H_3O^+) on the GO-PVA-PEG gel, and the adsorbent was further washed with deionized water. The regenerated GO-PVA-PEG gel was reused in the next cycle of the sorption experiment. All the experimental work was performed in duplicates and the relative standard deviation was less than 5%.

RESULT AND DISCUSSION

Synthesizing graphene oxide from graphite was achieved by the modified Hummer's method which demonstrated a less hazardous and more efficient way for graphite oxidation [30, 31]. Graphene oxide individual sheets can be viewed as they are decorated with different types of oxygen functional groups on both sides of the plane and around the edges [32, 33]. The ionization of carboxyl groups presented at the edges of graphene oxide sheet can electrostatically stabilize the graphene oxide in water, alcohols and certain organic solvents to form a colloidal suspension without surfactants (Fig. 1) [34-36].

The newly formed double network hydrogel may be considered a good adsorbent mainly due to the following advantages: (1) GO greatly improves the strength of the hydrogel, which makes it easy to regenerate and (2) the existence of oxygen-containing groups in the hydrogel network provides plenty of active sites able to adsorb different types of heavy metal ions.

The satisfactory heavy metals adsorption performance, good reusability, and relatively low cost reveal the potential of PVA-PEG-GO gel for practical application. GO-PEG-PVA is a novel network nanocomposite and these polymers have very good ability to swell and adsorb of heavy metals as well as they are biodegradable and nontoxic.

Characterization

The FTIR spectrum of GO (Fig. 3a) shows a broad peak between 3353 cm^{-1} in the high-frequency area together with a sharp peak at 1629 cm^{-1}, this corresponds to the stretching and bending vibrations of OH groups, and of water molecules adsorbed on the graphene oxide. Therefore, it can be concluded that GO has strong hydrophilicity [37]. Absorption peak observed at 1715 cm^{-1} represents C=O carbonyl group, while the stretching at 1130 and 1038 cm^{-1} represents the C–O epoxide group [38]. Finally, the peak observed at 877 cm^{-1} can be attributed to the stretching vibration of C=C. The presence of these peaks corresponding to the presence of oxygen-containing function groups, prove the oxidation of the graphite to graphene

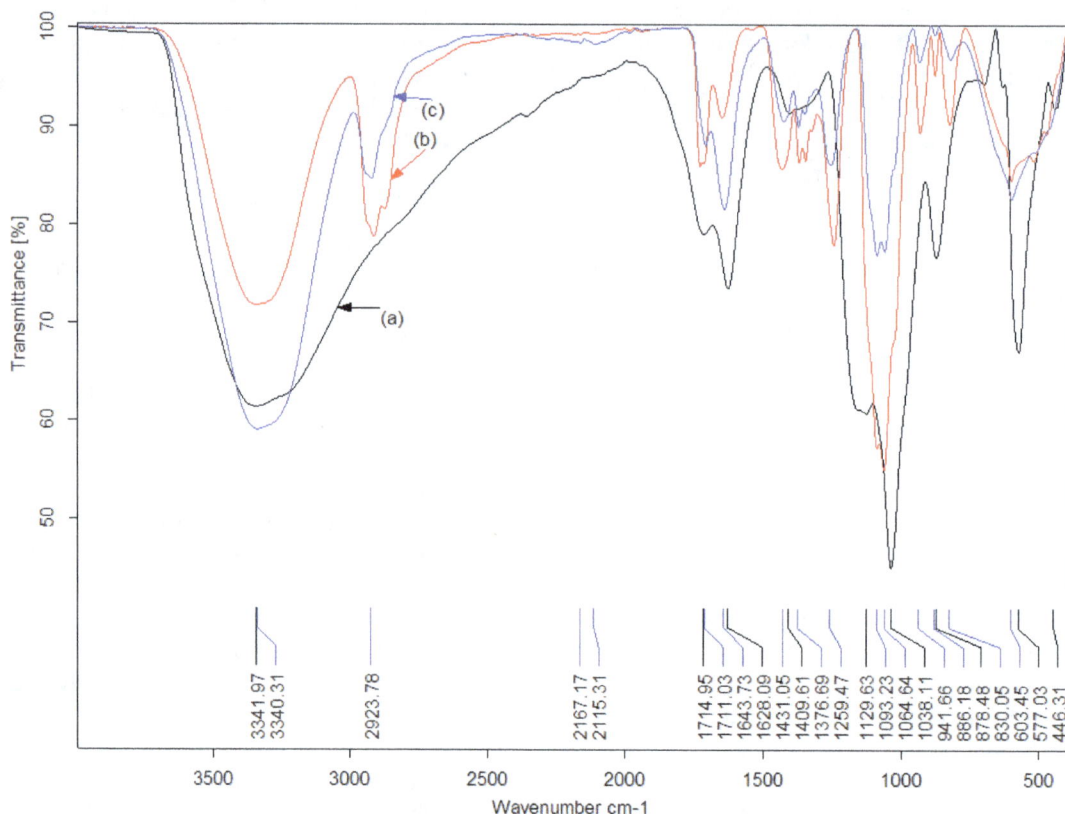

Fig. 3: FTIR Spectrum of (a) Graphene Oxide, (b) GO-PEG-PVA hydrogel and (C) PVA-PEG hydrogel.

oxide. The presence of polar groups, such as surface hydroxyl, led to the formation of hydrogen bonds between GO and water molecules; this further explains the hydrophilic nature of graphene oxide [39]. FTIR analysis is one of the most powerful techniques to investigate polymeric systems because it provides information on both polymer-polymer interactions and blends compositions using those vibrational modes attributed to the free and hydrogen bonded carbonyl and hydroxyl groups [40]. Theoretically, hydrogen bonding and/or other secondary interactions occurring between the functional groups on dissimilar polymers should cause a shift in the peak position of these participating functional groups. This behavior is shown by blending miscible polymers that exhibit extensive phase mixing. Stretching frequencies of the participating functional groups usually shifted with the hydrogen bonding interactions towards a lower wave-number; they showed an increase in peak intensity and broadening. The shift in peak position also depends on the strength of the interaction [40]

Fig. 3b,c show FTIR of synthesized hydrogels. The stretching band of hydroxyl groups in PVA in the composite hydrogels is observed at 3340 cm^{-1} as a broadband which indicates the presence of a high degree of intramolecular and intermolecular hydrogen bonding in the hydrogel network [40, 41]. The peak at 2914 cm^{-1} in Figs. 4b,c is the characteristic band of alkyl (CH$_2$) groups of both PVA and PEG. The stretching vibrational band of C=O group at 1730 cm^{-1} is attributed to the aldehyde group, while C=C groups appeared at 1650 cm^{-1}. The absorption band at 1436 cm^{-1} is assigned as CH$_2$ bending vibration while the deformation vibration of C–CH$_3$ is associated with the absorption band at 1351 cm^{-1} [40]. The peak at 1247 cm^{-1} is due to the C–O stretching mode, which is a contribution from both the PVA and PEG. The stretching at 1065 cm^{-1} represent the epoxide groups; it was also observed that the intensity of the peak in the case of a hydrogel containing GO, is higher than that of the hydrogel without GO. The vibrational band at 1093 cm^{-1} is mostly attributed to the crystallinity of the PVA, where it is related to carboxyl stretching band (C–O) [42].

Fig. 4 shows the X-ray diffraction pattern (XRD) of GO that had been synthesized by the modified Hummer's method. The sharp diffraction peak observed at $2\theta = 11.02°$ corresponds to the (001) diffraction peak of GO. The distance of GO's interlayer was calculated using Bragg's Law (Eq. 2).

$$d_{spacing} = \frac{\lambda}{2Sin\theta} \tag{2}$$

θ is Bragg's angle in radians and λ is X-ray's wavelength (0.1542 nm) [43]. The calculated interlayer spacing of GO was 0.71 nm, which is a large interlayer distance and might be an indication to the formation of functional groups, such as hydroxyl, epoxy and carboxyl groups [44]. The XRD curves of the PVA-PEG and PVA-PEG-GO nano-composites

are presented in Fig. 4b. One of the major factors that affect the polymer mechanical properties is the crystallinity. The XRD pattern of the pure PVA membranes revealed strong crystalline reflections at around $2\theta = 19.92°$ and a shoulder at 22.74. That represented reflections at 101 and 200 at the monoclinic unit cell, results that are characteristic of PVA [45]. In the XRD profile of PVA-PEG-GO nano-composite hydrogel, the intensity of PVA diffraction peaks decreased (Fig. 4b).

The SEM micrographs of synthesized GO with different scale bars are given in Fig. 5a, which shows that the graphene oxide has a layered structure that can afford ultrathin and homogeneous graphene films. Such films are continuous or folded at times and it is possible to see the individual sheet edges, including the

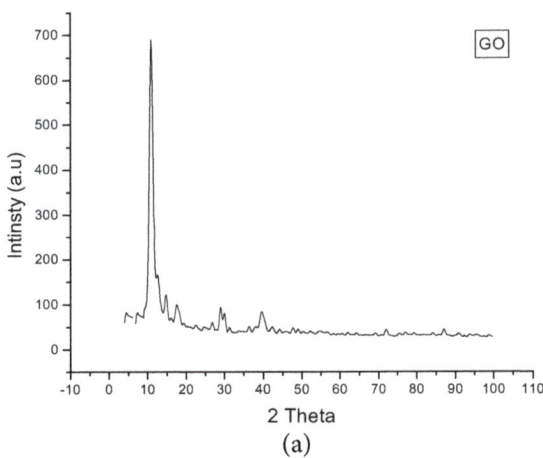

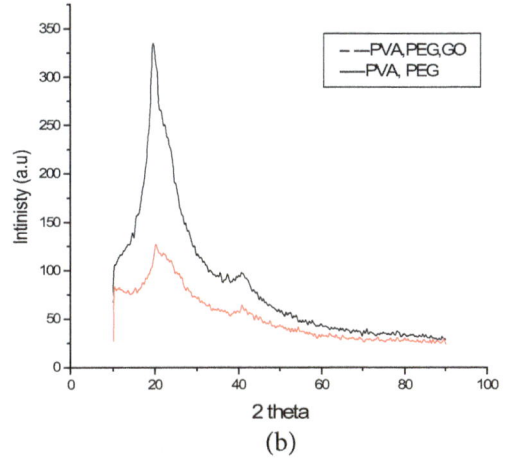

Fig. 4: XRD of GO (a) and (b) GO-PEG-PVA, PVA-PEG hydrogel.

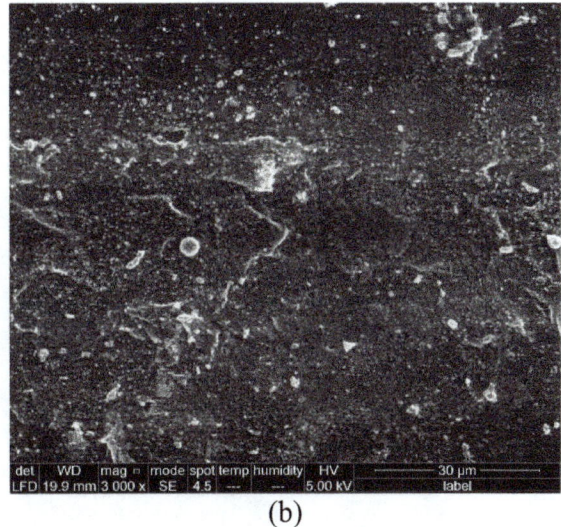

Fig. 5: (a) SEM micrograph of graphene oxide, (b) GO-PEG-PVA composite hydrogel.

wrinkled and kinked areas. On the other hand, the SEM image of PVA-PEG-GO composite hydrogel (Fig. 5b) shows the uniform dispersion of GO in the polymer matrices.

Batch Experiment
Effect of pH

It is well known that adsorption of Cu^{2+} ions from its water solution onto adsorbents depends on the pH value of the solution. The adsorption characteristics of Cu^{2+} ions on graphene oxide with various pH values ranging 1.0–7.0 were studied using an initial Cu^{2+} ions concentration of 500 mg l^{-1}. The results are reported in Fig. 6, which shows that the uptake of Cu^{2+} ions increased when the pH increased from 3.0 to 5.0.

The removal efficiency of GO, hydrogel with GO and hydrogel without GO increases with the increase of the pH value from 1 to reach a maximum of pH 5 and a decrease when the pH increases to 7. The results in case of GO can be explained based on the competition between Cu^{2+} ions and H_3O^+ for adsorption sites on GO. At low pH levels, an excess H_3O^+ competes with Cu^{2+} ions resulting in a low level of adsorbed Cu^{2+} ions. When the pH value increased, the covered H_3O^+ ions left the GO surface and made the sites available to Cu^{2+} ions. This condition suggests that an ion-exchange mechanism (H^+/Cu^{2+}) may be included in the adsorption of Cu^{2+} ions [46,47]. In addition, the negative charge on the surface of GO increased because the oxygen-containing functional groups became deprotonated with the increase in pH value. Hence, the electrostatic attraction between GO and Cu^{2+} ions was enhanced, hence further increasing the adsorption amount of Cu^{2+} ions.

While in the case of hydrogels, the adsorption of heavy metals was inhibited at low pH values. When the pH value was above 3.0, the adsorption efficiencies reached the adsorption balance points of Cu^{2+} ions, the optimum value was reached at pH 5. It is reasonable to say that at low pH values, there weren't any charges on the surface of the hydrogel. Therefore, the positive metal ions are difficult to adsorb on the surface of hydrogel–GO or the surface may be coated with H_3O^+ ions, so the positive metal ions are difficult to adsorb on the positively charged surface of the electrostatic repulsion [26]. However, the charged surface of the hydrogel (negative charge) increased with the increase of the pH value. So the combination of metal ions and hydroxyl groups on the PVA, PEG–GO gel became dominant and thereby resulted in the increase of metal ion sorption.

Effect of contact time

The relationship between the adsorption behavior of Cu^{2+} ions on adsorbents GO, PVA-PEG-GO and PVA-PEG hydrogels and the contact time were investigated using an initial Cu^{2+} ions concentration of 500 mg l^{-1} at optimum pH 5.0, the result is represented in Fig. 7. It was found that the adsorption capacity of GO and both hydrogels increased with increasing the time due to the existence of many binding sites for Cu^{2+} ions. The adsorption capacity gradually increased and finally remained constant, and the adsorption equilibrium occurred after stirring

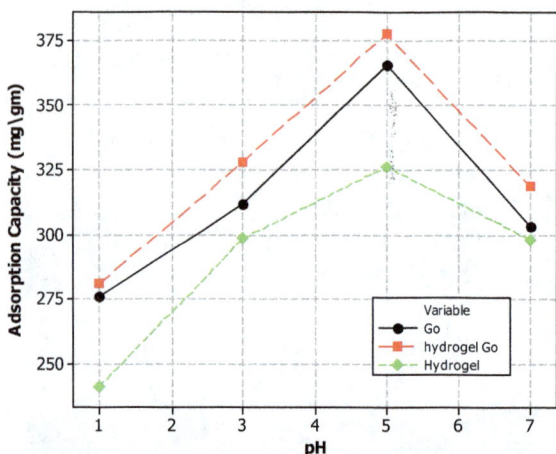

Fig. 6: Effect of initial solution pH on adsorption uptake of Cu(II) (initial Cu(II) concentration 500 mg l^{-1}, dosage of GO 100 mg l^{-1}, contact time 60 min)

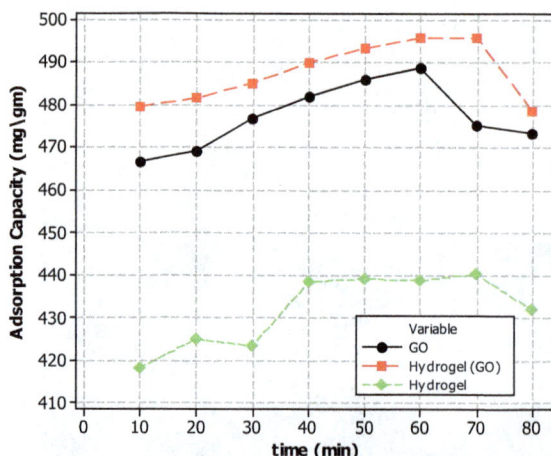

Fig. 7: Effect of contact time on adsorption of Cu(II) (initial Cu(II) mass concentration 500 mg l^{-1}, dosage of GO 100 mg l^{-1}, initial solution pH 5.0)

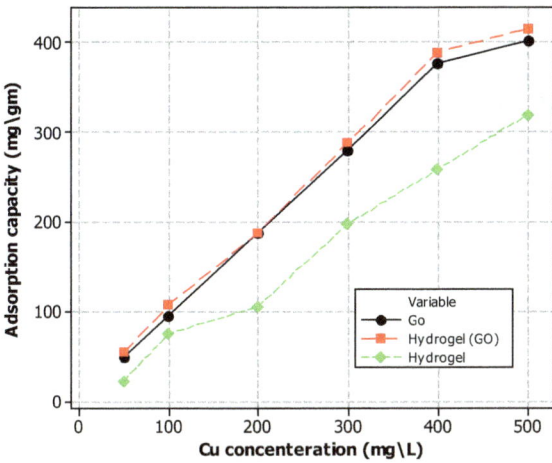

Fig. 8: Effect of Cu Concentration (initial dosage of GO 0.2 mg ml^{-1}, initial solution pH 5.0).

for 60 minutes. The delay in time of desorption may attribute to the weakening of the driving force resulting in the decrease of available adsorption sites and pH values. These decreased as a consequence of releasing H$^+$ from the oxygen-containing functional groups (COOH or OH) on the surface of GO and hydrogels to the solution, hence hindering adsorption. Therefore, the adsorption time was fixed at 60 minutes in subsequent adsorption experiments to make sure that the equilibrium is reached [48].

Effect of initial concentration on Cu^{2+}ions adsorption

The uptake of Cu^{2+} ions highly depended on the initial concentration of Cu^{2+} ions (Fig. 8). It is obvious that the adsorption capacity increased with the increase of initial Cu^{2+} ions' concentration. The adsorption capacity of GO-PVA-PEG hydrogel is higher than GO only and PVA-PEG hydrogel. The adsorption capacity of GO-PVA-PEG hydrogel increased from 55.4 to 415 mg g^{-1} when the initial Cu^{2+} ions' concentration increased from 50 to 500 mg g^{-1},while the adsorption capacity for GO and the PVA-PEG hydrogel increased from 50 to 400mg g^{-1} and 23.46 to 325 mg g^{-1}, respectively.

Adsorption kinetic studies

Studies the kinetic data were analyzed using two commonly kinetic models, the pseudo-first-order [44] and the pseudo-second-order [45], which can be expressed in the linear forms as follows:

$$Log(q_e - q_t) = \log (q_e) - \frac{k_1}{2.303}(t) \quad (3)$$

$$\frac{t}{q_t} = \frac{1}{k_2 q_e^2} + \frac{1}{q_e}(t) \quad (4)$$

where q_e and q_t are the adsorption capacity at equilibrium and time t, respectively (mg g^{-1}), k_1(min^{-1}) is rate constant of pseudo-first order adsorption, while k_2 (g/mg min) is rate constant of pseudo-second order adsorption. The adsorption kinetic plots are shown in Fig. 9(a,b) and the related parameters calculated from the two models are listed in Table 1. The values of the correlation coefficients indicate that the adsorption kinetics fits better to the pseudo-second-order model rather than to the pseudo-first-order model.

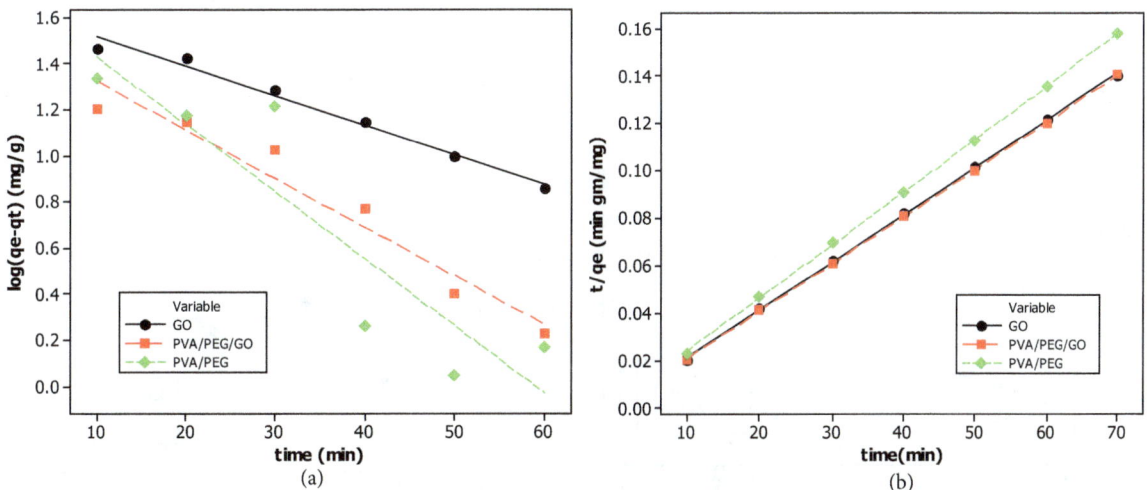

Fig. 9: (a) Pseudo first order model, (b) pseudo second order model for GO, PVA/PEG/GO, PVA/PEG at initial concentration of Cu^{2+} = 500 mg l^{-1}, dosage of GO 100 mg l^{-1}, initial solution pH 5.0).

Adsorption isotherms

An adsorption isotherm expresses the relationship between the amount of adsorbate adsorbed per unit weight of adsorbent (Q_e, mg g^{-1}) and the concentrations of adsorbate in the bulk solution (C_e, mg l^{-1}) at a given temperature under equilibrium conditions. As well as it shows the quality of an adsorbent by how much adsorbate it can sorb and keep in an immobilized form [46]. The experimental data were analyzed using the Langmuir and Freundlich adsorption isotherm models. The Langmuir isotherm model describes an ideal adsorption process, which contains the formation of a uniform single layer adsorbate on the outer surface of the adsorbent [20]. The Langmuir model was based on monolayer adsorption and could be described as the following equation:

$$\frac{C_e}{q_e} = \frac{1}{k_L Q_m} + \frac{1}{Q_m} C_e \qquad (5)$$

where C_e (mg l^{-1}) is the equilibrium concentration of adsorbate, q_e (mg g^{-1}) is the amount of metal adsorbed per gram of the adsorbent at equilibrium, Q_m (mg g^{-1}) is the maximum monolayer coverage capacity and k_L (l mg^{-1}) is the Langmuir isotherm constant determined by plotting C_e/q_e versus C_e.

The Freundlich isotherm model describes the adsorption on a heterogeneous surface with un-uniform energy. The Freundlich isotherm can be described as the following linear equation:

$$\text{Log } q_e = \log k_F + \frac{1}{\text{n}} \log C_e \qquad (6)$$

where k_F (mg g^{-1}) is the Freundlich isotherm constant and n is the adsorption intensity. Fig. 10 (a,b) show Langmuir and Freundlich adsorption isotherm models. The parameter values of Langmuir and Freundlich model could be determined by the linear relationship of $C_e/q_e - C_e$, and log q_e – log C_e, respectively and the results are presented in Table 2. The Higher correlation coefficient of Langmuir isotherm model for GO implied that monolayer adsorption was more readily to occur on the homogeneous surface of graphene oxide sheets. While Freundlich model was more fit to PVA-PEG-GO and PVA-PEG hydrogel. The maximum adsorption capacity (Q_m) was in order PVA-PEG-GO (917 mg g^{-1}) > GO (900 mg g^{-1}) > PVA-PEG (412 mg g^{-1}) (Table 2) and they are compared with another adsorbent in Table 3. The adsorption obeys Freundlich isotherm models for PVA-PEG-GO and n value > 1 so the adsorption is physical adsorption.

Table 1: Parameters of kinetic models

Adsorbent	q_e exp mg g^{-1}	Pseudo-first order			Pseudo-second order		
		K_1 (l min^{-1})	q_e Cal.	R^2	K_2	q_e Cal. mg g^{-1}	R^2
GO	496.46	0.027	44.4	0.98	2.79×10^{-3}	500	0.999
PVA-PEG-GO	495.16	0.048	34.5	0.94	4×10^{-3}	502	0.999
PVA-PEG	440	0.066	52.9	0.823	2.78×10^{-3}	444.5	0.999

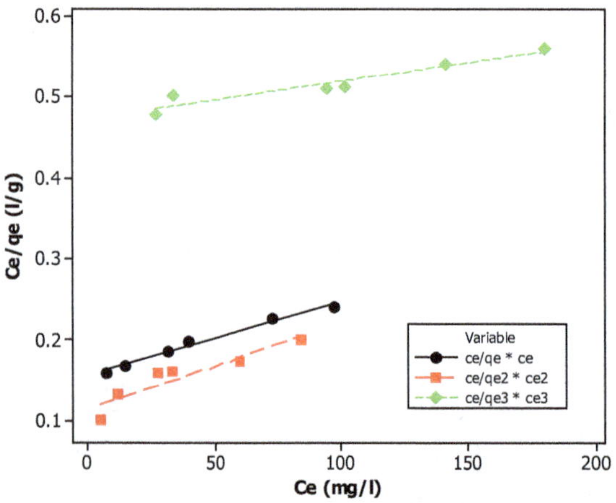

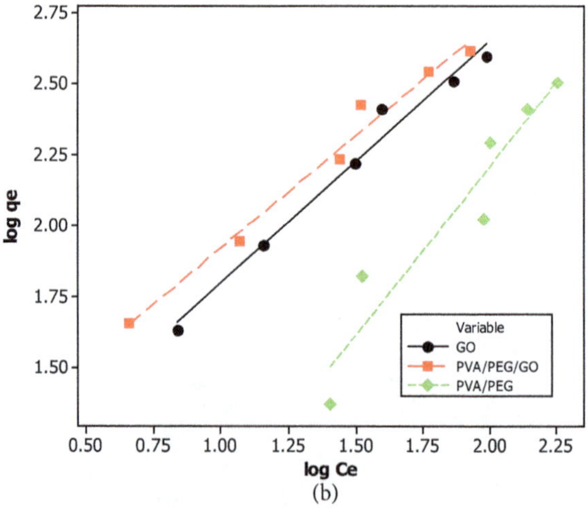

Fig. 10: (a) Langmuir and (b) Freundlich adsorption isotherm models of Cu (II) on GO, PVA-PEG-GO, and PVA-PEG.

Table 2: Fitting parameters for Langmuir and Freundlich isotherms

Model	GO	PVA-PEG-GO	PVA-PEG
Linear Langmuir			
Q_m (mg/g)	900	917	423
ka (l/mg)	5.76×10^{-3}	10.23×10^{-3}	0.9×10^{-3}
R_l	0.257	0.163	0.68
R^2	0.992	0.92	0.959
Linear Freundlich			
n	1.169	1.26	0.839
k_F	8.74	13.45	0.671
R^2	0.969	0.990	0.993

Table 3: Comparison of maximum adsorption capacity of Cu (II) on various adsorptions

Adsorbent	Maximum Capacity (mg g^{-1})	Reaction condition	Ref
Sulfonated magnetic GO	62.73	At intial Cu(II) conc. 60 mg/l, pH=6, time 30 min, and Temp=25±1	46
Graphene oxide	117.5	At intial Cu(II) conc. 20 mg/l, pH=5, time 150 min, and Temp=25±1	47
EDTA-MGO	301.2	At intial Cu(II) conc. 100 mg/l, pH=5.1, time 90 min, and Temp=25±1	48
graphene oxide-ethylenediamine triacetic acid	108.7	At intial Cu(II) conc. 20 mg/l, pH=5, time 90 min, and Temp=25±1	49
GO	909.0	At intial Cu(II) conc. 500 mg/l, pH=5, time 60 min, and Temp=25±1	This work
PVA-PEG-GO hydrogel	917	At initial Cu(II) conc. 500 mg/l, pH=5, time 60 min, and Temp=25±1	This work

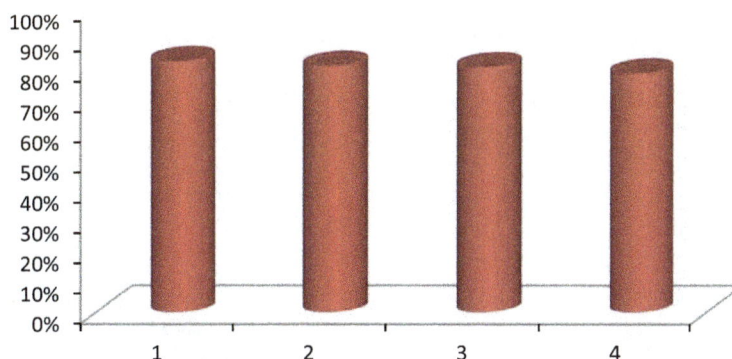

Fig. 11: Effect of recycle times on the Cu^{2+} ions removal efficiency of GO-PEG-PVA hydrogel.

Recyclability test

The industrial applicability of adsorbents depends not only on the adsorption capacity but also on the desorption regeneration. Treatment of the GO-PVA-PEG hydrogel by diluted HCl lead to large amounts of H$^+$ which caused protonation of hydroxyl groups of polyvinyl alcohol and polyethylene glycol. Therefore, metal ions can be replaced by H$^+$. After experiencing each adsorption

cycle, the metal-adsorbed by GO-PVA-PEG hydrogel was eluted with 0.1 M HCl solution, then neutralized by 0.1 M NaOH solution and finally washed with double distilled water for the next cycle [26]. Fig. 11 exhibits the reusability of GO-PVA-PEG hydrogel in removing Cu^{2+} ions. The removal efficiencies are 83% for the first cycle and about 81.4, 81, and 79 % for the second, third and fourth cycle, respectively. That proves the efficient reusability of the newly prepared hydrogel. Recycling with HCl decreased the pH value which leads to enhancing polymers cross-linking and decreased the hydrogel degradation.

CONCLUSIONS

The graphene oxide was prepared by oxidizing purified natural flake graphite via the modified Hummer's method; the PVA-PEG-GO nano-composites were prepared by a simple and environmentally friendly process that used water as the proceeding medium. The results of FTIR showed that graphite is oxidized by strong oxidants and that oxygen atoms are introduced into the graphite layers to form C=O, C-H, COOH and C-O-C bonds. FTIR of the hydrogel showed the interaction of GO with both polymers, PVA and PEG. The XRD results of GO show 2θ of $11.02°$ with interlayer spacing equal to 0.71 nm which attributed to good functionalization with hydroxyl groups. On the other hand, XRD of the hydrogel with GO showed the disappearance of the characteristic peak of GO and this emphasized the good dispersion of GO into the hydrogel. In addition to that, the characteristic peak of PVA decreased in the case of GO-PVA-PEG hydrogel due to a reduction in crystallinity. SEM image of GO-PVA-PEG hydrogel indicated the good dispersion of GO into the polymers. The adsorption obeys Freundlich isotherm models for PVA-PEG-GO, and the values of the correlation coefficients indicate that the adsorption kinetics fits better to the pseudo-second-order model rather than to the pseudo-first-order model. The present study clearly shows that GO-PVA-PEG hydrogel is effective in the removal of Cu^{2+} ions from water without leaving any toxic residual. It also proved how GO-PVA-PEG hydrogel is of good reusability, revealing its potential in practical application.

CONFLICT OF INTEREST

The authors declare that there is no conflict of interests regarding the publication of this manuscript.

REFERENCES

1. El-Nemr A. Impact, Monitoring, and Management of Environmental Pollution (Pollution Science, Technology & Abatement Series): Nova Science Publishers Incorporated; 2010.
2. Fayyaz F, Rahimi R, Rassa M, Maleki A. Efficient photo-oxidation of phenol and photo-inactivation of bacteria by cationic tetrakis(trimethylanilinium)porphyrins. Water Science and Technology: Water Supply. 2015;15(5):1099-105.
3. Khaled A, El Nemr A, El Sikaily A. Heavy Metals Concentrations in Biota of the Mediterranean Sea: A Review, Part I. Blue Biotechnology Journal. 2013;2(1):79.
4. Khaled A, El Nemr A, El Sikaily A. HEAVY METAL CONCENTRATIONS IN BIOTA OF THE MEDITERRANEAN SEA: A REVIEW, PART II. Blue Biotechnology Journal. 2013;2(2):191.
5. Maleki A, Rahimi R, Maleki S. Efficient oxidation and epoxidation using a chromium(VI)-based magnetic nanocomposite. Environ Chem Lett. 2016;14(2):195-9.
6. El-Nemr A. Non-conventional Textile Waste Water Treatment: Nova Science Publishers; 2012.
7. Abdelwahab O, El Sikaily A, Khaled A, El Nemr A. Mass-transfer processes of chromium(VI) adsorption onto guava seeds. Chem Ecol. 2007;23(1):73-85.
8. El Nemr A, El Sikaily A, Khaled A, Abdelwahab O. Removal of toxic chromium(VI) from aqueous solution by activated carbon using Casuarina equisetifolia. Chem Ecol. 2007;23(2):119-29.
9. El-Sikaily A, Nemr AE, Khaled A, Abdelwehab O. Removal of toxic chromium from wastewater using green alga Ulva lactuca and its activated carbon. J Hazard Mater. 2007;148(1):216-28.
10. El Nemr A. Pomegranate husk as an adsorbent in the removal of toxic chromium from wastewater. Chem Ecol. 2007;23(5):409-25.
11. Nemr AE. Potential of pomegranate husk carbon for Cr(VI) removal from wastewater: Kinetic and isotherm studies. J Hazard Mater. 2009;161(1):132-41.
12. Sun L, Fugetsu B. Mass production of graphene oxide from expanded graphite. Mater Lett. 2013;109:207-10.
13. Sun L, Fugetsu B. Effect of encapsulated graphene oxide on alginate-based bead adsorption to remove acridine orange from aqueous solutions. arXiv preprint arXiv:13070223. 2013.
14. Zhao G, Li J, Ren X, Chen C, Wang X. Few-Layered Graphene Oxide Nanosheets As Superior Sorbents for Heavy Metal Ion Pollution Management. Environ Sci Technol. 2011;45(24):10454-62.
15. Maleki A, Rahimi R, Maleki S. Efficient oxidation and epoxidation using a chromium(VI)-based magnetic nanocomposite. Environ Chem Lett. 2016;14(2):195-9.
16. Maleki A, Paydar R. Bionanostructure-catalyzed one-pot three-component synthesis of 3,4-dihydropyrimidin-2(1H)-one derivatives under solvent-free conditions. React Funct Polym. 2016;109(Supplement C):120-4.
17. Najafian A, Rabbani M, Rahimi R, Deilamkamar M, Maleki A. Synthesis and characterization of copper porphyrin into SBA-16 through "ship in a bottle" method: A catalyst for photo oxidation reaction under visible light. Solid State Sci. 2015;46(Supplement C):7-13.
18. Mahboubeh Rabbani, Hamideh Bathaee, Rahmatollah Rahimi, Maleki A. Photocatalytic degradation of p-nitrophenol and methylene blue using Zn-TCPP/Ag doped mesoporous TiO2 under UV and visible light irradiation. Desalin Water Treat. 2016;57(53):25848-56.
19. Zhao G, Ren X, Gao X, Tan X, Li J, Chen C, et al. Removal of Pb(ii) ions from aqueous solutions on few-layered graphene oxide nanosheets. Dalton Trans. 2011;40(41):10945-52.
20. Travlou NA, Kyzas GZ, Lazaridis NK, Deliyanni EA. Graphite oxide/chitosan composite for reactive dye removal. CHEM ENG J. 2013;217(Supplement C):256-65.

21. Zhu J, Wang Y, Liu J, Zhang Y. Facile One-Pot Synthesis of Novel Spherical Zeolite–Reduced Graphene Oxide Composites for Cationic Dye Adsorption. IND ENG CHEM RES. 2014;53(35):13711-7.

22. González JA, Villanueva ME, Piehl LL, Copello GJ. Development of a chitin/graphene oxide hybrid composite for the removal of pollutant dyes: Adsorption and desorption study. CHEM ENG J. 2015;280(Supplement C):41-8.

23. Ma Z, Liu D, Zhu Y, Li Z, Li Z, Tian H, et al. Graphene oxide/chitin nanofibril composite foams as column adsorbents for aqueous pollutants. Carbohydr Polym. 2016;144(Supplement C):230-7.

24. Jamnongkan T, Kantarot K, Niemtang K, Pansila PP, Wattanakornsiri A. Kinetics and mechanism of adsorptive removal of copper from aqueous solution with poly(vinyl alcohol) hydrogel. T NONFERR METAL SOC. 2014;24(10):3386-93.

25. Fan L, Luo C, Li X, Lu F, Qiu H, Sun M. Fabrication of novel magnetic chitosan grafted with graphene oxide to enhance adsorption properties for methyl blue. J Hazard Mater. 2012;215(Supplement C):272-9.

26. Kong X-b, Tang Q-y, Chen X-y, Tu Y, Sun S-z, Sun Z-l. Polyethylene glycol as a promising synthetic material for repair of spinal cord injury. Neural Regener Res. 2017;12(6):1003-8.

27. Xu R, Zhou G, Tang Y, Chu L, Liu C, Zeng Z, et al. New double network hydrogel adsorbent: Highly efficient removal of Cd(II) and Mn(II) ions in aqueous solution. CHEM ENG J. 2015;275(Supplement C):179-88.

28. Abraham TN, Kumar R, Misra RK, Jain SK. Poly(vinyl alcohol)-based MWCNT hydrogel for lead ion removal from contaminated water. J Appl Polym Sci. 2012;125(S1):E670-E4.

29. Khokan Chandra Sarker, Ullaha R. Determination of Trace Amount of Cu(II) Using UV-Vis. Spectrophotometric Method. International Journal of Chemical Studies. 2013;1(1):5-14.

30. Niyogi S, Bekyarova E, Itkis ME, McWilliams JL, Hamon MA, Haddon RC. Solution Properties of Graphite and Graphene. J Am Chem Soc. 2006;128(24):7720-1.

31. Hirata M, Gotou T, Horiuchi S, Fujiwara M, Ohba M. Thin-film particles of graphite oxide 1:: High-yield synthesis and flexibility of the particles. CARBON. 2004;42(14):2929-37.

32. Lerf A, He H, Forster M, Klinowski J. Structure of Graphite Oxide Revisited. The Journal of Physical Chemistry B. 1998;102(23):4477-82.

33. He H, Klinowski J, Forster M, Lerf A. A new structural model for graphite oxide. Chem Phys Lett. 1998;287(1):53-6.

34. Cote LJ, Kim F, Huang J. Langmuir–Blodgett Assembly of Graphite Oxide Single Layers. J Am Chem Soc. 2009;131(3):1043-9.

35. Cai D, Song M. Preparation of fully exfoliated graphite oxide nanoplatelets in organic solvents. J Mater Chem. 2007;17(35):3678-80.

36. Paredes JI, Villar-Rodil S, Martínez-Alonso A, Tascón JMD. Graphene Oxide Dispersions in Organic Solvents. LANGMUIR. 2008;24(19):10560-4.

37. Song J, Wang X, Chang C-T. Preparation and Characterization of Graphene Oxide. Journal of Nanomaterials. 2014;2014:6.

38. Thakur S, Karak N. Green reduction of graphene oxide by aqueous phytoextracts. CARBON. 2012;50(14):5331-9.

39. Du Q, Zheng M, Zhang L, Wang Y, Chen J, Xue L, et al. Preparation of functionalized graphene sheets by a low-temperature thermal exfoliation approach and their electrochemical supercapacitive behaviors. Electrochim Acta. 2010;55(12):3897-903.

40. Ali ZI, Eisa WH. Characterization of Electron Beam Irradiated Poly Vinyl Alcohol/Poly Ethylene Glycol Blends. Journal of Scientific Research. 2013;6(1):29-42.

41. Gadea JL, Cesteros LC, Katime I. Chemical–physical behavior of hydrogels of poly(vinyl alcohol) and poly(ethylene glycol). Eur Polym J. 2013;49(11):3582-9.

42. Mansur HS, Oréfice RL, Mansur AAP. Characterization of poly(vinyl alcohol)/poly(ethylene glycol) hydrogels and PVA-derived hybrids by small-angle X-ray scattering and FTIR spectroscopy. POLYMER. 2004;45(21):7193-202.

43. Fu C, Zhao G, Zhang H, Li S. Evaluation and characterization of reduced graphene oxide nanosheets as anode materials for lithium-ion batteries. Int J Electrochem Sci. 2013;8:6269-80.

44. Tang C-M, Tian Y-H, Hsu S-H. Poly(vinyl alcohol) Nanocomposites Reinforced with Bamboo Charcoal Nanoparticles: Mineralization Behavior and Characterization. Materials. 2015;8(8):4895-911.

45. Rengaraj S, Joo CK, Kim Y, Yi J. Kinetics of removal of chromium from water and electronic process wastewater by ion exchange resins: 1200H, 1500H and IRN97H. J Hazard Mater. 2003;102(2):257-75.

46. Mi X, Huang G, Xie W, Wang W, Liu Y, Gao J. Preparation of graphene oxide aerogel and its adsorption for Cu2+ ions. CARBON. 2012;50(13):4856-64.

47. Yang Y, Wu W-q, Zhou H-h, Huang Z-y, Ye T-t, Liu R, et al. Adsorption behavior of cross-linked chitosan modified by graphene oxide for Cu(II) removal. Journal of Central South University. 2014;21(7):2826-31.

48. Cui L, Wang Y, Gao L, Hu L, Yan L, Wei Q, et al. EDTA functionalized magnetic graphene oxide for removal of Pb(II), Hg(II) and Cu(II) in water treatment: Adsorption mechanism and separation property. CHEM ENG J. 2015;281(Supplement C):1-10.

49. Mejias Carpio IE, Mangadlao JD, Nguyen HN, Advincula RC, Rodrigues DF. Graphene oxide functionalized with ethylenediamine triacetic acid for heavy metal adsorption and anti-microbial applications. CARBON. 2014;77(Supplement C):289-301.

18

Kinetic and Isotherm Studies of Cadmium Adsorption on Polypyrrole/Titanium dioxide Nanocomposite

Marjan Tanzifi[1],*, Marzieh Kolbadi Nezhad[2], Kianoush Karimipour[1]

[1]*Department of Chemical Engineering, Faculty of Engineering, University of Ilam, Ilam, Iran*
[2]*School of Chemical Gas and Petroleum Engineering, Semnan University, Semnan, Iran*

ABSTRACT

The present work seeks to investigate the ability of polypyrrole/titanium dioxide nanocomposite to adsorb cadmium ions from aqueous solution. The impact of various experimental conditions, including solution pH, adsorbent dosage, adsorption time and initial concentration on the uptake of cadmium were studied. The adsorption kinetic was studied with the first-order, second-order, pseudo-first-order, pseudo-second-order and Morris–Weber models. The results revealed that adsorption process is controlled by pseudo-second-order model which illustrated that the adsorption process of cadmium is chemisorption-controlled. The adsorption capacity obtained from this model is 20.49 mg/g which close to the experimental value. The study yielded the result that when the initial concentration of the solution changed from 20 mg/l to 120 mg/l, the adsorption capacity increased from 0.99 to 24.52 mg/g. Further, Langmuir, Freundlich and Temkin isotherm models were applied to investigate the adsorption isotherm. Based on the results of the adsorption isotherm, Freundlich isotherm proved to be the best fit with the experimental data. Also, the morphology, chemical structure and thermal stability of adsorbent were studied by using SEM, EDX, FTIR, and TGA.

Keywords: *Adsorption; Cadmium; Isotherm; Kinetic; Polypyrrole; Titanium Dioxide*

INTRODUCTION

Heavy metals due to be toxic and non-biodegradable are a serious threat to health and the environment. These metals are dissolved in water so can be transmitted by water and accumulate in soil and living organisms [1]. Among heavy metals, cadmium is a toxic one, which exists in natural, industrial and agricultural environments [2]. Exposure to cadmium can lead to the destruction of human organs (kidney, liver, and lung), immune system, and cardiovascular disease [3-5]. So, removal of cadmium from water is essential and necessary.

Various processes are presented for the removal of heavy metals from water and wastewater such as sedimentation, oxidation/reduction, membrane filtration, osmosis, ion exchange as well as adsorption. Among them, ion exchange/adsorption process is a convenient and economical way to produce high-quality water [6-8].

Today, nanotechnology is one of the most important trends in materials science. The term of nanotechnology refers to materials that one or more of their dimensions (length, width, and thickness) are in the range of nanometers (1-100 nm) [9]. Large specific surface, high reactivity and high adsorption and desorption capacity are main features of nanomaterials [9-12]. Due to these features, use of nanomaterials has increased in

* Corresponding Author Email: *m.tanzifi@ilam.ac.ir*

environmental applications [13,14]. Various nano-adsorbents such as Copper oxide nano blades [15], nanoscale zero-valent iron particles [16], CdS nanoparticles [17], nano magnetite/chitosan films [18] and nano zeolites [19] have been studied for the removal of cadmium.

Since the detection of conducting polymers, a lot of researches has been done in relation to the features and applications of these materials. Conductive polymers are used in different fields such as composite materials, biosensors as well as adsorption [20-22]. These materials have ion exchange capacity, which is dependent on conditions of polymerization and type of dopant and polymer [23]. Among the various conductive polymers, polypyrrole is considered due to its ease of polymerization, high chemical stability and biocompatibility [24-27]. Combination of polypyrrole with some materials can improve adsorption capacity. Ghorbani et al (2011), showed that polypyrrole is suitable adsorbent for removing zinc from aqueous solutions [28]. In a study conducted by John Chen et al (2015), polypyrrole/montmorillonite nanocomposite showed high adsorption capacity for the adsorption of chromium [29]. Mthombeni et al (2013), used magnetized natural zeolite-polypyrrole composite as a potential adsorbent for vanadium [30].

Titanium dioxide nanoparticles are used in catalysts [31], gas sensors [32], batteries [33] and environmental cleanup [34,35]. These particles, used as nano-fillers, lead to the modification of the polymeric network, which consequently, increases the surface area of the adsorbent. Fabrication of conductive polymer composites through titanium dioxide have been reported [36,37].

In this study, polypyrrole/titanium dioxide nanocomposite was considered as an effective nano-adsorbent for removing cadmium ions from aqueous solution. The impacts of four parameters, including adsorbent dosage, pH of the solution, adsorption time and initial concentration were scrutinized on the efficiency of cadmium adsorption. Also, kinetics and isotherms of cadmium adsorption on polypyrrole/titanium dioxide nanocomposite were studied.

EXPERIMENTAL
Materials and Instrumentation
In this study, pyrrole monomer, polyvinyl alcohol (PVA), ferric chloride (III) ($FeCl_3$), sulfuric acid, sodium hydroxide and cadmium salt purchased from Merck company of Germany and titanium dioxide nanoparticles (TiO_2) purchased from Kronos company of Germany. With the dissolution of $Cd(NO_3)_2.4H_2O$ in distilled water, 100 ppm cadmium solution was prepared. By using this solution, solutions with different concentrations were prepared. 1 M solution of sulfuric acid was used to change the pH. Pyrrole monomer was distilled for use in polymerization.

The equipment used in this research include: magnetic stirrer (model MK20), digital scale (model FR 200), Fourier transform infrared spectrometer (FTIR) (model VERTEX 70; BRUKER) and thermogravimetric analyzer (TGA) (model Perkin Elmer). Also, to describe the morphology of the polypyrrole/titanium dioxide and determine the concentration of cadmium ions in solution, the scanning electron microscope (SEM) (model XL30 and model KYKY-EM 3200) and atomic adsorption spectrophotometer (model Analytic Jena) were used, respectively. Atomic adsorption spectrophotometer calibrated using standard solutions (0.5-10 ppm) then the concentration of cadmium ions in each solution was measured.

Preparation of nano-adsorbent
In this method, 0.1 g of titanium dioxide nanoparticles was added to 50 ml of distilled water. Then 1 ml of pyrrole monomer and 0.3 g of polyvinyl alcohol as a stabilizer was added to the solution. After 30 minutes, a solution of ferric chloride as oxidant was added drop by drop to the above solution. By adding oxidant, the solution was initially green and then to black. This color change is a sign of conversion of pyrrole monomer to polypyrrole. Polymerization was carried out for 5 hours at ambient temperature. The polymer is then filtered. In order to remove oligomers and impurities, the polymer was washed several times with distilled water. The final product was dried for 48 hours in an oven at 45 °C. Then, the resulting polymer was milled to convert the powder. Also for comparison, pure polypyrrole and polypyrrole/titanium dioxide composite (without stabilizer) were synthesized [38].

Cadmium adsorption experiments
In the present study, the effects of different experimental conditions such as pH, initial concentration, adsorbent dose and adsorption time on removal of cadmium and also kinetic and isotherm results were studied. A Certain amount of

polypyrrole/titanium dioxide nanocomposite was added to 50 ml of cadmium solution with a certain initial concentration. Then, the solution was filtered and the concentration of cadmium ions was measured by atomic adsorption spectrophotometer. The adsorption efficiency is calculated by the following equation:

$$Removal(\%) = \frac{C_i - C_f}{C_i} \times 100 \qquad (1)$$

In this equation, C_i and C_f represent the initial and final concentration of cadmium (mg/L), respectively. Also, q_e and q_t (mg/g) are adsorption capacity at equilibrium and time t, which calculated as follows:

$$q_e = (C_i - C_e) \times \frac{V}{m} \qquad (2)$$

$$q_t = (C_i - C_t) \times \frac{V}{m} \qquad (3)$$

Where Ce(mg/L) and C_t(mg/L) are the cadmium concentration at equilibrium and time t, respectively. Also, V is the volume of cadmium solution (L), and m is a mass of adsorbent (g).

RESULT AND DISCUSSION
Characterization of Adsorbent

Morphology of adsorbent was studied using scanning electron microscopy (SEM). Fig. 1 shows scanning electron microscope image of titanium dioxide nanoparticles. As can be seen, the particles have nearly spherical shape with an average diameter of 46 nm. Also, the particle size distribution is uniform. Fig. 2 shows SEM images of polypyrrole/titanium dioxide composite and polypyrrole/titanium dioxide nanocomposite made with polyvinyl alcohol. As can be seen in

Fig. 2a, polypyrrole/titanium dioxide composite without stabilizer has large particle size. The morphology and particle size of products depend on the presence of stabilizer that reduces the size of the polymer particles. PVA affects on the physical and chemical features of the solution, the rate of formation of polypyrrole, morphology, homogenization, particle size and size distribution. The presence of PVA as a stabilizer in the polypyrrole/titanium dioxide synthesis media has several advantages. First, increase the solubility of the composite in aqueous solvents and then prevents the accumulation of particles in during polymerization. PVA is leading to a better quality of the final product [39]. As can be seen in Fig. 2b, by adding polyvinyl alcohol to the polypyrrole/titanium dioxide synthesis environment, smaller, spherical and uniform particles are produced, because polyvinyl alcohol affects the viscosity of the solution and polymerization rate. In fact, the use of stabilizer led to the formation of nano-adsorbent which has a positive effect on its performance in adsorption.

Fig. 3 shows the analysis of energy dispersive X-ray spectroscopy (EDX) of products. This

Fig. 1: Scanning electron micrograph of titanium dioxide nanoparticles.

Fig. 2: Scanning electron micrograph of (a) polypyrrole/titanium dioxide composite and (b) polypyrrole/titanium dioxide nanocomposite (with PVA).

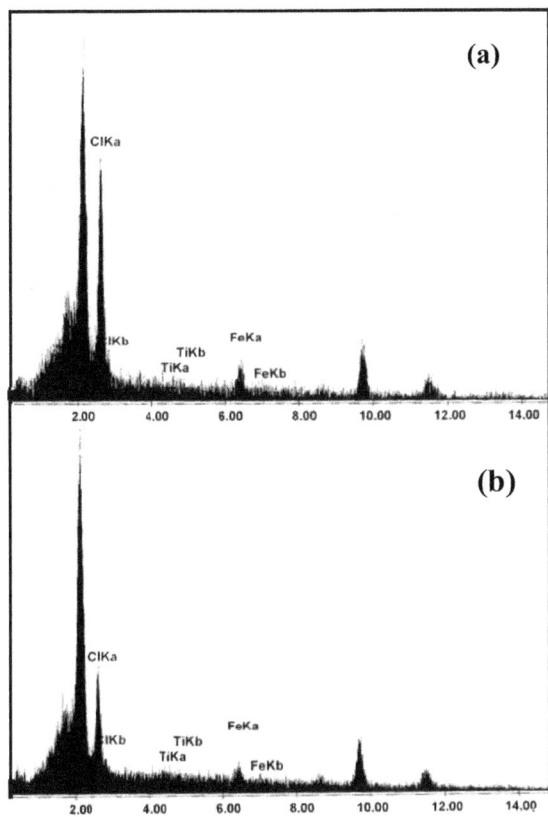

Fig.3: EDX analysis of (a) polypyrrole/titanium dioxide composite and (b) polypyrrole/titanium dioxide nanocomposite (with PVA).

analysis illustrates the presence of titanium dioxide nanoparticles in the synthesized nanocomposite. Further, Fe and Cl elements are related to ferric chloride (III), which was used as the oxidant. The presence of Fe and Cl ions in EDX spectrum shows that polypyrrole chain is doped by chloride anions. Table 1 presents the elemental compositions of synthesized polypyrrole/titanium dioxide composites with and without PVA which obtained from EDX analysis. Information obtained from SEM and EDX demonstrate the formation of polypyrrole in the presence of titanium dioxide nanoparticles.

The structure of synthesized products was specified by Fourier transform infrared (FTIR). Figs. 4a-4d are related to titanium dioxide nanoparticles, polypyrrole/titanium dioxide composite, pure polyvinyl alcohol and polypyrrole/titanium dioxide nanocomposite in presence of polyvinyl alcohol. The large peak at the wavelength 669 Cm^{-1} was ascribed to Ti-O bond (Fig. 4a). In Fig. 4b (FTIR spectra of polypyrrole/titanium dioxide composite), peak at 1547 Cm^{-1} is attributed to polypyrrole rings, which illustrates the polymer is generated. The other peaks are at 1313 cm^{-1} (C-N stretching vibration), 1175 cm^{-1} (C-H inplane deformation), 1043 cm^{-1} (N-H in-plane deformation), and 904 cm^{-1} (C-H out-of-plane deformation) [40]. The FTIR spectra of polypyrrole/titanium dioxide nanocomposite (Fig. 4d) shows all peaks of both PVA and polypyrrole nanocomposite have characteristic bands at 3457, 2973, 1844, 1533, 1445, 1293, 1159, 1039 cm^{-1} and 893 cm^{-1}, which are related to O-H band of PVA, C-H band of PVA, C=O band of PVA, pyrrole rings, CH_2 band of PVA, C-N band of polypyrrole, C-H band of polypyrrole, N-H group of polypyrrole and C-H band of polypyrrole, respectively. Also, a peak indicating the presence of titanium dioxide nanoparticles in the polymer can be seen in 667 Cm^{-1}. Results have been given in Table 2.

Table 1: EDX analysis data of polypyrrole/titanium dioxide with and without stabilizer

Sample	Elements (atoms%)		
	Fe	Ti	Cl
polypyrrole/titanium dioxide	41.31	6.81	51.88
polypyrrole/titanium dioxide (with PVA)	34.86	4.12	61.02

Table 2: FTIR spectra of polypyrrole/titanium dioxide with and without PVA

FTIR Spectra	Composite		
	Polypyrrole /titanium dioxide	Pure PVA	Polypyrrole /titanium dioxide (with PVA)
O-H	-	3392	3457
C-H	-	2941	2973
C=O	-	1717	1844
CH_2	-	1432	1445
C=C	1547	-	1533
C-N	1313	-	1293
C-H	1175	-	1159
N-H	1043	-	1039
C-H	904	-	893

In order to evaluate the impact of titanium dioxide nanoparticles on the thermal stability of polypyrrole, thermogravimetric analyzer (TGA) was used. Figs. 5a-5d illustrate TGA analysis of pure polypyrrole and its composite and nanocomposite. This analysis was performed under a nitrogen atmosphere in the temperature range 25-600 °C and the rate 10°C/min. TGA diagram of pure polypyrrole is shown in Fig. 5a. In this diagram, three stages of mass loss occur. The graph shows

that mass loss begins at 30°C and continues until 140°C. The mass loss remains constant until 320°C and then fast mass loss happens from 320 to 600. At first, the mass loss causes by the elimination of water molecules to be in the polypyrrole structure and then increases by removing of oligomers. The next rapid mass loss can be related to the destruction of the polypyrrole chain. In Fig. 5c, removing the retained water in the polypyrrole network, which is ascribed to water molecules adsorbed due to the

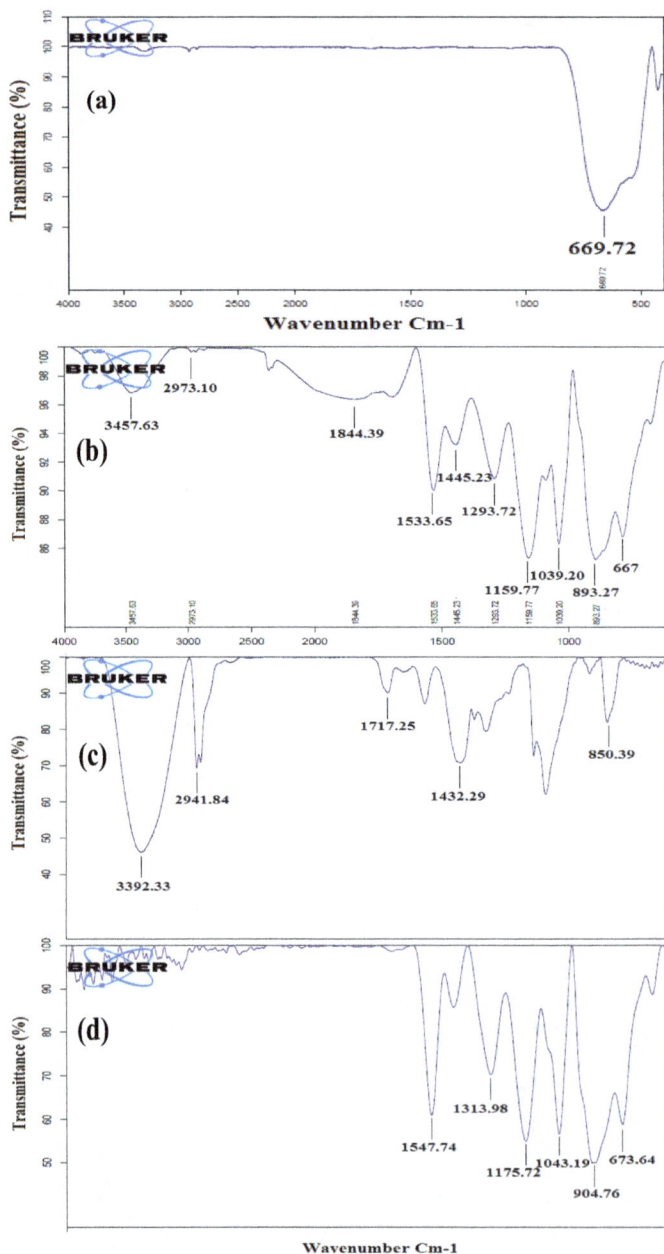

Fig. 4: FTIR spectra of (a) titanium dioxide nanoparticles (b) polypyrrole/titanium dioxide, (c) pure PVA and (d) polypyrrole/titanium dioxide (with PVA).

porous structure of polypyrrole [41], creates the first step of mass loss for the polypyrrole/titanium dioxide composite (about 7 wt% from 27°C to 148°C). The second mass loss stage (between 140°C to 380°C) has two reasons: The first reason is related to the loss of water molecules which form stronger bond with composite network and the second reason is related to the removal of volatile elements bound to polypyrrole chain and the omission of interaction between polypyrrole and titanium dioxide nanoparticles. Polypyrrole destruction and combustion at temperatures from 400-600°C, causes the third step on the mass loss. Mass loss in the temperature range 160-400°C (in Figs. 5b and 5d) can be ascribed to evaporation and degradation of polyvinyl alcohol. As can be seen, the thermal degradation rate of the nanocomposite, at temperatures above 350°C, is less than pure polypyrrole. Also according to Figs. 5a and 5d, mass loss of pure polypyrrole are 48% more than polypyrrole/titanium dioxide nanocomposite.

Effect of PH

The pH of the solution is an important parameter in the adsorption process of adsorbates. In this study, the effect of pH on adsorption efficiency of cadmium ions was investigated. The experiments were performed with the initial concentration of cadmium 50 mg/L, adsorption time 90 min and temperature 20°C. The experiments illustrated that at pH higher than 7 precipitate is formed (pH=7: sediment threshold). This shows that at the high pH, in addition of adsorption, deposit formation is also involved in the removal of cadmium. Therefore, cadmium uptake was studied in pH lower than 7. Results are shown in Fig. 6. As can be seen in the figure, in the range of pH=2.5 to 4.5, adsorption efficiency is maximum. By reducing the pH (less than 2.5), adsorption efficiency decreases. In low pH, the concentration of H^+ ions in the environment

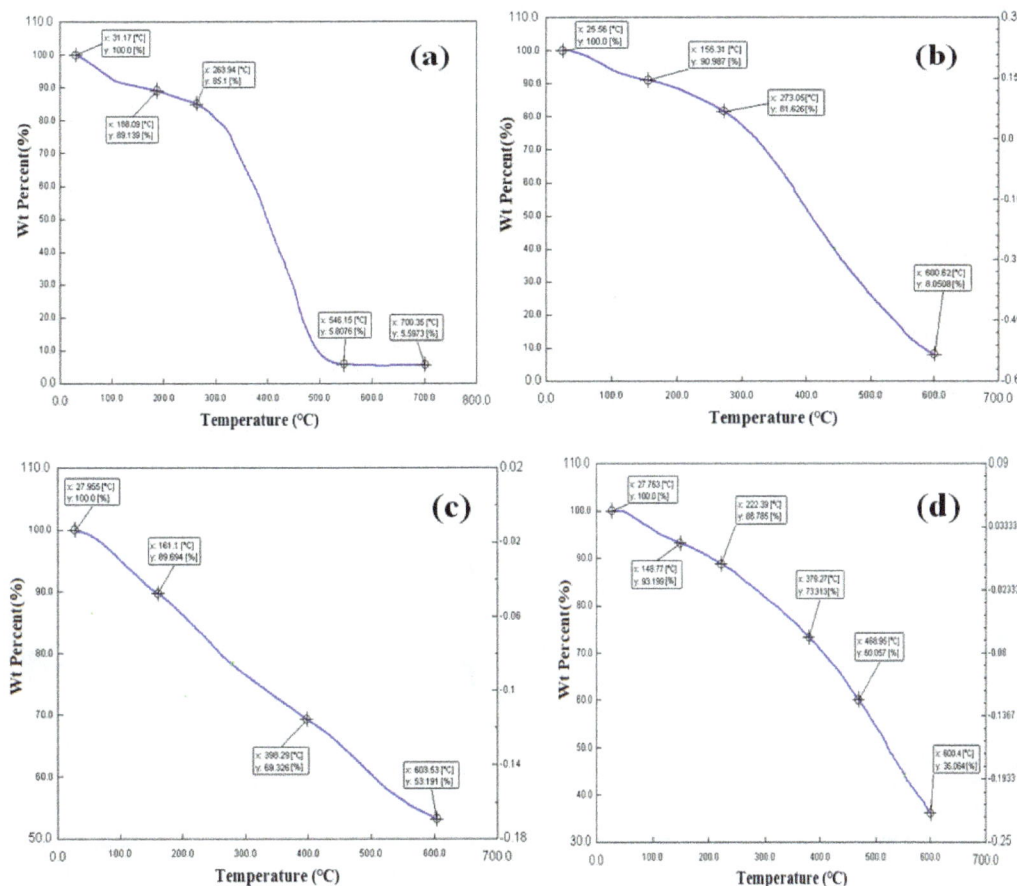

Fig. 5: TGA curve of (a) pure polypyrrole, (b) polypyrrole (with PVA), (c) polypyrrole/titanium dioxide and (d) polypyrrole/titanium dioxide (with PVA).

increases and these ions and metal ions compete on adsorption on the adsorbent surface, so the adsorption efficiency of metal ions is reduced. Polypyrrole/titanium dioxide nanocomposite has good performance in the range of pH=2.5-6.5, so it is suitable adsorbent in this pH range. Javadian and *et al.* showed that the effective adsorption of cadmium onto zeolite-based geopolymer occurs in the initial pH range of 2–5 [42].

Effect of amount of adsorbent

Adsorbent dosage is one of the important parameters in determining the adsorbed cadmium ions. In order to evaluate the impact of adsorbent dosage, varying amounts of polypyrrole/titanium dioxide nanocomposite (0.1 to 0.8 gr) were added to the 50 ml of cadmium solution (50 mg/l, pH=2.5). The effect of nanocomposite dosage on

the uptake of cadmium is shown in Fig. 7. As can be seen, by increasing the amount of nanocomposite from 0.1 to 0.2 gr, cadmium adsorption efficiency increases from 60.26% to 67.76%, because of available adsorption places for a constant amount of cadmium increase. For higher adsorbent values, curves increase with a gentle slope, and this suggests that cadmium ions uptake efficiency increased slightly. The reason is that, by increasing the nanocomposite dosage, nanomaterials due to a large specific surface and high reactivity, stick together and form a lump. So, adsorption efficiency does not change much. The optimum adsorbent dosage was considered 0.2 g.

Effect of initial concentration and cadmium adsorption Isotherm

Solutions with an initial concentration of 20-

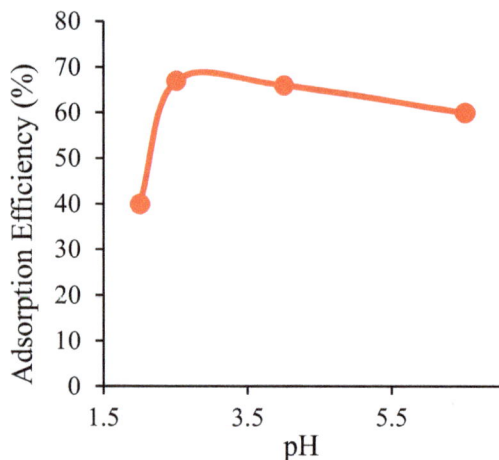

Fig. 6: Effect of pH on the adsorption efficiency of cadmium.

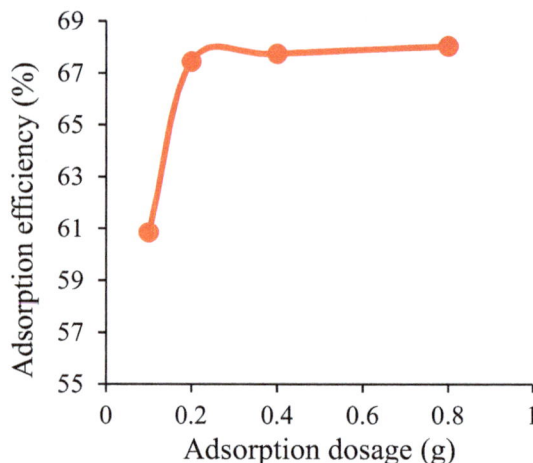

Fig. 7: Effect of adsorbent dosage on the adsorption efficiency of cadmium.

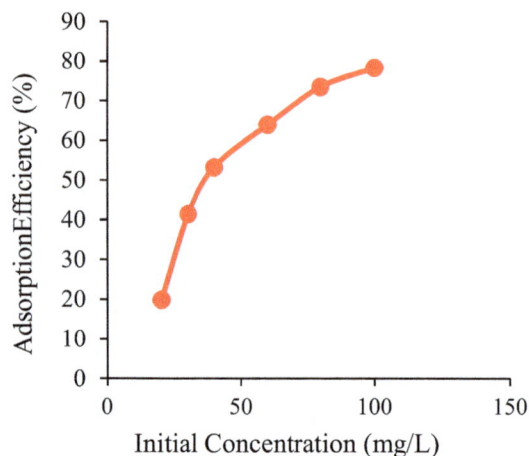

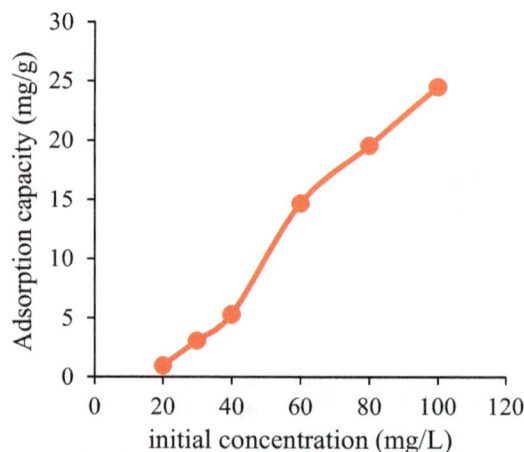

Fig. 8: Effect of initial concentration on the adsorption efficiency and capacity.

120 ppm, was prepared to investigate the impact of initial concentration of cadmium ions on adsorption efficiency and capacity. Nanocomposite dosage, PH, adsorption time and volume solution were considered 0.2 g, 2.5, 90 min and 50 ml, respectively. As shown in Fig. 8, the adsorption efficiency and capacity increase with increasing initial concentration of cadmium ions. When initial concentration increases from 20 to 120 mg/L, adsorption efficiency and capacity change from 19.9 to 81.74% and 0.995 to 24.52 mg/g, respectively.

Adsorption isotherm models illustrate the relationship between the cadmium concentration in solution and the quantity of cadmium ions adsorbed onto the specified amount of nanocomposite at a steady temperature. In this study, three adsorption isotherm models, namely; Langmuir, Freundlich, and Temkin were investigated.

In the Langmuir isotherm model, the adsorption energy is independent of surface coverage. All adsorption places are similar to each other and each place can adsorb only one type. This model is expressed by the following equation [43]:

$$q_e = \frac{q_m K_L C_e}{1 + K_L C_e} \quad (4)$$

where q_e, C_e and q_m are the quantity of adsorbed cadmium for the specified amount of nanocomposite (mg/g), equilibrium cadmium concentration (mg/L) and maximum adsorption capacity of nanocomposite (mg/g), respectively. Also, K_L is Langmuir constant. The Linear form of this equation presented as:

$$\frac{C_e}{q_e} = \frac{1}{q_m K_L} + \frac{1}{q_m} C_e \quad (5)$$

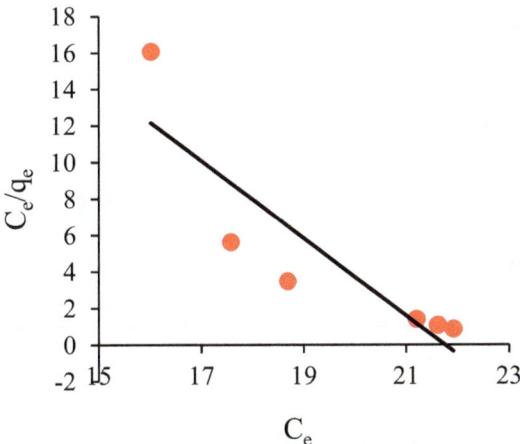

Separation factor (R_L) is a dimensionless constant that represents the essential specifications of the Langmuir isotherm:

$$R_L = \frac{1}{1 + K_L c_i} \quad (6)$$

where Ci is the initial concentration of cadmium solution and K_L is Langmuir constant. As the R_L obtained between 0-1, cadmium adsorption process onto nanocomposite is *favorable*. Langmuir plot is shown in Fig. 9. The negative slope of curve demonstrated that Langmuir equation is not appropriate for this data [28].

Freundlich model is an empiric equation that applied for adsorption on heterogeneous surfaces which is expressed as follows [44]:

$$q_e = K_F (C_e)^{1/n} \quad (7)$$

This equation can be written in linear form as follow:

$$log(q_e) = log(K_F) + \frac{1}{n} log(c_e) \quad (8)$$

K_f and $1/n$ are Freundlich constant and adsorption intensity, respectively, which calculated from intercept and slope of plot of $log(q_e)$ versus $log(C_e)$. Freundlich model plot is shown in Fig. 10. Temkin Isotherm has a factor that reflects interactions between adsorbate particles (cadmium) and adsorbent (nanocomposite). The equation of the isotherm can be expressed as follows [45]:

$$q_e = \frac{RT}{b_T} \ln(K_T C_e) \quad (9)$$

The above equation can be expressed in linear form as follows:

$$q_e = B \ln(K_T) + B \ln(C_e) \quad (10)$$

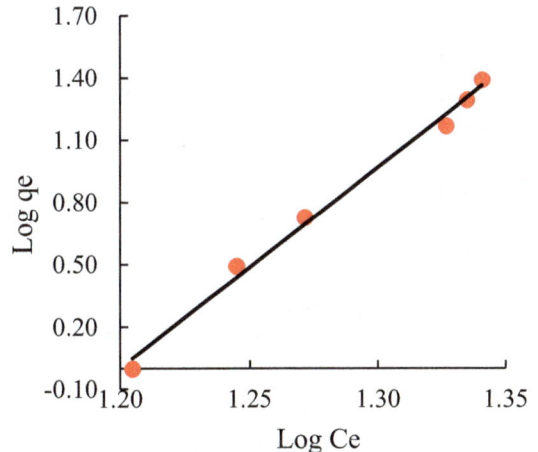

Fig. 9: Langmuir adsorption isotherm of cadmium ions onto polypyrrole/titanium dioxide nanocomposite.

Fig. 10: Freundlich adsorption isotherm of cadmium ions onto polypyrrole/titanium dioxide nanocomposite.

$$B = \frac{RT}{b_T} \qquad (11)$$

Where K_T, B, R, T, and b_T are equilibrium binding constant of Temkin (L/g), Temkin isotherm constant, universal gas constant (8.314 J/mol.K), absolute temperature (298.15 K) and constant appertained to adsorption heat (J/mol), respectively. Fig. 11 shows the Temkin isotherm plot.

The constants and correlation coefficient (R^2) for all isotherm models are illustrated in Table 3. Results show that Freundlich isotherm is the best model that represents the behavior of cadmium adsorption process onto polypyrrole/titanium dioxide nanocomposite because this model offers the highest correlation coefficient compared to other isotherms.

Effect of adsorption time and cadmium adsorption kinetics

Fig. 12 illustrates the effect of adsorption time on the adsorption of cadmium ions onto polypyrrole/titanium dioxide nanocomposite. The initial concentration of cadmium and pH were considered 100 ppm and 2.5. Also, adsorbent dosage was 0.2 g per 50 ml of cadmium solution. The impact of adsorption time on cadmium adsorption efficiency was conducted with the change in time from 2 to 100 min. The results showed that with increasing adsorption time, adsorption efficiency of cadmium increased. Adsorption efficiency reached from 78.4% to 82.01% with increasing the time from 2

to 100 min. As can be seen, removal of cadmium by polypyrrole/titanium dioxide nanocomposite, reached 81.8% at adsorption time 5 min. This demonstrated that in the short time, a large amount of cadmium is adsorbed by the polypyrrole/titanium dioxide nano-adsorbent.

Different kinetic models including the first-order, second-order, pseudo-first-order, pseudo-second and Morris–Weber were used to investigate the cadmium adsorption onto polypyrrole/titanium dioxide nanocomposite adsorbent. Kinetic models explain the adsorption rate of cadmium ions and it is clear that it controls the adsorption time in the intersection of the solution and solid.

The first-order equation is expressed in the linear form as follows:

$$\ln\left(\frac{C_0}{C_t}\right) = K_1 t \qquad (12)$$

Where C_0, C_t, and K_1 are initial concentration, cadmium concentration at time t and rate constant of the first-order model. Fig. 13 shows $\ln\left(\frac{C_0}{C_t}\right)$ versus t plot.

Table 3: Isotherm parameters for cadmium adsorption

Isotherm	parameters
Langmuir	y = -2.1186x+46.092
	$R^2 = 0.7817$
	R_L: $(-2.63) - (-0.278)$
	$K = 2.6 \times 10^{-12}$
Freundlich	n = 0.103
	$R^2 = 0.9925$
	B= 70.77
Temkin	$K_T = 0.699$
	$R^2 = 0.887$

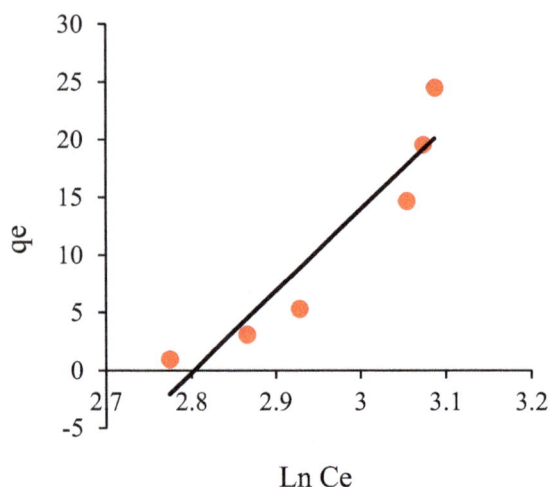

Fig. 11: Temkin adsorption isotherm of cadmium ions onto polypyrrole/titanium dioxide nanocomposite.

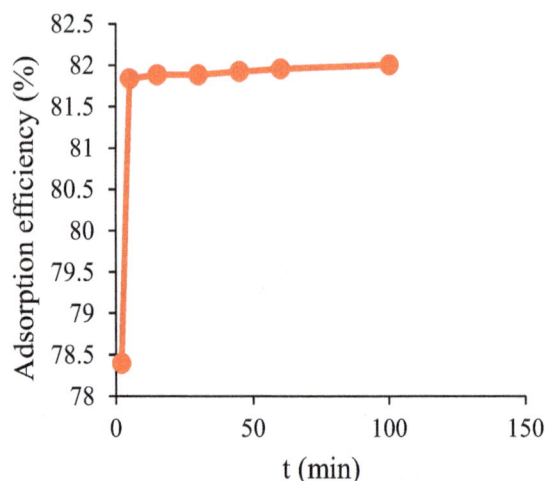

Fig. 12: Effect of adsorption time on adsorption efficiency.

The linear form of the second-order model is presented as follows:

$$\frac{1}{C_t} = K_2 t + \frac{1}{C_0} \tag{13}$$

where K_2 is rate constant of the second-order model which gained from slop of $\frac{1}{C_t} - \frac{1}{C_0}$ versus plot (Fig. 14).

In 1898, lagergren [46] presented the pseudo-first-order kinetic model for solid/liquid adsorption system based on adsorbent capacity. It presumes that the variation rate of adsorbate removal with time is straightly proportional to the difference in the saturation concentration and the quantity of solid adsorption with time. Pseudo-first-order equation is expressed as follows:

$$log(q_e - q_t) = log q_e - \left(\frac{K}{2.303}\right) t \tag{14}$$

where q_e and q_t are cadmium adsorption capacity at equilibrium and at time t. K is a constant rate of pseudo-first-order model. Fig. 15 shows the pseudo-first-order plot for adsorption cadmium

adsorbed onto polypyrrole/titanium dioxide nanocomposite.

Kinetic data of cadmium adsorption onto polypyrrole/titanium dioxide nanocomposite were also investigated by the pseudo-second-order equation. This kinetic model assumes that cadmium adsorption process is chemisorption. It is presented as follow [47]:

$$q_t = \frac{K q_e^2 t}{1 + K q_e t} \tag{15}$$

The linear form of this equation can express as:

$$\frac{t}{qt} = \frac{1}{K_2 q e^2} + \frac{1}{qe} t \tag{16}$$

where K_2 is the equilibrium rate constant of pseudo-second-order model (g/mg.min). The values of parameters of this model and regression coefficient (R^2) are summarized in Table 4. As shown in Fig. 16, $\frac{t}{qt}$ versus plot was a direct line with slope $\frac{1}{qe}$ and intercept $\frac{1}{K_2 q e^2}$.

Morris-Weber model illustrates that the intra-particle diffusion is a stage of speed controller

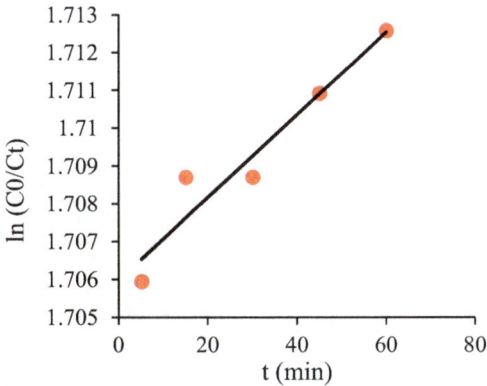

Fig. 13: First-order plot of cadmium adsorption onto polypyrrole/titanium dioxide nanocomposite.

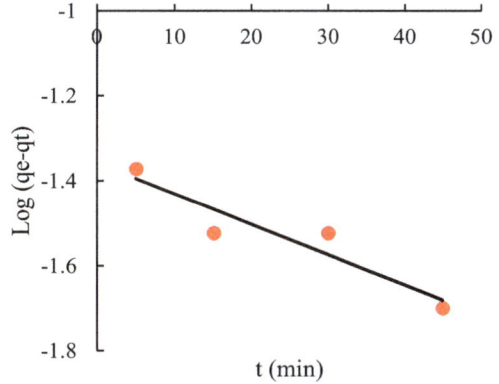

Fig. 15: Pseudo-first-order plot of cadmium adsorption onto polypyrrole/titanium dioxide nanocomposite.

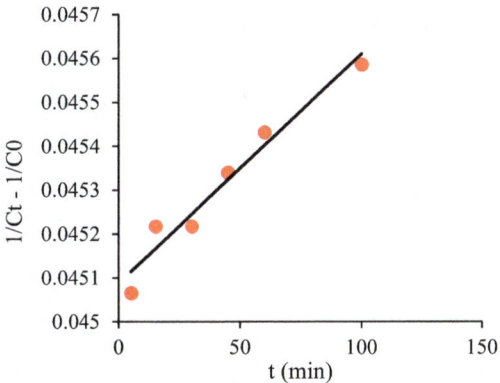

Fig. 14: Second-order plot of cadmium adsorption onto polypyrrole/titanium dioxide nanocomposite.

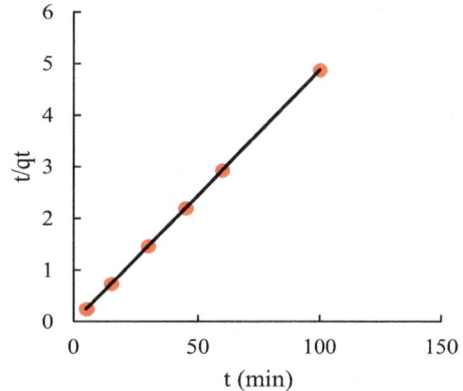

Fig. 16: Pseudo-second-order plot of cadmium adsorption onto polypyrrole/titanium dioxide nanocomposite.

in adsorption process [48]. This equation was expressed as follows:

$$q_t = K_{id}(t)^{0.5} + C \qquad (17)$$

In this equation, K_{id} is rate constant of intra-particle diffusion and C is a parameter depended on boundary layer thickness. If the Morris-Weber plot is linear, the process of cadmium adsorption is controlled by diffusion resistance. When the line passes through the origin, it indicates that intra-particle diffusion is the only rate-controlling step. The plot of qt versus $t^{0.5}$ is shown in Fig. 17. K_{id} and C constants are calculated from the slope and intercept of this plot.

Information gained from five kinetic equations is presented in Table 4. The correlation coefficient for first-order, second-order, pseudo-first-order, pseudo-second-order and Morris-Weber models were obtained 0.9279, 0.9554, 0.8765, 1 and 0.9641, respectively. As a result, cadmium adsorption

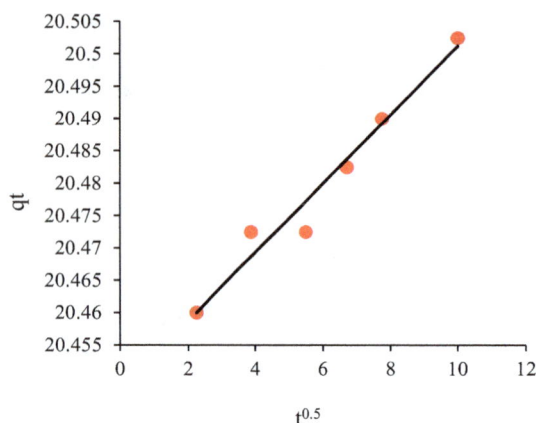

Fig. 17: Morris-Weber plot of cadmium adsorption onto polypyrrole/titanium dioxide nanocomposite

Table 4: Kinetic parameters for cadmium adsorption

Kinetic model	parameters
First-order	K_1=0.0001 R^2 =0.9279
Second-order	K_2=5×10^{-6} R^2 =0.9554
pseudo-first-order	q_e (experimental) =19.59 q_e (calculated) =22.85 K =0.016 R^2 = 0.8765
pseudo-second-order	q_e (experimental) =19.59 q_e (calculated) =20.4918 K =1.4884 R^2 = 1
Morris-Weber	K_{id} =0.0053 C= 20.448 R^2 =0.9641

process was controlled by pseudo-second-order model, which demonstrated that adsorption of cadmium onto polypyrrole/titanium dioxide nanocomposite is a chemical adsorption process. In studies conducted by Yang et al (2015) and Javadian et al (2016), adsorption of cadmium onto zeolite and carbon/aluminum composite were also controlled by pseudo-second-order kinetic model [41,49].

CONCLUSION

In this research, polypyrrole/titanium dioxide nanocomposite was prepared using polymerization of pyrrole monomer in the presence of ferric chloride (III) oxidant and stabilizer of polyvinyl alcohol at an ambient temperature and applied as a cadmium adsorbent from the water. Polypyrrole/titanium dioxide nanocomposite with the nano-sized particles has a high adsorption efficiency. The optimized conditions for adsorption cadmium were achieved adsorbent dosage 0.2 gr, pH 2.5 and adsorption time 60 min. Adsorption data matched well with the Freundlich model. The research yielded the result that cadmium adsorption process was controlled by pseudo-second-order model which demonstrates the chemical adsorption. Also, the adsorption capacity of nanocomposite obtained from this model was 20.49 mg/g, which is close to the empirical value. In addition, adsorbent characteristics such as thermal stability, chemical structure, and morphology were studied. EDX Analysis proved the presence of titanium dioxide nanoparticles in the nanocomposite structure. Also, the chemical structure of the synthesized nanocomposite was analyzed by FTIR. TGA results illustrated that the produced composite and nanocomposite of polypyrrole showed considerable thermal stability compared to pure polypyrrole.

CONFLICT OF INTEREST

The authors declare that there are no conflicts of interest regarding the publication of this manuscript.

REFERENCES

1. Kurniawan TA, Chan GYS, Lo W-H, Babel S. Physico-chemical treatment techniques for wastewater laden with heavy metals. CHEM ENG J. 2006;118(1):83-98.
2. Järup L. Hazards of heavy metal contamination. BRIT MED BULL. 2003;68(1):167-82.
3. Fowler BA. Monitoring of human populations for early markers of cadmium toxicity: A review. Toxicol Appl Pharmacol. 2009;238(3):294-300.
4. Järup L, Åkesson A. Current status of cadmium as an

environmental health problem. Toxicol Appl Pharmacol. 2009;238(3):201-8.

5. Thompson J, Bannigan J. Cadmium: Toxic effects on the reproductive system and the embryo. Reprod Toxicol. 2008;25(3):304-15.

6. Ghiloufi I, Ghoul JE, Modwi A, Mir LE. Ga-doped ZnO for adsorption of heavy metals from aqueous solution. Mater Sci Semicond Process. 2016;42(Part 1):102-6.

7. Liao B, Sun W-y, Guo N, Ding S-l, Su S-j. Equilibriums and kinetics studies for adsorption of Ni(II) ion on chitosan and its triethylenetetramine derivative. Colloids Surf, A. 2016;501(Supplement C):32-41.

8. Zareie C, Najafpour G. Preparation of nanochitosan as an effective sorbent for the removal of copper ions from aqueous solutions. International Journal of Engineering-Transactions B: Applications. 2013;26(8):829-36.

9. Tian J, Xu J, Zhu F, Lu T, Su C, Ouyang G. Application of nanomaterials in sample preparation. J Chromatogr A. 2013;1300(Supplement C):2-16.

10. El-Nahhal IM, Zourab SM, Kodeh FS, Elmanama AA, Selmane M, Genois I, et al. Nano-structured zinc oxide–cotton fibers: synthesis, characterization and applications. Journal of Materials Science: Materials in Electronics. 2013;24(10):3970-5.

11. Yan XM, Shi BY, Lu JJ, Feng CH, Wang DS, Tang HX. Adsorption and desorption of atrazine on carbon nanotubes. J Colloid Interface Sci. 2008;321(1):30-8.

12. Yusan S, Korzhynbayeva K, Aytas S, Tazhibayeva S, Musabekov K. Preparation and investigation of structural properties of magnetic diatomite nanocomposites formed with different iron content. J Alloys Compd. 2014;608(Supplement C):8-13.

13. Teimouri A, Nasab SG, Vahdatpoor N, Habibollahi S, Salavati H, Chermahini AN. Chitosan /Zeolite Y/Nano ZrO2 nanocomposite as an adsorbent for the removal of nitrate from the aqueous solution. Int J Biol Macromol. 2016;93(Part A):254-66.

14. Shojaei S, Khammarnia S, Shojaei S, Sasani M. Removal of Reactive Red 198 by Nanoparticle Zero Valent Iron in the Presence of Hydrogen Peroxide. Journal of Water and Environmental Nanotechnology. 2017;2(2):129-35.

15. Bhanjana G, Dilbaghi N, Singhal NK, Kim K-H, Kumar S. Copper oxide nanoblades as novel adsorbent material for cadmium removal. Ceram Int. 2017;43(8):6075-81.

16. Azzam AM, El-Wakeel ST, Mostafa BB, El-Shahat MF. Removal of Pb, Cd, Cu and Ni from aqueous solution using nano scale zero valent iron particles. J Environ Chem Eng. 2016;4(2):2196-206.

17. Raj R, Dalei K, Chakraborty J, Das S. Extracellular polymeric substances of a marine bacterium mediated synthesis of CdS nanoparticles for removal of cadmium from aqueous solution. J Colloid Interface Sci. 2016;462(Supplement C):166-75.

18. Lasheen MR, El-Sherif IY, Tawfik ME, El-Wakeel ST, El-Shahat MF. Preparation and adsorption properties of nano magnetite chitosan films for heavy metal ions from aqueous solution. Mater Res Bull. 2016;80(Supplement C):344-50.

19. Yousefi T, Torab-Mostaedi M, Charkhi A, Aghaei A. Cd(II) Sorption on Iranian nano zeolites: Kinetic and Thermodynamic Studies. Journal of Water and Environmental Nanotechnology. 2016;1(2):75-83.

20. Zuo X, Zhang Y, Si L, Zhou B, Zhao B, Zhu L, et al. One-step electrochemical preparation of sulfonated graphene/polypyrrole composite and its application to supercapacitor. J Alloys Compd. 2016;688(Part B):140-8.

21. Tanzifi M, Mansouri M, Heidarzadeh M, Gheibi K. Study of the Adsorption of Amido Black 10B Dye from Aqueous Solution Using Polyaniline Nano-adsorbent: Kinetic and Isotherm Studies. Journal of Water and Environmental Nanotechnology. 2016;1(2):124-34.

22. Tanzifi M, Karimipour K, Najafifard M, Mirchenari S. REMOVAL OF CONGO RED ANIONIC DYE FROM AQUEOUS SOLUTION USING POLYANILINE/TIO2 AND POLYPYRROLE/TIO2 NANOCOMPOSITES: ISOTHERM, KINETIC, AND THERMODYNAMIC STUDIES. International Journal of Engineering-Transactions C: Aspects. 2016;29(12):1659.

23. Weidlich C, Mangold KM, Jüttner K. Conducting polymers as ion-exchangers for water purification. Electrochim Acta. 2001;47(5):741-5.

24. Waltman RJ, Bargon J. Reactivity/structure correlations for the electropolymerization of pyrrole: An INDO/CNDO study of the reactive sites of oligomeric radical cations. TETRAHEDRON. 1984;40(20):3963-70.

25. Waltman RJ, Bargon J. Electrically conducting polymers: a review of the electropolymerization reaction, of the effects of chemical structure on polymer film properties, and of applications towards technology. Can J Chem. 1986;64(1):76-95.

26. Ruckenstein E, Chen J-H. Polypyrrole conductive composites prepared by coprecipitation. POLYMER. 1991;32(7):1230-5.

27. Diaz A. Electrochemical preparation and characterization of conducting polymers. CHEM SCRIPTA. 1981;17(1-5):145-8.

28. Omraei M, Esfandian H, Katal R, Ghorbani M. Study of the removal of Zn(II) from aqueous solution using polypyrrole nanocomposite. DESALINATION. 2011;271(1):248-56.

29. Chen J, Hong X, Xie Q, Tian M, Li K, Zhang Q. Exfoliated polypyrrole/montmorillonite nanocomposite with flake-like structure for Cr(VI) removal from aqueous solution. Res Chem Intermed. 2015;41(12):9655-71.

30. Mthombeni NH, Mbakop S, Ochieng A, Onyango MS. Vanadium (V) adsorption isotherms and kinetics using polypyrrole coated magnetized natural zeolite. J Taiwan Inst Chem Eng. 2016;66(Supplement C):172-80.

31. Zhong S, Ou Q, Shao L. PHOSPHORUS PROMOTED SO4 (2-)/TIO2 SOLID ACID CATALYST FOR ACETALIZATION REACTION. J Chil Chem Soc. 2015;60(3):3005-6.

32. Alev O, Şennik E, Kılınç N, Öztürk ZZ. Gas Sensor Application of Hydrothermally Growth TiO2 Nanorods. Procedia Eng. 2015;120(Supplement C):1162-5.

33. Kashale AA, Gattu KP, Ghule K, Ingole VH, Dhanayat S, Sharma R, et al. Biomediated green synthesis of TiO2 nanoparticles for lithium ion battery application. Composites Part B: Engineering. 2016;99(Supplement C):297-304.

34. Niehues E, Scarminio IS, TAKASHIMA K. Optimization of photocatalytic decolorization of the azo dye direct orange 34 by statistical experimental design. J Chil Chem Soc. 2010;55(3):320-4.

35. Kamal T, Anwar Y, Khan SB, Chani MTS, Asiri AM. Dye adsorption and bactericidal properties of TiO2/chitosan coating layer. Carbohydr Polym. 2016;148(Supplement C):153-60.

36. Gao F, Hou X, Wang A, Chu G, Wu W, Chen J, et al. Preparation of polypyrrole/TiO2 nanocomposites with enhanced photocatalytic performance. Particuology. 2016;26(Supplement C):73-8.

37. Zou Y, Wang Q, Jiang D, Xiang C, Chu H, Qiu S, et al. Pd-doped TiO2@polypyrrole core-shell composites as hydrogen-sensing materials. Ceram Int. 2016;42(7):8257-62.

38. Eisazadeh H, Ghorbani M. Copolymerization of pyrrole and vinyl acetate in aqueous and aqueous/nonaqueous media. J Vinyl Add Tech. 2009;15(3):204-10.

39. Rinaldi AW, Kunita MH, Santos MJL, Radovanovic E, Rubira AF, Girotto EM. Solid phase photopolymerization of pyrrole in poly(vinylchloride) matrix. Eur Polym J. 2005;41(11):2711-7.

40. Ghorbani M, Eisazadeh H. Fixed bed column study for Zn, Cu, Fe and Mn removal from wastewater using nanometer size polypyrrole coated on rice husk ash. Synth Met. 2012;162(15):1429-33.

41. Zhang H, Li GR, An LP, Yan TY, Gao XP, Zhu HY. Electrochemical Lithium Storage of Titanate and Titania Nanotubes and Nanorods. The Journal of Physical Chemistry C. 2007;111(16):6143-8.

42. Javadian H, Ghorbani F, Tayebi H-a, Asl SH. Study of the adsorption of Cd (II) from aqueous solution using zeolite-based geopolymer, synthesized from coal fly ash; kinetic, isotherm and thermodynamic studies. Arabian J Chem. 2015;8(6):837-49.

43. Langmuir I. The constitution and fundamental properties of solids and liquids. J Franklin Inst. 1917;183(1):102-5.

44. Freundlich H. Over the adsorption in solution. J Phys Chem. 1906;57(385471):1100-7.

45. Temkin M, Pyzhev V. Kinetics of ammonia synthesis on promoted iron catalysts. Acta physiochim URSS. 1940;12(3):217-22.

46. Lagergren S. About the theory of so-called adsorption of soluble substances. KUNGLIGA SVENSKA VETENSKAPSAKADEMIENS HANDLINGAR. 1898;24(4):1-39.

47. Ho YS, McKay G. Pseudo-second order model for sorption processes. Process Biochem. 1999;34(5):451-65.

48. Weber WJ, Morris JC. Kinetics of adsorption on carbon from solution. Journal of the Sanitary Engineering Division. 1963;89(2):31-60.

49. Yang W, Tang Q, Wei J, Ran Y, Chai L, Wang H. Enhanced removal of Cd(II) and Pb(II) by composites of mesoporous carbon stabilized alumina. Appl Surf Sci. 2016;369(Supplement C):215-23.

19

Photocatalytic Treatment of Synthetic Wastewater Containing 2,4 dichlorophenol by Ternary MWCNTs /Co-TiO$_2$ Nanocomposite Under Visible Light

Shahryar Nazarpour Laghani [1], Azadeh Ebrahimian Pirbazari [1,2,]*

[1] *Caspian Faculty of Engineering, College of Engineering, University of Tehran, Rezvanshahr, Iran.*
[2] *Fouman Faculty of Engineering, College of Engineering, University of Tehran ,Fouman, Iran.*

ABSTRACT

In this work, multi-walled carbon nanotubes (MWCNTs)/Co-TiO$_2$ nanocomposites were synthesized and investigated for photocatalytic degradation of 2,4-dichlorophenol (2,4-DCP) under visible light. Characterization of photocatalysts was done by means of XRD, FT-IR and SEM/EDX techniques. Obtained results showed cobalt doping can inhibit phase transformation from anatase to rutile and eliminate the recombination of electron-hole pairs. The presence of MWCNTs can both increase the photoactivity and change surface properties to achieve sensitivity to visible light. The optimum mass ratio of MWCNTs and cobalt (Co) dopant in TiO$_2$ was the prominent factor to harvest MWCNTs/Co-TiO$_2$ photocatalyst. The sample containing 3.13 wt% cobalt exhibited the highest activity under visible light for 2,4-DCP degradation, which was completed within 180 min using a 0.1 g/L dose of this photocatalyst in a 40 mg/L solution of the 2,4-DCP. The reactions follow the first-order kinetics. The reaction intermediates were identified by GC–MS technique. GC–MS analysis showed the major intermediates of 2,4-DCP degradation are simple acids like oxalic acid, acetic acid, etc. as the final products.

Keywords: *Carbon Nanotubes, Cobalt, 2,4-dichlorophenol, Degradation.*

INTRODUCTION

With the rapid development of industrialization, certain hazardous effects on the environment and human survival have emerged beside its benefit. Effluents from the use of pesticides, the textile, petrochemical, dyeing, plastic, and paper industries are highly toxic, carcinogenic and recalcitrant [1,2], and yet not readily degradable. Chlorophenols are toxic chemicals which are used in many industrial applications such as petrochemicals, pesticide, dye intermediates and paint [3]. Especially, 2,4-dichlorophenol (2,4-DCP) is an important chemical precursor for the manufacture of a widely used herbicide, 2,4-dichlorophenoxy acetic acid (2,4-D) [4]. However, 2,4-DCP may cause some pathological symptoms and changes to human endocrine systems. Their mode of exposure is through the skin and gastrointestinal tract. In recent years, concerns have been raised because of chlorophenols persistence and bioaccumulation both in animals and in humans [5-7].Therefore, it is important to find innovative and effective ways to minimize the harm of chlorophenols in the environment. Heterogeneous semiconductor photocatalysts have received significant attention owing to their potential application in a wide range

* Corresponding Author Email: *aebrahimian@ut.ac.ir*

of photoinduced reactions, notably photocatalytic hydrogen production, removal of organic contaminants and air pollutants, and electricity production using solar cells [8–11]. TiO_2 mediated photocatalytic degradation is a successful and convenient alternative to the conventional methods for the treatment of wastewater containing organic pollutants. TiO_2 has the advantage of good chemical stability, the absence of toxicity and relatively low cost, but a serious disadvantage is its wide band gap (Eg = 3.2 eV) that requires that UV radiation is used to trigger this attractive photocatalyst, which would greatly hinder the commercialization of TiO_2 photocatalysis. Photocatalytic degradation of organic contaminants using abundant natural solar radiation can be highly economical compared with the processes using artificial UV radiation, which require substantial electrical power input. In regard to this, various attempts have been made to extend the spectral response of TiO_2 into the visible region of the solar spectrum and enhance its photocatalytic activity [12]. Lots of attempts were tried to improve the photocatalytic performance of TiO_2 and doping with transition metal ions, such as silver, nickel, and iron was found to be a useful method [13-18]. Among all the available transition metals, cobalt was proved to be one of the most effective dopants to enhance the light response and photoactivity of TiO_2. Ebrahimian and et al [19] prepared cobalt doped TiO_2 nanoparticles which shows a wide absorption range extended into the visible region. Iwasaki [20] synthesized cobalt doped TiO_2 and found that the introduction of Co^{2+} could apparently shift the light absorption edge of anatase TiO_2 to the visible region and enhance photoactivity under both UV and visible light irradiation. Also, researchers found that the addition of co-sorbent carbon materials can enhance the photocatalytic efficiency of TiO_2 [21-23]. As a new member of the carbon family, carbon nanotubes (CNTs) with one-dimensional and hollow structure have received considerable interest since their discovery [24] due to their outstanding structural characters, e.g., mechanical strength [25], excellent thermal conductivity, unique electronic properties [26] and thermal stability [27]. CNTs can be used as a promising material for environmental cleaning. The collection process for MWCNTs with TiO_2 in new composite depends on many causes: one of them suggests that carbon nanotubes behave as a semiconductor supports because of their combination of physiochemical properties which include excellent conductance, high abilities for adsorption [28]. A similar literature has shown that the activities of TiO_2 increased due to abilities of MWCNTs to decrease TiO_2 crystalline grain and particle sizes [29], or increased in the activity of the particles because the direct interaction between MWCNTs and TiO_2 reduces the recombination of electron and hole (h^+/e^-) [30]. Generally, It is believed that the change on activities can be related to the TiO_2–CNTs bonding which can be formed through some physic/chemical interactions such as Vander Walls interaction. The above finding stimulated the advance improvement of an efficient photocatalyst in visible light region. To the best of our knowledge, MWCNTs /Co-TiO_2 nanocomposites prepared using modified sol-gel process have not yet been reported. In the present work, these nanocomposites were prepared using titanium isopropoxide (TIP) as titanium precursor. The performance of the resultant photocatalysts was evaluated by photocatalytic treatment of synthetic wastewater containing 2,4 dichlorophenol (2,4-DCP) under visible light. The textural properties of the resulting photocatalysts were investigated using XRD, FTIR and SEM/EDX. The reaction intermediates were identified by gas chromatography-mass spectrometry (GC–MS) technique.

EXPERIMENTAL
Materials and reagents
Cobalt (II) chloride hexahydrate ($CoCl_2.6H_2O$) was supplied by (Merck, No.102539). Titanium isopropoxide (TIP), (Merck No. 8.21895), ethanol (Merck No. 818760), deionized water and multi-walled carbon nanotubes functionalized by carboxylic groups (MWCNTs) were provided by Neutrino Corporation (Iran), (The average diameter of the MWCNTs was 10-20 nm, and the length was 0.5-2μm), were used for photocatalyst synthesis. High-purity 2,4-DCP, 98%, (Merck No. 803774) was used as a probe molecule for photocatalytic tests.

Preparation of MWCNTs/Co-TiO$_2$ nanocomposite
MWCNTs/Co-TiO_2 nanocomposite was prepared by a modified sol-gel method. An appropriate amount of $CoCl_2.6H_2O$ (Table 1), 10 mL TIP and 30 mL ethanol were stirred for 2 h (Solution A). Then, solution B, 20 mL ethanol, 5 mL deionized water and 2 mL hydrochloric acid and an amount of MWCNTs (Table 1) was

Table 1: Nomenclatures of the prepared samples

Samples	Materials	Nomenclatures
TiO$_2$	TiO$_2$	TiO$_2$
MWCNTs	MWCNTs	MWCNTs
MWCNTs/TiO$_2$	0.55* g MWCNTs + TiO$_2$	MWCNTs/TiO$_2$
Co-TiO$_2$	0.48 g CoCl$_2$.6H$_2$O + TiO$_2$	Co-TiO$_2$
MWCNTs/ Co-TiO$_2$	0.03 g MWCNTs + 0.48 g CoCl$_2$.6H$_2$O + TiO$_2$	MWCNTs/ Co-TiO$_2$ (3.03)**
MWCNTs/ Co-TiO$_2$	0.04 g MWCNTs + 0.48 g CoCl$_2$.6H$_2$O + TiO$_2$	MWCNTs/ Co-TiO$_2$ (3.13)
MWCNTs/ Co-TiO$_2$	0.05 g MWCNTs + 0.48 g CoCl$_2$.6H$_2$O + TiO$_2$	MWCNTs/ Co-TiO$_2$ (3.52)
MWCNTs/ Co-TiO$_2$	0.10 g MWCNTs + 0.48 g CoCl$_2$.6H$_2$O + TiO$_2$	MWCNTs/ Co-TiO$_2$ (4.14)

*The average amount of MWCNTs that used for synthesis of MWCNTs/ Co-TiO$_2$ samples.

**The number in parenthesis is the weight percent of cobalt in the final solid that obtained by EDX analysis.

added into the Solution A and stirred for 12 h at room temperature. The sol was formed after 12 h of stirring followed by aging at room temperature for 24 h and evaporated at 80 °C for 8 h. Finally, the dried powder was calcined at 450 °C under air for 2 h to get an MWCNTs/Co-TiO$_2$ sample. For comparison, four MWCNTs/Co-TiO$_2$ samples with different amounts of MWCNTs, pure TiO$_2$, Co-TiO$_2$ and MWCNTs/TiO$_2$ samples were synthesized by the same route. From now on, the prepared samples will be shown according to Table 1.

Characterization

Fourier transforms infrared (FTIR) analysis was applied to determine the surface, functional groups, using FTIR spectroscopy (FTIR-2000, Bruker), where the spectra were recorded from 4000 to 400 cm^{-1}. The XRD patterns were recorded on a Siemens, D5000 (Germany). X-ray diffractometer using Cu K$_\alpha$ radiation as the X-ray source. The diffractograms were recorded in the 2θ range of 20-80°. The morphology of the prepared samples was characterized using scanning electron microscope (SEM) (Vegall-Tescan Company) equipped with an energy dispersive X-ray (EDX).

Photocatalytic degradation of 2,4-DCP

In a typical run, the suspension containing 10 mg photocatalyst and 100 mL aqueous solution of 2,4-DCP (40 mg/L) was stirred first in the dark for 10 min to establish adsorption/desorption equilibrium. Irradiation experiments were carried out in a self-built reactor. A visible (Halogen, ECO OSRAM, 500W) lamp was used as irradiation source (its emitting wavelength ranges from 350 nm to 800 nm with the predominant peak at 575 nm).At certain intervals, small aliquots (2 mL) were withdrawn and filtered to remove the photocatalyst particles. These aliquots were used for monitoring the degradation progress, with Rayleigh UV-2601 UV/VIS spectrophotometer (λ_{max}= 227nm).

Statistical analysis

All experiments were performed in triplicate and the average values were presented. The data were analyzed by one-way analysis of variance (ANOVA) using SPSS 11.5 for Windows. The data were considered statistically different from control at P < 0.05.

Identification of degradation intermediates

The photocatalytic reaction intermediates were identified by GC–MS in an Agilent 190915-433 instrument equipped with an HP-5MS capillary column (30 m × 0.25 mm). The column temperature was programmed at 50 °C for 2 min, and from 50 to 250 °C at a rate of 10 °C min^{-1}. The sample used for GC–MS analysis was prepared according to the following procedure: The obtained degradation product was acidified to pH 1 and subsequently extracted with dichloromethane. After dichloromethane was evaporated to dryness under vacuum, 10 mL methanol was added to dissolve the residue. Then, 1 mL concentrated sulfuric acid was added and the combined solution was refluxed for about 3 h. The solution was further extracted with dichloromethane followed by concentrating to about 1 mL under reduced pressure. The released chloride ions originating from the degradation of 2,4-DCP were identified and determined by the AgNO$_3$ method.

RESULT AND DISCUSSION

X-ray diffraction analysis

Fig. 1a shows the XRD pattern of the prepared TiO$_2$. The diffractions found at 2θ= 27.4°, 36.1°, and 41.2° belonged to the rutile crystalline phase of TiO$_2$ [31] and the diffractions at 2θ= 25.28°, 37.80°, 48.18°, and 54.09 are the main diffractions for the anatase crystalline phase of TiO$_2$ (JCPDS 21-1272). The XRD pattern revealed the prepared TiO$_2$ containing predominant anatase crystalline phase and a few of rutile crystalline phase. The

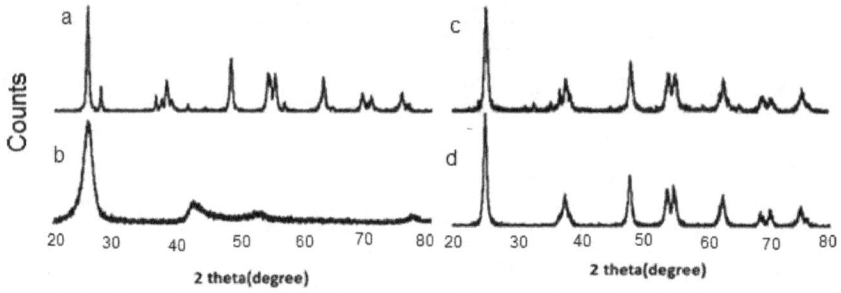

Fig. 1: The XRD patterns of a) TiO$_2$, b) MWCNTs, c) MWCNTs/TiO$_2$ and d) Co-TiO$_2$

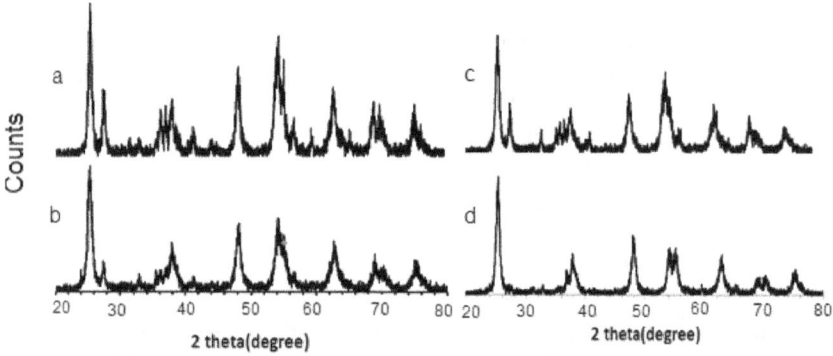

Fig. 2: The XRD patterns of a) MWCNTs/ Co-TiO$_2$ (3.03), b) MWCNTs/ Co-TiO$_2$ (3.13), c) MWCNTs/ Co-TiO$_2$ (3.52) and d) MWCNTs/ Co-TiO$_2$ (4.14)

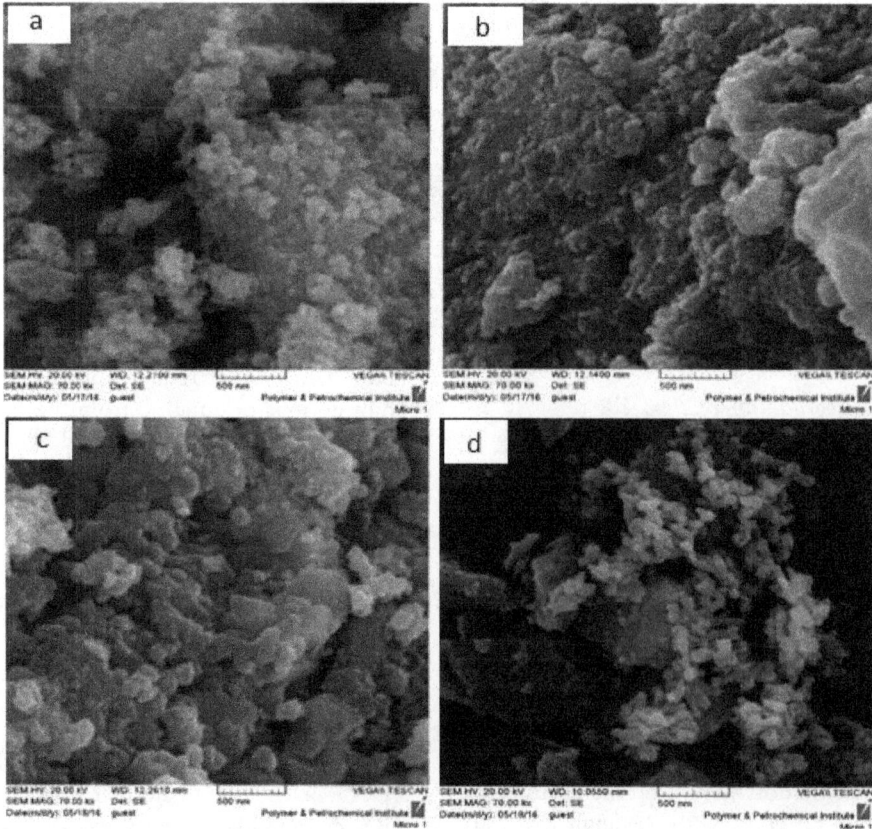

Fig. 3: SEM micrographs of a) MWCNTs/ Co-TiO$_2$ (3.03), b) MWCNTs/ Co-TiO$_2$ (3.13), c) MWCNTs/ Co-TiO$_2$ (3.52) and d) MWCNTs/ Co-TiO$_2$ (4.14)

Table 2: Phase and average crystal size of the prepared samples

Sample	Phase	Average crystal size (nm)
TiO$_2$	Anatase-Rutile	18.32
MWCNTs/TiO$_2$	Anatase	15.00
Co-TiO$_2$	Anatase	15.49
MWCNTs/ Co-TiO$_2$ (3.03)	Anatase-Rutile	12.32
MWCNTs/ Co-TiO$_2$ (3.13)	Anatase-Rutile	10.11
MWCNTs/ Co-TiO$_2$ (3.52)	Anatase-Rutile	11.19
MWCNTs/ Co-TiO$_2$ (4.14)	Anatase	11.52

Table 3: Elemental chemical analysis of the MWCNTs/Co-TiO$_2$ samples

Sample	C wt%	Ti wt%	O wt%	Co wt%
MWCNTs/Co-TiO$_2$(3.03)	9.01	38.27	49.68	3.03
MWCNTs/Co-TiO$_2$(3.13)	7.80	37.87	51.19	3.13
MWCNTs/Co-TiO$_2$(3.52)	7.01	44.13	45.35	3.52
MWCNTs/Co-TiO$_2$(4.14)	8.46	42.92	44.48	4.14

XRD pattern of MWCNTs (Fig. 1b) shows a broad crystalline diffraction around $2\theta=25.5°$, which represents the characteristic diffraction of MWCNTs [32]. In XRD pattern of MWCNTs/ TiO$_2$ (Fig. 1c), we didn't observe the strong and main diffraction of MWCNTs at $2\theta=25.5°$, which was overlapped with the main diffraction of anatase TiO$_2$ at $2\theta=25.3°$ and the relatively large difference between the mass percent of MWCNTs and TiO$_2$ and low crystallinity of MWCNTs could be the reasons for the MWCNTs diffraction not to be detectable [33]. Also, the XRD pattern of Co-TiO$_2$ sample (Fig. 1d) didn't show any cobalt phase indicating that cobalt ions uniformly dispersed among the anatase crystallites and showed only pure anatase phase for TiO$_2$ [34-37]. The XRD patterns of the ternary MWCNTs/Co-TiO$_2$ nanocomposites (Fig. 2) revealed the main diffractions of anatase and rutile crystalline phase for TiO$_2$. Also, the XRD patterns (Fig. 2), showed, increasing of cobalt doping inhibited the phase transformation from anatase to rutile in the ternary MWCNTs/Co-TiO$_2$ nanocomposites [38] and we observed pure anatase crystalline phase for the sample MWCNTs/Co-TiO$_2$ (4.14) (containing the highest cobalt doping in our synthesized samples) (Fig. 2d).

The diffraction patterns of the prepared samples show considerable line width, indicating the samples containing small crystal. The average crystal size of each sample is calculated from the full width at half maximum (FWHM) of the (101) diffraction peak using Scherrer's equation [39].

$$D = \frac{K\lambda}{\beta \cos\theta} \qquad (1)$$

Where D is the average crystal size of the sample, λ the X-ray wavelength (1.54056 Å), β

the full width at half maximum (FWHM) of the diffraction peak (radian), K is a coefficient (0.89) and θ is the diffraction angle at the peak maximum. All the prepared samples are in nano-size range (Table. 2), from 10.11 to 18.32 nm, and all the samples showed smaller crystal size compared to pure TiO$_2$. In the case of ternary MWCNTs/ Co-TiO$_2$ nanocomposites, it can be concluded that the addition of cobalt to titania hinders the growth of TiO$_2$ nanoparticles. This may be due to the formation of Co–O bond on the surface of the doped TiO$_2$, which restricted the crystallite growth of TiO$_2$ [40].

SEM/EDX analysis

SEM images of the ternary MWCNTs/Co-TiO$_2$ nanocomposites are shown in Fig. 3. Some aggregation can be observed that probably happened during the synthesis process. The high viscosity of the sol might be one of the reasons to induce this phenomenon [41]. The EDX patterns of the ternary MWCNTs/Co-TiO$_2$ nanocomposites in Fig. 4 show two peaks around 0.2 and 4.5 keV. The intense peak is assigned to the bulk TiO$_2$ and the less intense one to the surface TiO$_2$. The peaks of cobalt are distinct in Fig. 4 at 0.6, 6.9 and 7.5 keV. The less intense peak is assigned to cobalt in the TiO$_2$ lattice [42, 43]. These results confirmed the existence of cobalt atoms in the ternary MWCNTs/ Co-TiO$_2$ nanocomposites but the XRD patterns do not show any diffractions related to cobalt. Therefore, it may be concluded that cobalt ions are uniformly dispersed among the TiO$_2$ lattice during the synthesis process. EDX results are given in Table 3. Fig. 5 shows elemental mapping images of the ternary MWCNTs/Co-TiO$_2$ nanocomposites. From the elemental mapping mode, highly and

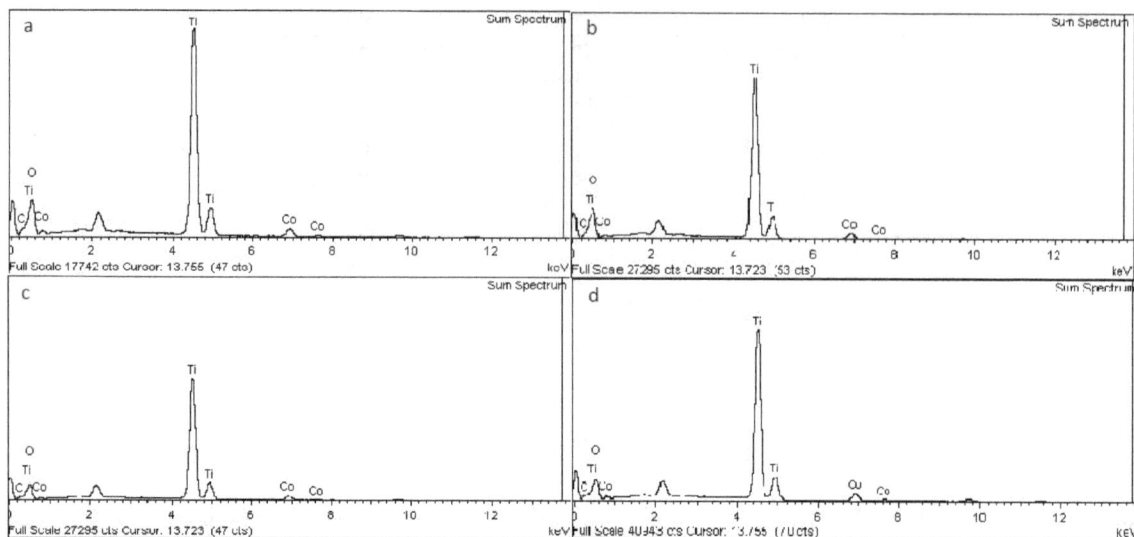

Fig. 4: EDX patterns of a) MWCNTs/ Co-TiO$_2$ (3.03), b) MWCNTs/ Co-TiO$_2$ (3.13), c) MWCNTs/ Co-TiO$_2$ (3.52) and d) MWCNTs/ Co-TiO$_2$ (4.14).

uniformly dispersion of cobalt was observed in the TiO$_2$ lattice especially for the sample MWCNTs/ Co-TiO$_2$ (3.13). This implies a good interaction between cobalt and TiO$_2$ in the preparation process using the sol-gel method.

FTIR analysis

The FTIR spectra of the ternary MWCNTs/ Co-TiO$_2$ nanocomposites are shown in Fig. 6. The vibrations observed at ~3400, 2930 and 2850 cm^{-1} are attributed to the Ti – OH bond [44]. The spectra show the relatively strong band at ~ 1630 cm^{-1} observed for all the samples which are due to the OH bending vibration of chemisorbed and/or physisorbed water molecule on the surface of the catalysts. The strong vibration in the range of 700-500 cm^{-1} is attributed to stretching vibrations of Ti –O–Ti bond [44]. The weak peak at about 514 cm^{-1} assigned to stretching vibrations of Co–O emerged a little [45], the Co–O vibration is not strong because of the broad spectrum of TiO$_2$ and a small amount of Co dopant. FT-IR results reminded the formation of a small part of Co–O bond. It was probably the existence of Co–O bond that hindered the recombination of generated photo holes and photoelectrons [40]. FTIR results showed the ternary MWCNTs/Co-TiO$_2$ nanocomposites contain MWCNTs, cobalt, and TiO$_2$.

Photocatalytic degradation of 2,4-DCP

The photocatalytic degradation efficiency of 2,4-DCP using the prepared samples under visible light showed in Fig.7 and Table 4. Among the ternary nanocomposites, the photocatalytic activity of the MWCNTs/Co-TiO$_2$ (3.13) sample was the highest and 82% degradation of 2,4-DCP obtained after 180min irradiation under visible light. We obtained the degradation percent of 2,4-DCP in the presence of pure TiO$_2$, MWCNTs/TiO$_2$ and Co-TiO$_2$ samples, 44%, 71% and 67%, respectively during 180min under visible light. The degradation percent of 2,4-DCP in the presence of the ternary MWCNTs/Co-TiO$_2$ nanocomposites is higher than pure TiO$_2$, it can be noticed that the introduction of MWCNTs and cobalt obviously caused a synergetic effect on 2,4-DCP degradation and led to a higher photocatalytic activity (Table 4). The higher photocatalytic activity of MWCNTs and Co co-modified TiO$_2$ in the ternary nanocomposites may be explained as firstly, an appropriate amount of the doped Co in TiO$_2$ could effectively capture the photo-induced electrons and holes, which inhibited the combination of photoinduced carriers and improved the photocatalytic activity. Secondly, MWCNTs/Co-TiO$_2$ samples had more surface hydroxyl groups than the pure TiO$_2$ sample which would be beneficial for the adsorption of 2,4-DCP.

Table 4: Degradation percent of 2,4-DCP under visible light after 180 min.

Sample	%Removal
TiO$_2$	44.01
MWCNTs/TiO$_2$	70.81
Co-TiO$_2$	67.00
MWCNTs/ Co-TiO$_2$ (3.03)	65.67
MWCNTs/ Co-TiO$_2$ (3.13)	82.03
MWCNTs/ Co-TiO$_2$ (3.52)	69.80
MWCNTs/ Co-TiO$_2$ (4.14)	60.91

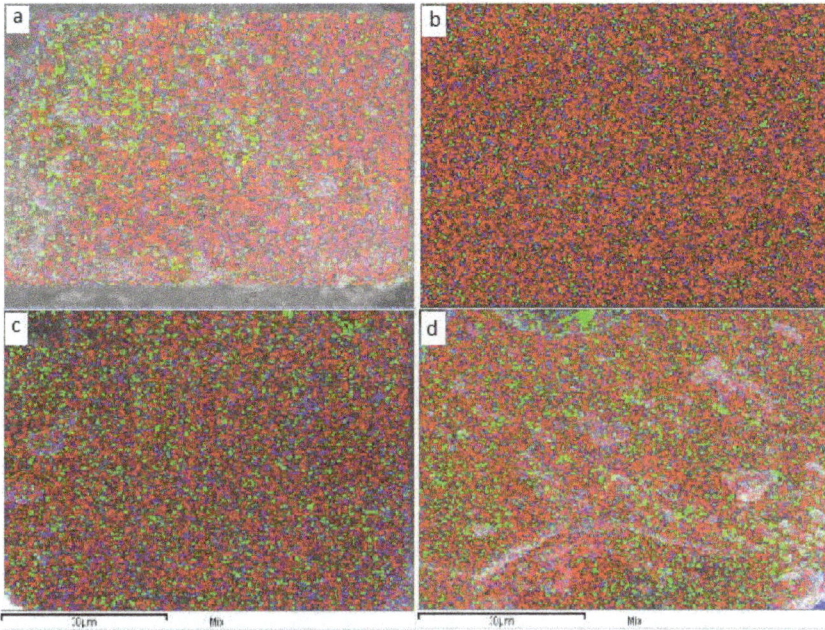

Fig. 5 : Elemental mapping images of a) MWCNTs/ Co-TiO$_2$ (3.03), b) MWCNTs/ Co-TiO$_2$ (3.13), c) MWCNTs/ Co-TiO$_2$ (3.52) and d) MWCNTs/ Co-TiO$_2$ (4.14).

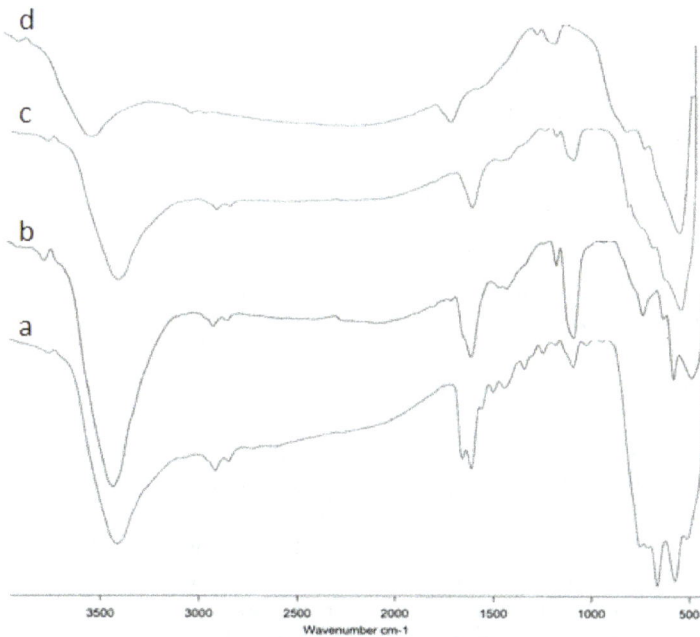

Fig. 6: FTIR spectra of of a) MWCNTs/ Co-TiO$_2$ (3.03), b) MWCNTs/ Co-TiO$_2$ (3.13), c) MWCNTs/ Co-TiO$_2$ (3.52) and d) MWCNTs/ Co-TiO$_2$ (4.14).

The abundant hydroxyl groups adsorbed on the surface of the catalyst could facilitate the formation of hydroxyl radicals which could optimize the degradation process of the adsorbed 2,4-DCP on the surface [46]. Thirdly, because of presence MWCNTs, the surface area of MWCNTs/Co-TiO$_2$ samples were slightly larger than that of pure TiO$_2$ which might favor the adsorption of 2,4-DCP and provide more possibly accessible active sites. Also, there is a synergetic effect between MWCNTs and TiO$_2$ and MWCNTs acting as a photosensitizer. MWCNTs can trap the photo-induced electrons and form superoxide radical ion and/or hydroxyl radical on the surface of TiO$_2$, which are responsible

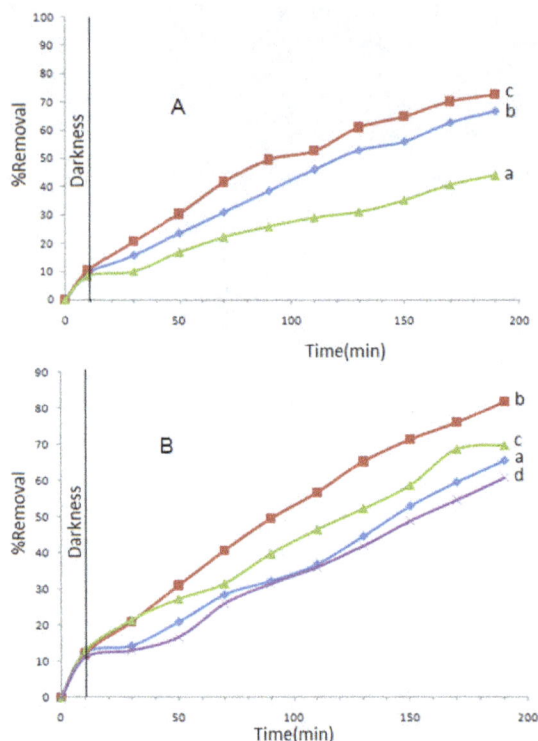

Fig. 7: Photocatalytic degradation of 2,4-DCP over A: (a) TiO_2, b) Co-TiO_2 and c) MWCNTs/TiO_2, B: a) MWCNTs/ Co-TiO_2 (3.03), b) MWCNTs/ Co-TiO_2 (3.13), c) MWCNTs/ Co-TiO_2 (3.52) and d) MWCNTs/Co-TiO_2 (4.14) (2,4-DCP Conc. 40 mg/L, 10 mg catalyst, 100 mL 2,4-DCP, visible light, irradiation time: 180 min).

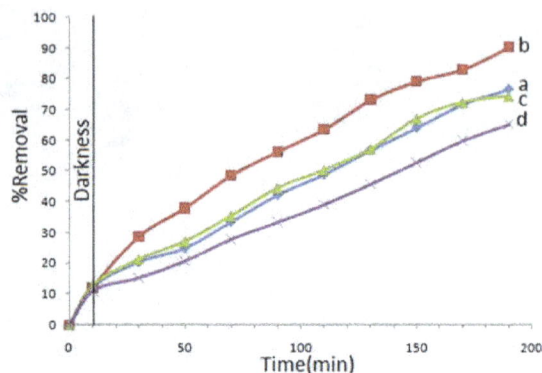

Fig. 8: Photocatalytic removal of 2,4-DCP over a) MWCNTs/ Co-TiO_2 (3.03), b) MWCNTs/ Co-TiO_2 (3.13), c) MWCNTs/ Co-TiO_2 (3.52) and d) MWCNTs/ Co-TiO_2 (4.14) (2,4-DCP Conc. 40 mg/L, 10 mg catalyst, 100 mL 2,4-DCP, visible light, irradiation time: 180 min and H_2O_2 conc. 0.01M).

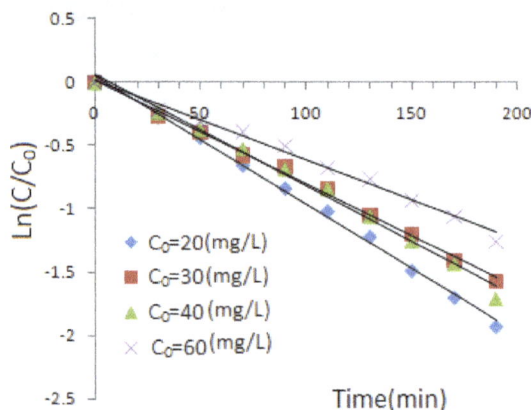

Fig. 9: Effect of initial concentration on the degradation of 2,4-DCP. Irradiation. source: visible lamp, 100mL 2,4-DCP, 10mg catalyst

for the degradation of the organic compound. Due to the introduction of the MWCNTs, an increase of surface charge on TiO_2 in the hybrid catalysts can be suggested. The surface charge may lead to modifications of the fundamental process of electron/hole pair formation while applying visible irradiation [47]. Consequently, it may be the unique interaction between TiO_2 and the MWCNTs that endows the ternary nanocomposite with a higher catalytic activity in the photocatalytic removal of MO compares to pure TiO_2.

Also from Table 4, while Co dopant content increased from 3.03 wt.% to 3.13 wt.%, the dominant Co^{2+} captured electrons which then moved to the absorbed O_2 to reach a higher photocatalytic reaction. As the literature reported [48], with the substitution for Ti^{4+} by Co^{2+} in crystal structure of TiO_2, the catalyst can introduce a new impurity level to the conduction band of TiO_2 and the electrons can be promoted from the valence band to these impurity levels, resulting in a narrowing of the band gap. This fact indicates that there are more photogenerated electrons and holes which can be introduced to participate in the photocatalytic

reactions [49]. However, recombination of photogenerated electrons and holes is one of the most significant factors that deteriorate the photoactivity of the TiO_2 catalyst. Any factor that suppresses the electron-hole recombination will, therefore, enhance the photocatalytic activity[50,51]. In general, if the size of doping metal ion is very similar to Ti^{4+}, it is very likely that metal ion enters into the interstitial site of the TiO_2 crystal. The doping metal ion located mainly on the shallow surface of TiO_2 can induce defects. The defects can become the centers of shallow electrons or holes traps, which would efficiently improve separation of an electron-hole pair. Hence the photocatalyst will have a high photocatalytic activity [52]. The radius of Co^{2+}(0.074 nm) is slightly bigger than Ti^{4+}(0.068 nm) [52]. When calcination was performed at high-temperature

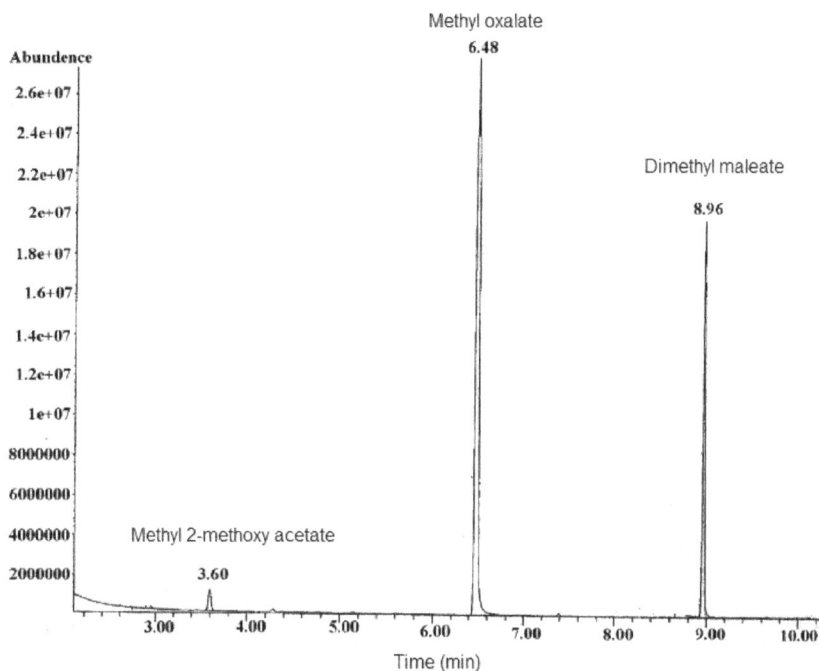

Fig. 10: GC chromatogram of the final products of the photocatalytic degradation of 2,4-DCP by MWCNTs/Co-TiO$_2$ (3.13).

Co^{2+} ions may enter the interstitial site of TiO$_2$ crystal and can create crystal defect. In this present work, FT-IR spectra revealed that there were Co–O bonds in the TiO$_2$/Co nanoparticles. The formation of the Co–O hindered the recombination of photogenerated electrons and holes, so the cobalt doping increased the photocatalytic activity of naked TiO$_2$. According to Fig. 7 and Table 4, 3.13 wt% was the optimal doping content. When the doping ratio is 3.52 or 4.14 wt%, the degradation rate became slower. This is because, at a low doping level of metal, photogenerated holes and electrons are well separated, increasing the efficiency of the photocatalyst. However, at high doping level, there is a considerable chance for multiple trapping which will reduce the efficiency of the catalyst. Moreover, the detrimental activity at high cobalt loading may be attributed to the blockage of many TiO$_2$ active sites due to the large amounts [40].

The activity of the samples for degradation of 2,4-DCP may even be better by addition of hydrogen peroxide to the reactor (Fig. 8). Almost 82% of 2,4-DCP is removed within 180 min in the presence of MWCNTs/Co-TiO$_2$ (3.13), when there is no H$_2$O$_2$ added. In the presence of H$_2$O$_2$, more hydroxyl radicals are produced compared to the conditions without using H$_2$O$_2$ and we obtained 91% removal for 2,4-DCP. Hydroxyl radicals may be generated by direct photolysis of hydrogen

peroxide [53,54], or by reaction of hydrogen peroxide with superoxide radical [55, 56]. Since in our study a source of visible light is employed, it is unlikely that direct photolysis of hydrogen peroxide is significant. Basically, wavelengths shorter than 300 nm provide enough energy for photocleavage of the H$_2$O$_2$ molecules. This can be explained according to the below reaction:

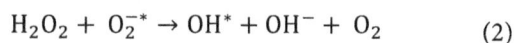

$$H_2O_2 + O_2^{-*} \rightarrow OH^* + OH^- + O_2 \qquad (2)$$

The faster hydroxyl radical formation is associated with the higher degradation rate. This photocatalyst did not exhibit any photoactivity in the dark, neither in the absence nor in the presence of H$_2$O$_2$, suggesting that it is necessary to photoexcite both TiO$_2$ and the cobalt in order to a obtain substantial improvement of the degradation of 2,4-DCP.

The kinetic of 2,4-DCP degradation

The photocatalytic degradation of 2,4-DCP is a first-order reaction and its kinetics may be expressed as $\ln(C/C_0) = -k_{obs}t$ (Fig. 9). In this equation k_{obs} (min^{-1}) is the apparent rate constant, C_0 and C are the initial concentration and concentration at reaction time t of 2,4-DCP, respectively. The k_{obs} are found from the slopes of the straight lines obtained by plotting $\ln(C/C_0)$ versus irradiation time (Fig. 9). The reaction rates, rate constants and half-lifes

Table 5: The rate, rate constant (k_{obs}) and half-time ($t_{1/2}$) for various concentrations of 2,4-DCP.after 180 min.

Concentration (mg/L)	R^2	Initial reaction rate (mg/L.(min^{-1}))	k_{obs}(min^{-1})	$t_{1/2}$ (min)
20	0.9964	0.202	0.0101	68.61
30	0.9958	0.264	0.0088	78.75
40	0.9916	0.328	0.0082	84.51
60	0.9906	0.378	0.0063	110.00

R^2 shows correlation coefficient for fit experimental data (Fig. 9)

($t_{1/2}$) at various initial concentrations of 2,4-DCP are given in Table 5. The results summarized in Table 5 show that the reaction rate of degradation of 2,4-DCP is faster at higher initial concentration. However, the rate constants decrease to some extent when the initial concentration increases.

Identification of intermediate products

The intermediate species formed during photocatalytic degradation of 2,4- DCP, were identified by GC–MS technique. The major reaction intermediates identified in an aliquot withdrawn after 200 min following a degradation condition specified as in Fig. 7. Presence of these intermediates (Fig. 10) supports our proposed mechanism which is based on OH radicals. The hydroxyl radicals attack 2,4-DCP converting it to chlorocatechol and then to chlorobenzoquinone. Subsequently, hydroxyl groups would break the aromatic rings of chlorobenzo-quinone transferring them into simple acids like oxalic acid, acetic acid, etc. as the final products [57,58]. In addition to identifying the organic intermediates, chloride ions were also detected and identified as one of the final products of the photocatalytic removal. The amount of Cl$^-$ in the reaction media at the end of the photocatalytic experiment almost equals the amount of chlorine present in 2,4-DCP indicating essentially complete degradation.

Photocatalyst reuse

The effectiveness of photocatalyst reuse was examined for degradation of 2,4-DCP during a four cycles experiment. Each experiment was carried out under identical conditions. 100mL 2,4-DCP with an initial concentration of 40 mg/L, 10mg MWCNTs/Co-TiO$_2$ (3.13) and 180 min irradiation time under visible light were used. After each experiment, the solution residue from the photocatalytic degradation was filtered, washed and the solid was dried. The dried catalyst samples were used again for the degradation of 2,4-DCP, employing similar experimental conditions. Recycling experiments showed, no obvious decrease of the photocatalytic removal efficiency (from 82% to 78%) was observed

after four cycles and indicating that our MWCNTs/Co-TiO$_2$ (3.13) is renewable for environmental applications. Possibly deactivation of the part of the photocatalyst surface, due to permanent adsorption of intermediate species, might be involved in the reduction of its activity.

CONCLUSIONS

In summary, ternary MWCNTs /Co-TiO$_2$ nanocomposites were successfully prepared by the modified sol-gel process and characterized by a different analysis. It was found that the ternary MWCNTs /Co-TiO$_2$ nanocomposites presented enhanced photocatalytic degradation of 2,4-dichlorophenol (2,4-DCP) and exhibited expansion in spectral response range shifted to the visible region. This was probably due to the affiliation of special properties by MWCNTs and Co dopant. The presence of MWCNTs could create many active sites and increase surface area while Co doping promoted the separation of photogenerated carriers. Furthermore, the major problem of TiO$_2$ photocatalyst was reduced by narrowing their band gap by the Co dopant. Therefore, adding a suitable amount of MWCNTs and Co into the TiO$_2$ lead to a great improvement in the photocatalytic degradation of 2,4-DCP and this new photocatalyst showed remarkable activity in the visible light region. The reactions follow a pseudo-first-order kinetics and the observed rate constant values change with 2,4-DCP concentrations. Oxalic acid and maleic acid were the major intermediate species at the final stage of the degradation process as identified by gas chromatography-mass spectrometry (GC–MS) technique.

ACKNOWLEDGEMENTS

The authors wish to acknowledge the financial support of the University of Tehran for supporting this research.

CONFLICT OF INTEREST

The authors declare that there is no conflict of interests regarding the publication of this manuscript.

REFERENCES

1. Anandan S. Photocatalytic effects of titania supported nanoporous MCM-41 on degradation of methyl orange in the presence of electron acceptors. Dyes and Pigments. 2008;76(2):535-41.

2. Lam S-M, Sin J-C, Mohamed AR. Parameter effect on photocatalytic degradation of phenol using TiO2-P25/ activated carbon (AC). Korean Journal of Chemical Engineering. 2010;27(4):1109-16.

3. Ahlborg UG, Thunberg TM, Spencer HC. Chlorinated Phenols: Occurrence, Toxicity, Metabolism, And Environmental Impact. CRC Critical Reviews in Toxicology. 1980;7(1):1-35.

4. Lee H-C, In J-H, Kim J-H, Hwang K-Y, Lee C-H. Kinetic analysis for decomposition of 2,4-dichlorophenol by supercritical water oxidation. Korean Journal of Chemical Engineering. 2005;22(6):882-8.

5. Dionysiou DD, Khodadoust AP, Kern AM, Suidan MT, Baudin I, Laîné J-M. Continuous-mode photocatalytic degradation of chlorinated phenols and pesticides in water using a bench-scale TiO2 rotating disk reactor. Applied Catalysis B: Environmental. 2000;24(3):139-55.

6. Arnoldsson K, Andersson PL, Haglund P. Formation of Environmentally Relevant Brominated Dioxins from 2,4,6,-Tribromophenol via Bromoperoxidase-Catalyzed Dimerization. Environmental Science & Technology. 2012;46(13):7239-44.

7. Bandara J, Mielczarski JA, Lopez A, Kiwi J. 2. Sensitized degradation of chlorophenols on iron oxides induced by visible light: Comparison with titanium oxide. Applied Catalysis B: Environmental. 2001;34(4):321-33.

8. White JL, Baruch MF, Pander JE, Hu Y, Fortmeyer IC, Park JE, et al. Light-Driven Heterogeneous Reduction of Carbon Dioxide: Photocatalysts and Photoelectrodes. Chemical Reviews. 2015;115(23):12888-935.

9. Liu J, Bai H, Wang Y, Liu Z, Zhang X, Sun DD. Self-Assembling TiO2 Nanorods on Large Graphene Oxide Sheets at a Two-Phase Interface and Their Anti-Recombination in Photocatalytic Applications. Advanced Functional Materials. 2010;20(23):4175-81.

10. Ashford DL, Gish MK, Vannucci AK, Brennaman MK, Templeton JL, Papanikolas JM, et al. Molecular Chromophore–Catalyst Assemblies for Solar Fuel Applications. Chemical Reviews. 2015;115(23):13006-49.

11. Ioannidou E, Ioannidi A, Frontistis Z, Antonopoulou M, Tselios C, Tsikritzis D, et al. Correlating the properties of hydrogenated titania to reaction kinetics and mechanism for the photocatalytic degradation of bisphenol A under solar irradiation. Applied Catalysis B: Environmental. 2016;188(Supplement C):65-76.

12. Wang H, Wang H-L, Jiang W-F, Li Z-Q. Photocatalytic degradation of 2,4-dinitrophenol (DNP) by multi-walled carbon nanotubes (MWCNTs)/TiO2 composite in aqueous solution under solar irradiation. Water Research. 2009;43(1):204-10.

13. Sarteep Z, Ebrahimian Pirbazari A, Aroon MA. Silver Doped TiO2 Nanoparticles: Preparation, Characterization and Efficient Degradation of 2,4-dichlorophenol Under Visible Light. Journal of Water and Environmental Nanotechnology.

2016;1(2):135-44.

14. Devi LG, Kottam N, Murthy BN, Kumar SG. Enhanced photocatalytic activity of transition metal ions Mn2+, Ni2+ and Zn2+ doped polycrystalline titania for the degradation of Aniline Blue under UV/solar light. Journal of Molecular Catalysis A: Chemical. 2010;328(1):44-52.

15. Gurkan YY, Kasapbasi E, Cinar Z. Enhanced solar photocatalytic activity of TiO2 by selenium(IV) ion-doping: Characterization and DFT modeling of the surface. Chemical Engineering Journal. 2013;214(Supplement C):34-44.

16. Zou X-X, Li G-D, Guo M-Y, Li X-H, Liu D-P, Su J, et al. Heterometal Alkoxides as Precursors for the Preparation of Porous Fe– and Mn–TiO2 Photocatalysts with High Efficiencies. Chemistry – A European Journal. 2008;14(35):11123-31.

17. Feng H, Zhang M-H, Yu LE. Hydrothermal synthesis and photocatalytic performance of metal-ions doped TiO2. Applied Catalysis A: General. 2012;413(Supplement C):238-44.

18. Van Nghia N, Ngoc Khoa Truong N, Phi Hung N. Hydrothermal synthesis of Fe-doped TiO 2 nanostructure photocatalyst. Advances in Natural Sciences: Nanoscience and Nanotechnology. 2011;2(3):035014-8.

19. Pirbazari AE, Monazzam P, Kisomi BF. Co/TiO2 Nanoparticles: Preparation, Characterization and Its Application for Photocatalytic Degradation of Methylene Blue. Desalination and Water Treatment. 2017;63:283-92.

20. Iwasaki M, Hara M, Kawada H, Tada H, Ito S. Cobalt Ion-Doped TiO2 Photocatalyst Response to Visible Light. Journal of Colloid and Interface Science. 2000;224(1):202-4.

21. Liu SX, Chen XY, Chen X. A TiO2/AC composite photocatalyst with high activity and easy separation prepared by a hydrothermal method. Journal of Hazardous Materials. 2007;143(1):257-63.

22. Sakthivel S, Kisch H. Daylight Photocatalysis by Carbon-Modified Titanium Dioxide. Angewandte Chemie International Edition. 2003;42(40):4908-11.

23. Zhang L-W, Fu H-B, Zhu Y-F. Efficient TiO2 Photocatalysts from Surface Hybridization of TiO2 Particles with Graphite-like Carbon. Advanced Functional Materials. 2008;18(15):2180-9.

24. Iijima S. Helical microtubules of graphitic carbon. Nature. 1991;354(6348):56-8.

25. Yu M-F, Files BS, Arepalli S, Ruoff RS. Tensile loading of ropes of single wall carbon nanotubes and their mechanical properties. Physical review letters. 2000;84(24):5552-3.

26. Cha SI, Kim KT, Lee KH, Mo CB, Jeong YJ, Hong SH. Mechanical and electrical properties of cross-linked carbon nanotubes. Carbon. 2008;46(3):482-8.

27. Serp P, Corrias M, Kalck P. Carbon nanotubes and nanofibers in catalysis. Applied Catalysis A: General. 2003;253(2):337-58.

28. Wang W, Serp P, Kalck P, Faria JL. Photocatalytic degradation of phenol on MWNT and titania composite catalysts prepared by a modified sol–gel method. Applied Catalysis B: Environmental. 2005;56(4):305-12.

29. Lee TY, Alegaonkar PS, Yoo J-B. Fabrication of dye sensitized solar cell using TiO2 coated carbon nanotubes. Thin Solid

Films. 2007;515(12):5131-5.

30. Yao Y, Li G, Ciston S, Lueptow RM, Gray KA. Photoreactive TiO2/Carbon Nanotube Composites: Synthesis and Reactivity. Environmental Science & Technology. 2008;42(13):4952-7.

31. Baiju KV, Shajesh P, Wunderlich W, Mukundan P, Kumar SR, Warrier KGK. Effect of tantalum addition on anatase phase stability and photoactivity of aqueous sol-gel derived mesoporous titania. Journal of Molecular Catalysis A: Chemical. 2007;276(1):41-6.

32. Ahmmad B, Kusumoto Y, Somekawa S, Ikeda M. Carbon nanotubes synergistically enhance photocatalytic activity of TiO2. Catalysis Communications. 2008;9(6):1410-3.

33. Wang W, Serp P, Kalck P, Faria JL. Visible light photodegradation of phenol on MWNT-TiO2 composite catalysts prepared by a modified sol-gel method. Journal of Molecular Catalysis A: Chemical. 2005;235(1):194-9.

34. Jiang P, Xiang W, Kuang J, Liu W, Cao W. Effect of cobalt doping on the electronic, optical and photocatalytic properties of TiO2. Solid State Sciences. 2015;46(Supplement C):27-32.

35. Chen Q, Ji F, Liu T, Yan P, Guan W, Xu X. Synergistic effect of bifunctional Co-TiO2 catalyst on degradation of Rhodamine B: Fenton-photo hybrid process. Chemical Engineering Journal. 2013;229(Supplement C):57-65.

36. Preethi T, Abarna B, Vidhya KN, Rajarajeswari GR. Sol-gel derived cobalt doped nano-titania photocatalytic system for solar light induced degradation of crystal violet. Ceramics International. 2014;40(8, Part B):13159-67.

37. Kaushik A, Dalela B, Kumar S, Alvi PA, Dalela S. Role of Co doping on structural, optical and magnetic properties of TiO2. Journal of Alloys and Compounds. 2013;552(Supplement C):274-8.

38. Fan C, Xue P, Sun Y. Preparation of Nano-TiO2 Doped with Cerium and Its Photocatalytic Activity. Journal of Rare Earths. 2006;24(3):309-13.

39. Khan M, Cao W. Cationic (V, Y)-codoped TiO2 with enhanced visible light induced photocatalytic activity: A combined experimental and theoretical study. Journal of Applied Physics. 2013;114(18):183514.

40. Vijayan P, Mahendiran C, Suresh C, Shanthi K. Photocatalytic activity of iron doped nanocrystalline titania for the oxidative degradation of 2,4,6-trichlorophenol. Catalysis Today. 2009;141(1):220-4.

41. Silva CG, Faria JL. Photocatalytic oxidation of benzene derivatives in aqueous suspensions: Synergic effect induced by the introduction of carbon nanotubes in a TiO2 matrix. Applied Catalysis B: Environmental. 2010;101(1):81-9.

42. Zhang K, Zhang FJ, Chen ML, Oh WC. Comparison of catalytic activities for photocatalytic and sonocatalytic degradation of methylene blue in present of anatase TiO2–CNT catalysts. Ultrasonics Sonochemistry. 2011;18(3):765-72.

43. Wang Y, Huang Y, Ho W, Zhang L, Zou Z, Lee S. Biomolecule-controlled hydrothermal synthesis of C–N–S-tridoped TiO2 nanocrystalline photocatalysts for NO removal under simulated solar light irradiation. Journal of Hazardous Materials. 2009;169(1):77-87.

44. Bae E, Choi W. Highly Enhanced Photoreductive Degradation of Perchlorinated Compounds on Dye-Sensitized Metal/TiO2 under Visible Light. Environmental Science & Technology. 2003;37(1):147-52.

45. Guo Q, Guo X, Tian Q. Optionally ultra-fast synthesis of CoO/Co3O4 particles using CoCl2 solution via a versatile spray roasting method. Advanced Powder Technology. 2010;21(5):529-33.

46. Yang Q, Deng Y, Hu W. Preparation of alumina/carbon nanotubes composites by chemical precipitation. Ceramics International. 2009;35(3):1305-10.

47. Wang W, Serp P, Kalck P, Faria JL. Visible light photodegradation of phenol on MWNT-TiO2 composite catalysts prepared by a modified sol–gel method. Journal of Molecular Catalysis A: Chemical. 2005;235(1):194-9.

48. Yang J, Cui S, Qiao J-q, Lian H-z. The photocatalytic dehalogenation of chlorophenols and bromophenols by cobalt doped nano TiO2. Journal of Molecular Catalysis A: Chemical. 2014;395(Supplement C):42-51.

49. Li C-J, Wang J-N, Wang B, Gong JR, Lin Z. A novel magnetically separable TiO2/CoFe2O4 nanofiber with high photocatalytic activity under UV–vis light. Materials Research Bulletin. 2012;47(2):333-7.

50. Sakthivel S, Shankar MV, Palanichamy M, Arabindoo B, Bahnemann DW, Murugesan V. Enhancement of photocatalytic activity by metal deposition: characterisation and photonic efficiency of Pt, Au and Pd deposited on TiO2 catalyst. Water Research. 2004;38(13):3001-8.

51. Nagaveni K, Hegde MS, Madras G. Structure and Photocatalytic Activity of Ti1-xMxO2±δ (M = W, V, Ce, Zr, Fe, and Cu) Synthesized by Solution Combustion Method. The Journal of Physical Chemistry B. 2004;108(52):20204-12.

52. Chand R, Obuchi E, Katoh K, Luitel HN, Nakano K. Effect of transition metal doping under reducing calcination atmosphere on photocatalytic property of TiO2 immobilized on SiO2 beads. Journal of Environmental Sciences. 2013;25(7):1419-23.

53. Peyton GR, Glaze WH. Destruction of pollutants in water with ozone in combination with ultraviolet radiation. 3. Photolysis of aqueous ozone. Environmental Science & Technology. 1988;22(7):761-7.

54. Ollis DF, Pelizzetti E, Serpone N. Photocatalyzed destruction of water contaminants. Environmental Science & Technology. 1991;25(9):1522-9.

55. Poulios I, Micropoulou E, Panou R, Kostopoulou E. Photooxidation of eosin Y in the presence of semiconducting oxides. Applied Catalysis B: Environmental. 2003;41(4):345-55.

56. Cornish BJPA, Lawton LA, Robertson PKJ. Hydrogen peroxide enhanced photocatalytic oxidation of microcystin-LR using titanium dioxide. Applied Catalysis B: Environmental. 2000;25(1):59-67.

57. Shibata T, Irie H, Tryk DA, Hashimoto K. Effect of Residual Stress on the Photochemical Properties of TiO2 Thin Films. The Journal of Physical Chemistry C. 2009;113(29):12811-7.

58. Chaliha S, Bhattacharyya KG. Fe(III)-, Co(II)- and Ni(II)-impregnated MCM41 for wet oxidative destruction of 2,4-dichlorophenol in water. Catalysis Today. 2009;141(1):225-33.

20

Removal of Cobalt Ions from Contaminated Water Using Magnetite Based Nanocomposites: Effects of Various Parameters on the Removal Efficiency

Saeed Tizro, Hadi Baseri*

School of Chemistry, Damghan University, Damghan, Iran.

ABSTRACT

Cobalt is one of the most hazardous heavy metals present in the environment. Magnetic based nanoadsorbents were used for removal of Co(II) ions in this work. The characteristics results of FT-IR, XRD, TGA, and FE-SEM show that applied coatings were modified magnetite nanoparticles efficiently. The results of TEM indicate that magnetic nanoadsorbents were produced on the nanoscale with average particle sizes of 60±10 nm. Batch experiments were carried out to determine the removal efficiency of the nanoadsorbents. pH, temperature, contact time, adsorbent dose, shaking rate and the initial concentration of analyte were the studied parameters. At optimized conditions of operation parameters, the maximum removal percentage of 92% was obtained by using magnetite-citric acid as an adsorbent. Equilibrium data for Co(II) ions adsorption onto magnetite-citric acid were fitted well by Langmuir isotherm model and the maximum adsorption capacity for Co(II)ions was obtained 43.292 mg/g at 313 K. Also, thermodynamic parameters reveal the spontaneity, feasibility and endothermic nature of the Co(II) ions adsorption process. In addition, the cobalt ions can be desorbed from magnetite-citric acid nanoadsorbent by using nitric acid solution with 95% desorption efficiency and the magnetite-citric acid nanoadsorbent exhibits good recyclability.

Keywords: *Adsorption, Cobalt ions, Contaminated Water, Magnetite*

INTRODUCTION

Cobalt is a very toxic heavy metal ion and it´s containing compounds are widely used in many industrial applications such as mining, electroplating, metallurgical, paints, pigments and electronic [1]. The permissible limits of cobalt in the livestock wastewater and irrigation water are 1.0 and 0.05 mg/L, respectively [2]. The presence of cobalt in the environment leads to several health troubles such as low blood pressure, vomiting, nausea, heart diseases, vision problems, sterility, thyroid damage, hair loss, bleeding, diarrhea, bone defects and may also cause mutations (genetic changes) in

*Corresponding Author Email: baseri@du.ac.ir

living cells [2,3]. So the fast and effective removal of Co(II) from the aqueous solution is necessary for the environmental protection. There are several different techniques for the removal of Co(II) ions, such as chemical precipitation, oxidation, ion-exchange, reverse osmosis, membrane electrolysis, coagulation and adsorption [5-9]. Among these available techniques, adsorption method has been used widely, because it is simple, adaptable, economical, and cost-effective. So many materials such as clay minerals, oxides, activated carbon, zeolites, ion exchange resin and cellulose have been used as adsorbents [10-14]. Low adsorption capacity and difficult separation are the problems that limit the application of these materials.

Recently, magnetite-based nanocomposite adsorbents have been fabricated and have shown good properties for drug delivery [15] energy storage [16,17] and water treatment[18]. Also, the use of magnetite-based nanoadsorbents in water treatment provides a suitable and comfortable approach for separating and removing the pollutants by applying an external magnetic field.

This present study pursues: (1) synthesis of different magnetite-based nanoadsorbents by using co-precipitation method; (2) characterization of these adsorbents by using Fourier transform infrared spectroscopy (FT-IR), X-ray diffractometer (XRD), Field emission scanning electron microscopy (FE-SEM),thermogravimetric (TGA) analysis and Transmission electron microscopy (TEM); (3) study the effects of main parameters that affect adsorption of Co(II) ions; (4) study isotherm, thermodynamic and kinetic parameters of adsorption process and (5) study reusability and stability of nanoadsorbents. The main object of this work is to develop cost efficient, biocompatible and easily available adsorbent for the environmental applications.

EXPERIMENTAL
Materials
Ferric chloride (FeCl$_3$·6H$_2$O), ferrous sulfate (FeSO$_4$·7H$_2$O), ammonia solution (with analytical grade, Merck Chemical Company) and double-distilled water were used for the preparation of magnetite nanoparticles. Natural zeolite clinoptilolite was prepared from the West Semnan, Iran and all of the coatings (citric acid (C$_6$H$_8$O$_7$), ascorbic acid (C$_6$H$_8$O$_6$), salicylic acid (C$_7$H$_6$O$_3$), starch (C$_6$H$_{10}$O$_5$)$_n$ and saccharose (C$_{12}$H$_{22}$O$_{11}$)) were purchased from the Sigma–Aldrich company.

Synthesis of magnetite-based nanoadsorbents
We used previously reported co-precipitation method [19] for the synthesis of bare magnetite nanoparticles. In order to synthesize Fe$_3$O$_4$ based nanoadsorbents (except magnetite-zeolite), 4.2 g of FeSO$_4$·7H$_2$O and 6.1 g of FeCl$_3$·6H$_2$O were dissolved in 100 mL of double-distilled water and heated at 100 °C for 1 h. After that 10 mL of ammonium hydroxide solution (25%) was added to the mixture. Then the temperature of the system decreased to the 80 °C and 0.5 g of the coating was dissolved in 50 mL of distilled water and was

added to the mixture. The mixture was stirred at this temperature for about 1 hour. The dark brown precipitate was collected by centrifugation and washed with deionized water and acetone for 3 times and dried at 70°C for 5 h.

Magnetite-clinoptilolite nanoadsorbent was prepared by the co-precipitation method as follows: First of all, for activation of natural zeolite, 10 mL of H$_2$SO$_4$-HCl (%10) solution was added to the zeolite and the mixture was relaxed for 3 h. Then acid was evacuated and the sample was washed with deionized water 2 times and dried at 200°C for about 3 h. We added the activation zeolite to the Fe$_3$O$_4$ sample. Centrifugation, washing and drying processes of the obtained sample were done just like the last samples and the final product's color was dark brown.

Characterization
The surface functional groups of samples were determined by Fourier transform infrared spectrum (Spectrum RXI).The crystal structures of the synthesize nanoadsorbents were determined by XRD analysis (D8-Advance, Bruker AXS, Cu K α1, λ=1.54 °A). The thermogravimetric analyses of the samples were performed with a thermal analysis system (STA 503). For these measurements, the weight losses of dried samples were monitored under N$_2$ from room temperature to 600 °C at a rate of 10°C/min. Surface study and size measurement of the nanostructures were done by Field emission scanning electron microscopy (Hitachi S4160) and transmission electron microscopy (Philips CM 30), respectively. Atomic adsorption spectroscopy (Chemtech analytical CTA-2000) was used for determination of Co(II) ions concentration in supernatant.

Batch adsorption experiments
In order to evaluate the adsorption ability of synthesized nanoadsorbents, batch experiments were carried out at room temperature. Hydrated Co(NO$_3$)$_2$ was used as the source of Co(II) ions and pH values were adjusted by using 0.1 M NaOH or 0.1 M HCl. In a typical removal experiment, 0.2 g of magnetite-citric acid nanoadsorbent was added into 100 mL of Co(II) ions solution (25 mg/L), sealed and shaken for 1 h(rpm was 500). Then as can be seen from Fig. 1, magnetite was easily separated by using an external magnetic field. Co(II) ions concentration in the supernatant solution was measured by using atomic absorption

spectroscopy. All of the experiments were carried out in triplicates and average values were used in the graphs. The adsorption capacity and removal percentage of Co(II) ions were calculated by using Eqs. (1) and (2), respectively [20,21].

$$q_e = \frac{(C_0 - C_e)}{W} \times V \tag{1}$$

$$\text{Removal efficiency (\%)} = \frac{(C_0 - C_e)}{C_0} \times 100 \tag{2}$$

Where q_e (mg/g) is the adsorption capacity at equilibrium, C_0 and C_e are the initial and equilibrium concentration of Co(II) ions (mg/L); V is the total volume of solution in (L) and W is the adsorbent mass(g).

RESULT AND DISCUSSION
Characterization of nanoadsorbents
The crystalline structures of differently synthesized nanoadsorbents were identified with XRD pattern (Fig. 2). Diffraction peaks with 2θ at 30.2°, 35.5°, 43.3°, 57.2°, and 62.8° that were observed for bare magnetite spectra, indicate a cubic spinel structure of this nanoparticle [22]. The same set of characteristic peaks were also observed for magnetite-citric acid, magnetite-salicylic acid and magnetite-starch samples that indicate the stability of the crystalline phase of magnetite nanoparticles during functionalization of the surface [23]. The magnetite-clinoptilolite pattern shows three crystalline peaks with 2θ at 35°, 58° and 62° related to the 311, 511 and 440 crystallographic planes of the face-centered cubic (FCC) iron oxide nanocrystals [24]. The average crystallite size of different nanoadsorbents was calculated by using the Scherrer's equation [25]:

$$D_c = \frac{0.9\,\lambda}{L.\,Cos\Theta_B} \tag{3}$$

Where D_c is the crystallite size (A°), λ is 1.54056

Fig. 1. Magnetic nanoadsorbent separation in Co(II) ions media (a) before using magnet, (b) after using magnet

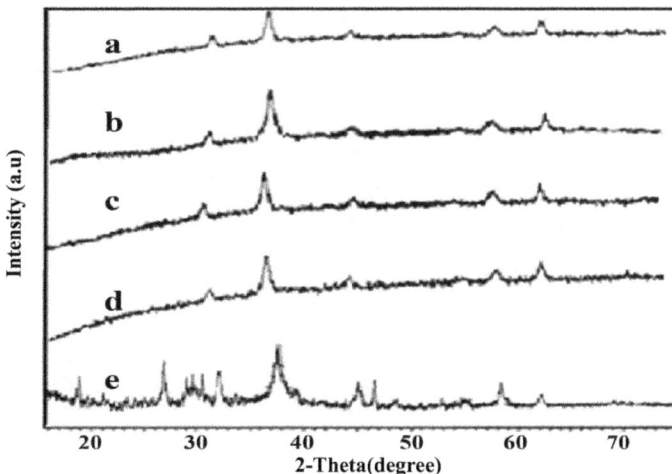

Fig. 2. XRD pattern of (a) bare magnetite, (b) magnetite-citric acid, (c) magnetite-salicylic acid, (d) magnetite-starch and (e) magnetite-clinoptilolite

A°, L is the full width at half-maximum (FWHM) and Θ_B is Bragg angle. Obtained results are summarized in Table 1.

The FT-IR spectra of bare magnetite nanoparticles, magnetite-citric acid, magnetite-salicylic acid, magnetite-starch, and magnetite-clinoptilolite are illustrated in Fig. 3. The low-intensity adsorption band at 570 cm^{-1} attributed to Fe-O stretching vibration of the magnetite nanoparticles was observed in all five samples [26,27]. The observed band at nearly 3500 cm^{-1} in bare magnetite spectra is due to –OH stretching vibrations. There is a large and intense band at

nearly 3450 cm^{-1} in magnetite-citric acid sample spectra that could be devoted to the structural OH groups of molecular water or citric acid. The intense band at nearly 1600 cm^{-1} for the magnetite-citric acid, revealed the binding of a citric acid radical to the magnetite surface (Fig. 3b). In the spectrum of magnetite-salicylic acid (Fig. 3c), the two peaks that were observed at 2850 and 2920 cm^{-1} were corresponded to the asymmetric and symmetric CH$_2$ stretching vibrations, respectively [28]. In the magnetite-starch spectra (Fig. 3d), the peak that was observed at 1090 concerned with the C-O stretching vibration in the C-O-H group, while the

Table 1. Average crystallite size of different nanoadsorbents.

Nanoadsorbents	Average crystallite size (nm)
Bare magnetite	16
Magnetite-citric acid	19
Magnetite-salicylic acid	18
Magnetite-starch	19
Magnetite-clinoptilolite	20

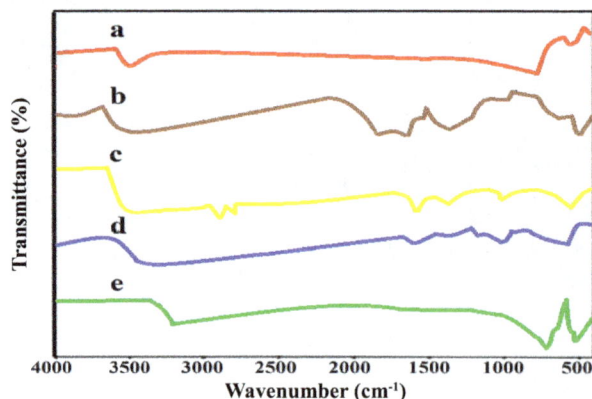

Fig. 3. FT-IR spectra of (a) bare magnetite, (b) magnetite-citric acid, (c) magnetite-salicylic acid, (d) magnetite-starch and (e) magnetite-clinoptilolite

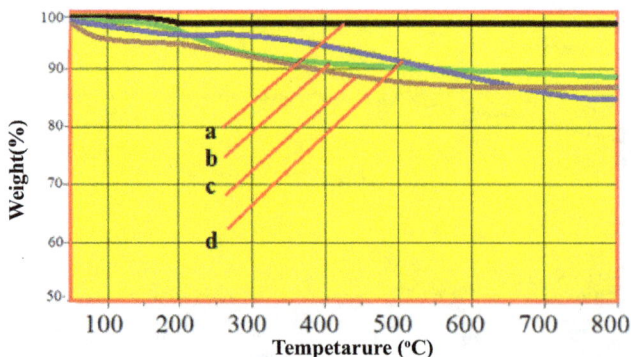

Fig. 4. TGA curve for(a) bare magnetite, (b) magnetite-citric acid, (c) magnetite-salicylic acid and (d) magnetite-starch

$1160~cm^{-1}$ peak corresponded to the C-O stretching vibration in the C-O-C group. Unlike the bare magnetite sample, the greatly intensified peaks were observed for the magnetite-starch sample due to the H-bonded OH groups of amylose and amylopectin of starch [29]. In comparison with magnetite, FTIR spectrum of magnetite-clinoptilolite in the range of 440-$580~cm^{-1}$ showed vibration bands that related to Fe-O functional groups (Fig. 3e).

TGA curves of bare magnetite, magnetite-citric acid, magnetite-salicylic acid and magnetite-starch were shown in Fig. 4. TGA curve of the bare magnetite nanoparticles shows a weight loss of about 2% below 200°Cwhich can be attributed to the loss of adsorbed water in the sample and no other weight loss was observed until 800 °C. TGA curve for Fe_3O_4-citric acid exhibits two steps of weight loss. The first weight loss can be attributed to the loss of residual water in the sample and the second is due to the loss of citric acid in the range of 200 to 600 °C. The total weight loss for this sample is about 12%. For the magnetite-salicylic acid sample, a weight loss of about 4% is observed between room temperature and 200 °C, which is attributed to the evaporation of adsorbed water. A secondary weight loss of about 10% in the temperature range of 200 to 600 °C is attributed to the decomposition of salicylic acid. For Fe_3O_4-starch sample, weight loss at temperatures below 250°C can be attributed to water desorption and a drastic weight loss of about 13% from 250 to 600 °C is due to the loss or decomposition of starch. The obtained TGA curves of these samples confirm the successful modifications of magnetite nanoparticles.

SEM images of the magnetite-ascorbic acid sample are shown in Figs. 5(a) and 5(b). MNPs are roughly spherical and tend to aggregate together because of their high surface energy and adhesion.

Fig. 5. SEM images of magnetite-ascorbic acid

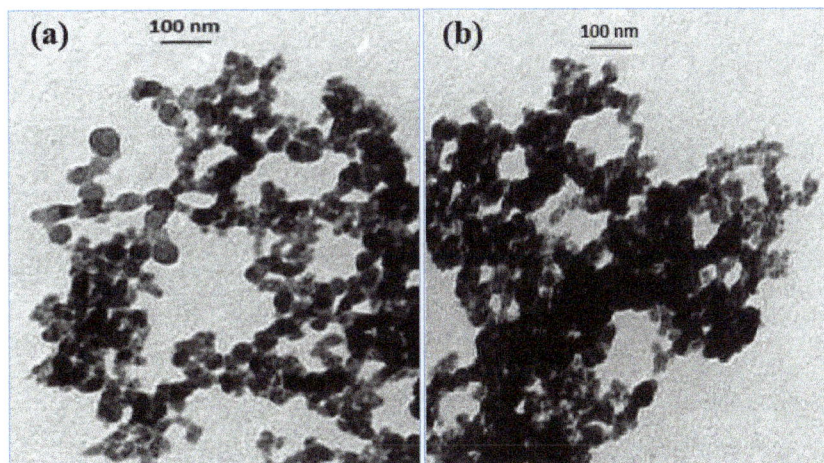

Fig. 6. TEM image of (a) bare magnetite nanoparticles, (b) ascorbic acid coated magnetic nanoparticles

TEM images of magnetite and magnetite-ascorbic acid are shown in Fig. 6a and 6b, respectively. Well-shaped spherical or ellipsoidal magnetic nanoparticles are observed in TEM images. Particle size distribution histogram is presented in Fig. 7 which is obtained by measuring 100 nanoparticles. An average size of 60 ± 10 nm was obtained for magnetite-ascorbic acid nanoparticles. Crystallite size obtained from XRD pattern is smaller than the particle size obtained from TEM, revealing the polycrystalline structure of most observed nanoparticles. Also, the similar size of magnetite with magnetite-ascorbic acid indicates that the binding process did not result in agglomeration and the change in the size of the nanoparticles.

Adsorption study using nanoadsorbents

Some of the main parameters affecting the Co(II) ions adsorption by using synthesized nanoadsorbents were investigated and optimized here. These parameters are pH, temperature, contact time, adsorbent dose, shaking rate and initial concentration of the analyte.

Effect of pH

pH is a key factor affecting the adsorption of heavy metal ions. So, the dependence of pH on the removal of Co(II) ions was studied at a constant Co(II) ions concentration (25 mg./L) using 0.2g of nanoadsorbent. As shown in Fig. 8, cobalt adsorption increased with increase in pH range from 2 to 8 and slowly decreased or fixed with increasing pH in the range of 8 to 12. Therefore, pH 8 was chosen for further experiments. The removal decreasing trend at pH above 8 can correspond to nanoadsorbents probable destroying in an alkaline environment and certain precipitation of $Co(OH)_2$ from the solution. At pH higher than 8, the dominance of OH^- ions in the solution creates a competition between negatively charged nanoadsorbent surface and OH^-

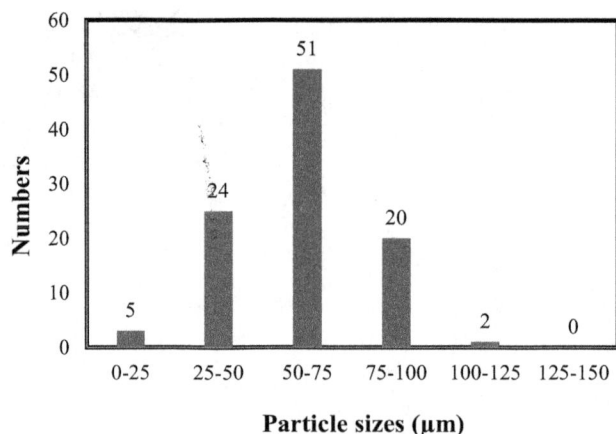

Fig. 7. Obtained histogram from SEM images of magnetite-ascorbic acid

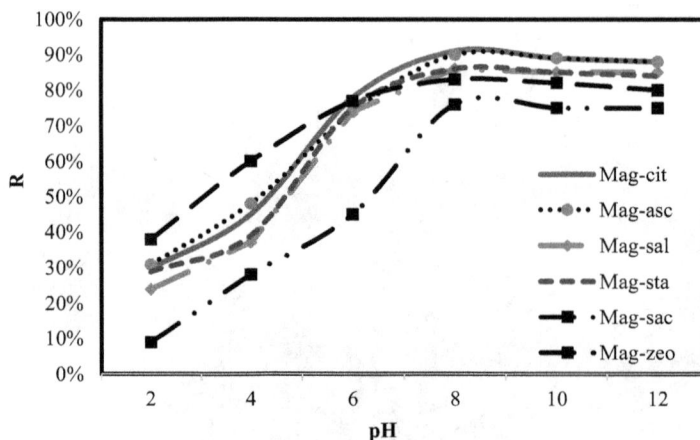

Fig. 8. Effect of pH on the removal of Co(II) (adsorbents dose: 0.2 g, Co(II) concentration: 25 mg/L)

ions which resulted in a decrease in the adsorption of Co(II) metal ions[31]. The suppressed adsorption of metal ions at lower pH implies that acid treatment is a possible and useful method to regenerate these nanoadsorbents [32,33].

Effect of temperature

The effects of temperature on the adsorption of Co(II) ions were explained in Fig. 9. It was observed that removal increased with increase in the temperature range from 293 to 313 K and further fixed or even slowly decreased at the higher temperatures. Probably the connivance decrease in the removal of Co(II) ions that observed at higher temperatures was due to the weakening of the

adsorptive forces between the active sites on the nanoadsorbents surfaces and Co(II) ions[24].

Effect of contact time

Fig. 10 shows the time-dependent behavior of Co(II) ions removal from aqueous solution by using different synthesized adsorbents. As shown, the removal amount of Co(II) ions increased with increase in contact time and the equilibrium time was reached within 50 min. The initial adsorption rate was very fast which was due to the existence of a greater number of available sites on nanoadsorbents surfaces for Co(II) ions adsorption. According to the obtained results contact time of 50 min selected for further experiments.

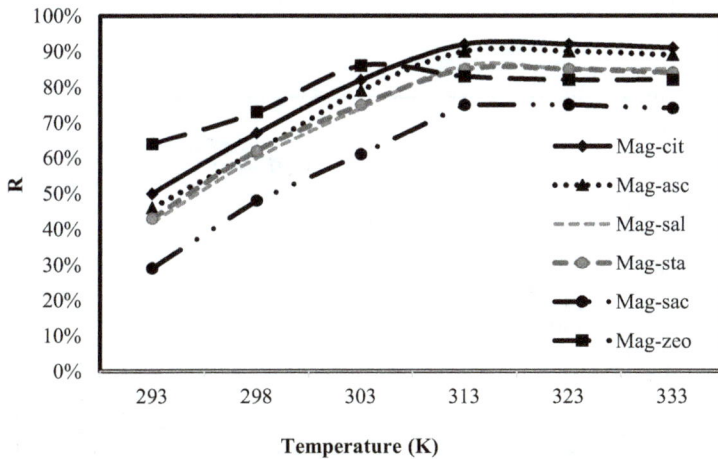

Fig. 9. Effect of temperature on the removal of Co(II) (adsorbents dose: 0.2 g, Co(II) concentration: 25 mg/L).

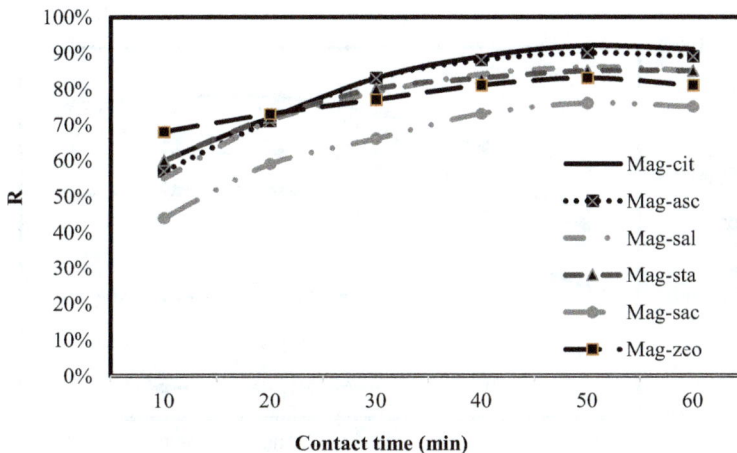

Fig. 10. Effect of contact time on the removal of Co(II) (adsorbents dose: 0.2 g, Co(II) concentration: 25 mg/L).

Effect of adsorbent dose

In order to study the adsorbent dose effect on the removal of Co(II) ions, the adsorption experiments were carried out with different concentrations of nanoadsorbents. 40, 80, 120, 160, 200 and 240 mg of each magnetite based nanoadsorbents were applied. As can be seen from Fig. 11, by increasing the dose of the nanoadsorbents, the number of adsorption sites available for adsorbent interaction is increased, thereby resulting in the increased percentage of Co(II) ions removal from the solution. In all cases, the optimal amount of adsorbent dose was 0.2 g and no significant increase was observed with further increases in adsorbent dose.

Effect of shaking rate

Since an optimum shaking rate is essentially needed to maximize the interactions between metal ions and adsorption sites of nanoadsorbents, the removal of Co(II) ions under various shaking rates was investigated and is shown in Fig. 12. It is found that Co(II) ions removal increased with increase in shaking rates from 100 to 500 rpm and with further increases in shaking rate from 500 to 750 rpm, the effect of shaking rate on the cobalt adsorption became comparatively negligible. This can be explained by the fact that, for a relatively lower shaking rate, the system is incompletely mixed; hence the poor dispersion of nanoparticles in solution resulted in only a small portion of surface area of adsorbent being exposed and reacted with the Co(II) ions. With further increasing the shaking rate from 500 to 750 rpm, the effect of shaking rate on the Co(II) adsorption became comparatively insignificant; since the system was well mixed under a comparatively higher shaking

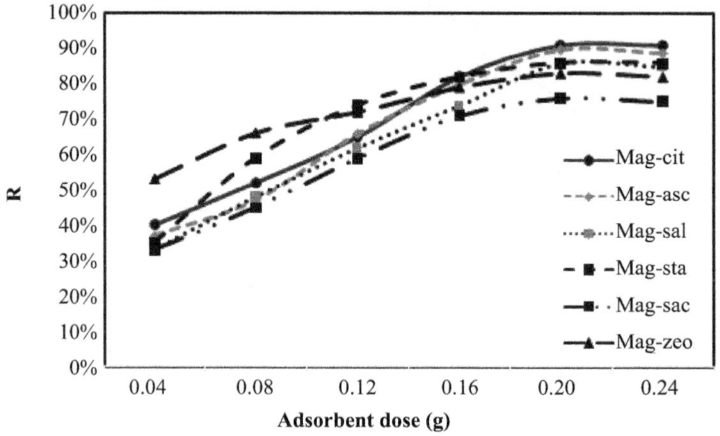

Fig. 11. Effect of adsorbent dosage on the removal of Co(II)

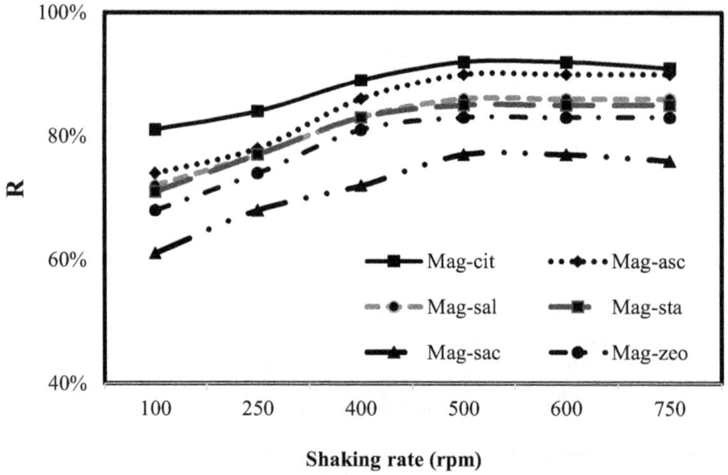

Fig. 12. Effect of shaking rate on the removal of Co(II) (pH: 8, Co(II) concentration: 25 mg/L).

rate, say 500 rpm[34]. As a consequence, a shaking rate of 500 rpm was selected as the optimum value and was used for further experiments.

Effect of initial concentration

The initial concentration of the analyte is also an important parameter for removal of heavy metals. As shown in Fig. 13, by increasing the initial concentration of Co(II) ions, removal amounts were decreased. Regarding maximum removal of Co(II) ions at 25 mg/L concentration of the analyte and a further significant decrease in removal amounts, 25 ppm was selected for further studies.

After obtaining optimum values of different parameters, we did the Co(II) ions adsorption by using bare magnetite nanoparticles. Removal efficiency was obtained 69% in this situation. From obtained results, it is completely clear that magnetite-citric acid is the most efficient adsorbent among the tested materials for the removal of Co(II) ions. So we did further studies just by using this nanoadsorbent.

Adsorption isotherms

Cobalt adsorption is significantly influenced by the initial concentration of Co(II) ions in aqueous solution. In this study, the initial concentration of Co(II) ions varied from 10 to 100 mg/L while the adsorbent dosage is 0.2 g/L, pH is 8 and contact time is 50 min. The experimental data were fitted to Langmuir [35] and Freundlich [36] isotherm model.

The Langmuir isotherm model was used to describe the chemisorptions and monolayer coverage of adsorbates onto nanoadsorbents and

it is linear form can be expressed by the following equation:

$$\frac{C_e}{q_e} = \frac{1}{K_L q_m} + \frac{1}{q_m} C_e \qquad (4)$$

Where q_e is the amount of Co(II) adsorbed at equilibrium in mg/g, C_e is the solute equilibrium concentration in mg/L, q_{max} and K_L is Langmuir constants indicating the saturated capacity of adsorbents and energy term, respectively.

The Freundlich isotherm model which indicates the exponential distribution of active sites and their energies and surface heterogeneity of the adsorbents are described by the following equation:

$$q_e = K_f C_n^{1/n} \qquad (5)$$

Where K_f and $1/n$ are the Freundlich constants related to the adsorption capacity and the adsorption intensity, respectively.

Table 2 shows that the adsorption behavior of Co(II) ions onto the magnetite-citric acid nanoadsorbent is better described by Langmuir isotherm model because this model yields higher correlation coefficients. This indicates that monolayer coverage of magnetite-citric acid adsorbent is the main sorption mechanism.

The maximum adsorption capacity determined by the Langmuir model was 43.292 mg/g which was compared with other reports in Table 3.

Thermodynamic parameters

The effect of temperature on the adsorption isotherm was investigated under isothermal conditions in the temperature range of 303to 323

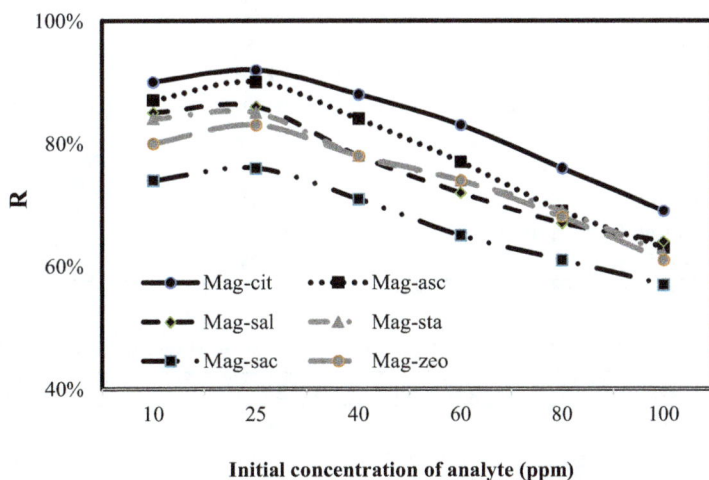

Initial concentration of analyte (ppm)
Fig. 13. Initial concentration effect on removal of Co(II).

K and optimal pH, contact time and adsorbent dose. Thermodynamic parameters such as change in free energy () (KJ/mol), enthalpy () (KJ/mol), and entropy () (J/mol.K) are calculated using the following thermodynamic functions [44,45]:

$$\Delta G^0 = -RT \ln K_d \tag{6}$$

$$\ln K_d = -\frac{\Delta H^0}{RT} + \frac{\Delta S^0}{R} \tag{7}$$

$$K_d = \frac{q_e}{C_e} \tag{8}$$

$$\Delta G^0 = \Delta H^0 - T\Delta S^0 \tag{9}$$

where R is the universal gas constant (8.314 J/mol.K), T is the absolute temperature in Kelvin, K_d is distribution coefficient, (mg/g) is the equilibrium concentration of Co(II) ions adsorbed onto magnetite-citric acid and (mg/L) is the remained concentration of Co(II) ions in the aqueous solution. The values of the different thermodynamic parameters are summarized in Table 4. value was obtained 47.684 kJ/mol and it means that adsorption process has endothermic nature [46]. The negative and positive values indicate the spontaneous nature of the adsorption process

and affinity of the Co(II) ions adsorbent [47,48], respectively. The value of decreases with increase in temperature suggesting that higher temperature makes the adsorption process favourable but this stop at temperature upper than 323 K (Fig. 9).The positive entropy change for the process is caused by the increase in degree of freedom or randomness at the adsorbent-adsorbate interface during the adsorption of Co(II) ions at different temperatures [49].

Kinetic parameters

The adsorption kinetics of Co(II) ions onto magnetite-citric acid is investigated with the help of the pseudo-first order [50] and pseudo-second order [51] model.

The linear form of pseudo-first order kinetic model is expressed by the following equation:

$$\ln(q_e - q_t) = \ln q_t - k_1 t \tag{10}$$

Where q_e and q_t refer to the adsorption capacity of Co(II) ions (mg/g) at equilibrium and at any time, respectively and k_1 is the rate constant of pseudo-first order adsorption (1/min).

The pseudo-second-order kinetic rate equation

Table 2. Isotherm parameters for the adsorption of cobalt on magnetite-citric acid adsorbent

Temperature (K)	Langmuir model			Freundlich model		
	K_L(L/mg)	q_m(mg/g)	R^2	N	K_F (mg/kg)(L/mg)^(1/n)	R^2
293	0.08	27.84	0.995	1.32	1.90	0.983
303	0.11	34.32	0.998	1.60	4.02	0.985
313	0.13	43.29	0.999	1.68	5.38	0.971

Table 3. Comparison of adsorption capacity (mg/g) of different adsorbents for removal of Co(II) ions.

Adsorbent	Adsorption capacity (mg/g)	Reference
Kaolinite	0.92	[37]
Coir pith	12.82	[38]
Natural zeolites	14.38	[39]
EDTA-modified silica gel	20.00	[40]
Synthetic hydroxyapatite	20.19	[41]
Lemon peel adsorbent	22.00	[42]
Fe_3O_4/GO	22.70	[43]
Al-pillared bentonite clay	38.60	[1]
Present study	43.29	-

Table 4. Thermodynamic parameters for the adsorption of cobalt on magnetite-citric acid adsorbent

Temperature(K)	ΔG^0(kJ/mol)	ΔH^0(kJ/mol)	ΔS^0(J/mol.K)
303	-6.25	47.68	0.18
313	-6.74		
323	-7.16		

Table 5. Kinetic parameters for the adsorption of cobalt on magnetite-citric acid adsorbent.

Pseudo-first order				Pseudo-second order		
k_1 (min^{-1})	$q_{e.cal}$ (mg/g)	R^2	$q_{e,exp}$	k_2 (min^{-1})	$q_{e.cal}$ (mg/g)	R^2
0.06	8.34	0.995	13.72	0.01	13.41	0.998

is expressed as follows:

$$\frac{t}{q_t} = \frac{1}{k_2 q_e^2} + \frac{1}{q_e} t \qquad (11)$$

Where k_2 is the rate constant of pseudo-second order adsorption (g/mg.min).

As can be seen from Table 5, the adsorption of Co(II) ions onto magnetite-citric acid is more favorably described by a pseudo-second-order kinetic model because of its greater correlation coefficient (0.998) and closer values of calculated and experimental q_e. The higher values of R^2 of pseudo-second order and applicability of Langmuir isotherm indicated that the process to be chemisorptions in nature [45,52,53].

Stability and reusability of nanoadsorbents

To evaluate desorption of Co(II) ions loaded magnetite-citric acid, various concentration of 0.05, 0.1, 0.15 and 0.2 M HNO$_3$ was used as desorption media and quantitative desorption efficiencies were 72%, 92%, 94% and 95%, respectively. The maximum value of desorption efficiency was 95% by using 0.20 M concentrate of HNO$_3$. To evaluate the reusability of adsorption–desorption process, five times desorption were carried out. In this study, the initial concentration of Co(II) ions was 25 mg/L, the adsorbent dose was 0.2 g, the reaction temperature was 298 K, pH was 3 and shaking rate was 500 rpm. In each experiment, the used magnetite-citric acid nanoadsorbent was collected by using an external magnet. Obtained results that are summarized in Table 6 show a wonderful reusability for selected nanoadsorbent. It was observed that with an increase in cycle numbers, uptake capacity of Co(II) ions was slowly decreased. Following reaction was used in order to determine the efficient recovery of Co(II) ions:

$$R(\%) = \frac{C_{des}}{C_{ads}} \times 100 \qquad (12)$$

CONCLUSION

The main idea of this study was to provide a simple, cost-effective, reproducible and inexpensive method to remove Co(II)ions from aqueous solution. To

Table 6. Reusability of magnetite-citric acid for adsorption/desorption of Co(II) ions during five cycles

Cycle numbers	Removal percentage
1	92
2	92
3	91
4	90
5	90

this end, different magnetic nanocomposites were synthesized by the simple co-precipitation method. Some of these nanoadsorbents weren´t synthesized until now. After determining and characterizing of different synthesized nanoadsorbents structure, several parameters affecting the removal of Co(II) ions were evaluated and optimized. The results show that in all cases, the maximum removal efficiency of Co(II) ions was obtained by using magnetite-citric acid as an adsorbent. Also, the results show that pH, temperature, contact time, adsorbent dose and initial concentration factors have a high effect on the removal of Co(II) ion. It is found that the adsorption behavior of Co(II) ions is better described by the Langmuir isotherm model and the kinetics of Co(II) ions adsorption follows the pseudo-second order model. Moreover, the results indicate that the method has a suitable repeatability and nanoadsorbents have high stability. Results of this work suggest that the magnetite-based nanoadsorbents are wonderful adsorbents for the removal of heavy metal ions from synthetic and industrial wastewater by using the technology of magnetic separation.

CONFLICT OF INTEREST

The authors declare that there are no conflicts of interest regarding the publication of this manuscript.

REFERENCES

1. Sayadi MH, Rezaei MR. Impact of land use on the distribution of toxic metals in surface soils in Birjand city, Iran. Proceedings of the International Academy of Ecology and Environmental Sciences. 2014;4(1):18-29.

2. Sayadi MH, Rezaei MR, Rezaei A. Fraction distribution and bioavailability of sediment heavy metals in the environment surrounding MSW landfill: a case study. Environmental Monitoring and Assessment. 2014;187(1):4110.

3. Sayadi MH, Rezaei MR, Rezaei A. Sediment Toxicity and Ecological Risk of Trace Metals from Streams Surrounding a Municipal Solid Waste Landfill. Bulletin of Environmental Contamination and Toxicology. 2015;94(5):559-63.

4. Sayadi MH, Shabani M, Ahmadpour N. Pollution Index and Ecological Risk of Heavy Metals in the Surface Soils of Amir-Abad Area in Birjand City, Iran. Health Scope. 2015;4(1):ee21137.

5. Ghaleno OR, Sayadi M, Rezaei M, Kumar CP, Somashekar R, Nagaraja B. Potential ecological risk assessment of heavy metals in sediments of water reservoir case study: Chah Nimeh of Sistan. Proc Int Acad Ecol Environ Sci. 2015;5(4):89-96.

6. Mohan D, Pittman Jr CU. Arsenic removal from water/wastewater using adsorbents—A critical review. Journal of Hazardous Materials. 2007;142(1–2):1-53.

7. Sayadi M, Torabi S. Geochemistry of soil and human health: A review. Pollution Research. 2009;28(2):257-62.

8. Smedley PL, Kinniburgh DG. A review of the source, behaviour and distribution of arsenic in natural waters. Applied Geochemistry. 2002;17(5):517-68.

9. Nordstrom DK. Worldwide Occurrences of Arsenic in Ground Water. Science. 2002;296(5576):2143-5.

10. Organization WH. Guidelines for drinking-water quality. Vol. 1, Recommendations. 3rd ed: World Health Organization; 2004.

11. Wang P, Sun G, Jia Y, Meharg AA, Zhu Y. A review on completing arsenic biogeochemical cycle: Microbial volatilization of arsines in environment. Journal of Environmental Sciences. 2014;26(2):371-81.

12. Nguyen TV, Vigneswaran S, Ngo HH, Pokhrel D, Viraraghavan T. Specific Treatment Technologies for Removing Arsenic from Water. Engineering in Life Sciences. 2006;6(1):86-90.

13. Bhargavi RJ, Maheshwari U, Gupta S. Synthesis and use of alumina nanoparticles as an adsorbent for the removal of Zn(II) and CBG dye from wastewater. International Journal of Industrial Chemistry. 2015;6(1):31-41.

14. Bissen M, Frimmel FH. Arsenic — a Review. Part II: Oxidation of Arsenic and its Removal in Water Treatment. Acta hydrochimica et hydrobiologica. 2003;31(2):97-107.

15. Akin I, Arslan G, Tor A, Ersoz M, Cengeloglu Y. Arsenic(V) removal from underground water by magnetic nanoparticles synthesized from waste red mud. Journal of Hazardous Materials. 2012;235–236:62-8.

16. Kanel SR, Manning B, Charlet L, Choi H. Removal of Arsenic(III) from Groundwater by Nanoscale Zero-Valent Iron. Environmental Science & Technology. 2005;39(5):1291-8.

17. Rahmani A, Ghaffari H, Samadi M. A comparative study on arsenic (III) removal from aqueous solution using nano and micro sized zero-valent iron. Iranian Journal of Environmental Health Science & Engineering. 2011;8(2):157-66.

18. Pena M, Meng X, Korfiatis GP, Jing C. Adsorption Mechanism of Arsenic on Nanocrystalline Titanium Dioxide. Environmental Science & Technology. 2006;40(4):1257-62.

19. Li Q, Easter NJ, Shang JK. As(III) Removal by Palladium-Modified Nitrogen-Doped Titanium Oxide Nanoparticle Photocatalyst. Environmental Science & Technology. 2009;43(5):1534-9.

20. Ghosh D, Luwang MN. Arsenic detection in water: YPO4:Eu3+ nanoparticles. Journal of Solid State Chemistry. 2015;232:83-90.

21. Martinson CA, Reddy KJ. Adsorption of arsenic(III) and arsenic(V) by cupric oxide nanoparticles. Journal of Colloid and Interface Science. 2009;336(2):406-11.

22. Shrivas K, Shankar R, Dewangan K. Gold nanoparticles as a localized surface plasmon resonance based chemical sensor for on-site colorimetric detection of arsenic in water samples. Sensors and Actuators B: Chemical. 2015;220:1376-83.

23. JANBAZ FM, KHOLGHI M, HORFAR A, HAGHSHENAS D. EXPERIMENTAL INVESTIGATION OF ARSENIC REMOVAL BY USING FE NANO PARTICLES IN BATCH EXPERIMENT. journal of Environmental Studies. 2014;39(4):149-56.

24. Koohpayehzadeh H, Torabian A, Nabi Bidhendi G, Habashi N. Nanoparticle Zere-valent Iron Affect on As (V) Removal from Drinking Water. Journal of Water and Wastewater(parallel title); Ab va Fazilab (in persian). 2012;23(3):60-7.

25. Olyaie E, Banejad H, Rahmani AR, Afkhami A, Khodaveisi J. Feasibility study of using Calcium Peroxide Nanoparticles in Arsenic Removal from Polluted Water in Agriculture and It's Effect on the Irrigation Quality Parameters Iranian Journal of Health and Environment. 2012;5(3):319-30.

26. Adlim M, Abu Bakar M, Liew KY, Ismail J. Synthesis of chitosan-stabilized platinum and palladium nanoparticles and their hydrogenation activity. Journal of Molecular Catalysis A: Chemical. 2004;212(1–2):141-9.

27. De Corte S, Hennebel T, Fitts JP, Sabbe T, Bliznuk V, Verschuere S, et al. Biosupported Bimetallic Pd–Au Nanocatalysts for Dechlorination of Environmental Contaminants. Environmental Science & Technology. 2011;45(19):8506-13.

28. Arsiya F, Sayadi MH, Sobhani S. Green synthesis of palladium nanoparticles using Chlorella vulgaris. Materials Letters. 2017;186:113-5.

29. Rahmani A, Ghaffari H, Samadi M. Removal of Arsenic (III) from Contaminated Waterby Synthetic Nano Size Zerovalent Iron. World Academy of Science, Engineering and Technology, International Journal of Environmental, Chemical, Ecological, Geological and Geophysical Engineering. 2010;4(2):96-9.

30. Goswami A, Raul PK, Purkait MK. Arsenic adsorption using copper (II) oxide nanoparticles. Chemical Engineering Research and Design. 2012;90(9):1387-96.

31. Sartape AS, Mandhare AM, Jadhav VV, Raut PD, Anuse MA, Kolekar SS. Removal of malachite green dye from aqueous solution with adsorption technique using Limonia acidissima (wood apple) shell as low cost adsorbent. Arabian Journal of Chemistry.

32. Sajadi F, Sayadi MH, Hajiani M. Study of optimizing the process of Cadmium adsorption by synthesized silver nanoparticles using Chlorella vulgaris. Journal of Birjand University of Medical Sciences. 2016;23(2):119-29.

21

Carboxymethyl-β-cyclodextrin Modified Magnetic Nanoparticles for Effective Removal of Arsenic from Drinking Water: Synthesis and Adsorption Studies

Sedigheh Zeinali [*,1], *Maryam Abdollahi* [1], *Samad Sabbaghi* [1,2]

[1]*Department of Nanochemical Engineering, Faculty of Advanced Technologies, Shiraz University, Shiraz, Iran*

[2]*Nanotechnology Research Institute, Shiraz University, Shiraz, Iran*

ABSTRACT

The β-cyclodextrin coated magnetic nanoparticles were prepared by the surface modification of Fe_3O_4 magnetic nanoparticles using carboxymethyl-β-cyclodextrin. Prepared nanoparticles were characterized by X-ray diffraction analysis, transmission electron microscope, Fourier transform infrared spectroscopy, dynamic light scattering and vibrating sample magnetometer. The β-cyclodextrin modified Fe_3O_4 nanoparticles have a narrow size distribution with mean diameter about 10 nm. They exhibit superparamagnetic properties at room temperature with saturation magnetization of 48 emu/g. Since, the most reported technologies for arsenic removal are more effective in removing As(V) rather than As(III), the adsorption ability of these nanoparticles was investigated for removing As (III) from aqueous solution. The adsorption behavior of this material can be influenced by various factors such as contact time, pH, adsorbent dosage and initial concentration of As(III), which their effects were studied. Equilibrium data were fitted by Langmuir isotherm and the maximum removal percentage was obtained about 85% at optimum conditions. Using these modified Fe_3O_4 nanoparticles, the arsenic concentrations can be reduced to the allowed limits declared by the World Health Organization.

KEYWORDS: *Arsenic removal; Magnetic nanoparticle; β-cyclodextrin; Adsorption; Drinking water treatment*

INTRODUCTION

Arsenic contamination, which is present mainly as oxyanion compounds in drinking water and groundwater, is a serious environmental worldwide problem because of its toxicity and health hazards. The ingestion of arsenic can result in both cancerous and non-cancerous health effects such as disturbance of cardiovascular and nervous system function, pigmentation, depigmentation, skin cancer and cancer of internal organs [1].

Arsenic exists in four oxidation states, +5 (arsenate), +3 (arsenite), 0 (arsenic), and -3 (arsine). The toxicity of arsenic depends on its speciation, For example, arsenite is significantly more toxic than arsenate. It is typically more difficult to remove arsenite than arsenate from contaminated water. Under normal pH conditions (in natural waters), arsenite is mostly found as an uncharged species (H_3AsO_3), and negatively charged species ($H_2AsO_3^-$

* Corresponding Author Email: zeinali@shirazu.ac.ir

, $HAsO_3^{2-}$ and AsO_3^{3-}) are found only at high pH [2].

Arsenic contamination of water results from both natural and anthropogenic activities. Arsenic is present in water due to dissolution of minerals as well as human activities such as mining, smelting of metal ores and use of pesticides causing arsenic pollution [3]. Reports have indicated arsenic pollution in many regions of countries around the world including USA, China, Chile, Bangladesh, Taiwan, Mexico, Argentina, Poland, Canada, Japan, India and Islamic Republic of Iran [4,5]. Due to the negative impacts of arsenic on human health, the US Environmental Protection Agency, the World Health Organization (WHO), and the European Commission have reduced the maximum contaminant level of arsenic in drinking water from 50 to 10 ppb. However, some countries have kept the earlier WHO guideline of 50 ppb as their standard [3,6]. Many arsenic removal technologies such as precipitation [7], membrane separation [8,9], nanofiltration [10,11], coagulation [12], ion exchange [13], microbial transformation [14,15], and adsorption [16,17,18] have been developed. Most of these techniques are only effective in removing arsenate, and they require a pretreatment step for oxidation of arsenite to arsenate. These pretreatment chemical methods can be used easily; however, their applications are limited by producing large amount of toxic sludge, and they need further treatment before disposal into the environment [19].

Among the removal methods, adsorption from solution has received more attention due to its low cost and high efficiency for arsenic treatment [20]. Many adsorbent materials have been used including activated alumina, activated carbon, red mud, bauxsol, etc [21]. Therefore, one of the key factors in adsorption-based technologies is the development of highly effective and inexpensive adsorbents [22]. Recently, adsorption through Fe_3O_4 magnetic nano-sized particles has been more popular because of the specific characteristics of these materials. Particular properties of them such as extreme small size, high surface-area-to-volume ratio and the absence of internal diffusion resistance, provides better kinetics for adsorption of arsenic from aqueous solution. Magnetic nano-adsorbents have the qualities of both magnetic separation techniques and nano-sized materials, thus they can be easily separated with an external magnetic field [23, 24]. However, these adsorbents are not stable and hardly can be recycled because Fe_3O_4 is highly susceptible to oxidation when it is exposed to the atmosphere due to its small size [25]. In order to improve the stability and functionality, the iron oxide nanoparticles are often modified with natural or synthetic polymers.

Cyclodextrins are polysaccharides produced through the degradation of starch by cyclodextrin glucanotransferase enzyme. β-cyclodextrin (β-CD) is a cyclic oligosaccharide consist of seven α-D-glucose units connected through α-(1,4) linkages. The structure of these molecules is like toroidal truncated cones containing a lipophilic cavity with two hydroxyl groups one lying on the outside and the other lying in the inside [26,25]. Cyclodextrins are available on a large scale. Their production is not costly, and most importantly they have the ability to form inclusion complexes (guest-host complexes) with a wide variety of organic and inorganic compounds in its hydrophobic cavity. In addition, metal ions can be complexed by cyclodextrins through hydroxyl groups [27].

In this study, a magnetic nano-adsorbent was fabricated by surface modification of Fe_3O_4 nanoparticles with β-CD for the adsorption of arsenic contamination. The size, morphology and properties of the β-CD modified magnetic nanoparticles were characterized using different analytical tools. The mean diameter of β-CD modified Fe_3O_4 magnetic nanoparticles were ~10 nm. These modified nanoparticles exhibit superparamagnetic properties at room temperature; therefore, they can be easily separated by applied magnetic field. They were evaluated as absorbents to remove arsenic, and the effects of several factors such as, pH, initial arsenic concentration, contact time, and mechanism of arsenic adsorption onto β-CD modified magnetic nanoparticles were investigated in this report. In Iran, in western and north-western provinces especially in Kurdistan province arsenic contamination has been reported in groundwater sources [5,28]. Real water samples from Kurdistan villages were collected, and the ability of modified magnetic nanoparticles in removing arsenic in real samples was examined. Overall, the major end of the current research is ascertaining the capability of β-CD modified magnetic nanoparticles to remove trivalent arsenic from water in a one-step operation with no pretreatment. The maximum removal efficiency of arsenic (III) at optimal conditions is about 85%.

EXPERIMENTAL

Materials

Iron (II) chloride tetrahydrate, iron (III) chloride hexahydrate, sodium hydroxide (NaOH), ammonium hydroxide (25%), β-cyclodextrin (β-CD), monochloroacetic acid, Carbodiimides (cyanamide, CH_2N_2), arsenic trioxide (As_2O_3) and methanol (99%) were purchased from Merck. All chemicals were the guaranteed or analytic grade reagents commercially available and used without further purification. Arsenic solutions that were used in different experiments were prepared by diluting a stock solution (1000 ppm of arsenic). The stock solution was prepared by solving 50 mg of As_2O_3 salt in 100 ml deionized water. For preparing working solution from 1 up to 100 ppm of arsenic, 0.01 ml to 1 ml of the stock solution were diluted in 10 ml deionized water.

Apparatus

In order to investigate the size and morphology of magnetic nanoparticles, transmission electron microscopy (CM10, Philips) was used. β-CD's grafting onto the surface of magnetic nanoparticles (MNPs) was monitored by Fourier transform infrared spectroscopy (RX1, Perkin-Elmer). The crystalline structure of Fe_3O_4 nanoparticles was characterized by X-ray diffraction technique (D8 Advance, Bruker). The magnetic properties of the Fe_3O_4 nanoparticles coated with β-cyclodextrin were investigated using a vibrating sample magnetometer (Megnatis Daghigh Kavir, Iran). The size and size distribution of nanoparticles were measured by particle size analyzer (L-550, HORIBA).

Preparation of Nacked Fe_3O_4 magnetic nanoparticles

Fe_3O_4 nanoparticles were prepared by chemical co-precipitation method. A complete precipitation of Fe_3O_4 was achieved under alkaline condition, while maintaining a molar ratio 1:2 of Fe^{2+} and Fe^{3+}. In a typical synthesis to obtain 1 g Fe_3O_4 precipitate, 0.86 g of $FeCl_2·4H_2O$ and 2.36 g of $FeCl_3·6H_2O$ were dissolved in 40 ml of distilled water with vigorous stirring at a speed of 1000 rpm. 5 ml of NH_4OH (25%) was added after the solution was heated to 80 °C. The reaction was continued for 30 min at 80 °C under constant stirring to ensure the complete growth of the nanoparticle crystals. The resulting particles were then washed with Distilled water at least 5 times to remove any unreacted chemicals and dried [24].

Preparation of Carboxymethyl β-cyclodextrin

Carboxymethyl-β-cyclodextrin (CM-βCD) was prepared following the previous procedure [29]. CM-βCD was synthesized in the alkaline condition by reacting monochloroacetic acid with β-CD. A mixture of β-CD (2 g) and NaOH (1.86g) in water (7.4 ml) was treated with a 16.3% monochloroacetic acid solution (5.4 ml) at 50 °C for 5 h. The temperature of the reaction mixture was reduced to room temperature, and then the pH values were adjusted in the range of 6–7. The obtained neutral solution was added to methanol, and the produced white precipitate, carboxymethylated β-cyclodextrin, was filtered and dried in oven at 50 °C.

Preparation of CM-βCD Modified Magnetic Nanoparticles

β-cyclodextrin modified magnetic nanoparticles can be prepared using one-step or two-step methods. By one-step co-precipitation method, in which iron precursors (Fe^{2+} and Fe^{3+}) and CM-βCD were mixed together, the binding of CM-βCD onto the Fe_3O_4 surface was conducted. Briefly, 0.57 g of $FeCl_2·4H_2O$, 1.57 g $FeCl_3·6H_2O$ and 1 g CM-βCD were dissolved in 26.7 ml of distilled water with vigorous stirring at a speed of 1200 rpm. When the temperature of the reaction mixture reached 90 °C, 3.5 ml of NH_4OH (25%) was added in drop. The reaction was maintained at 90 °C under constant stirring for 1 h. The resulting nanoparticles were then washed with distilled water to remove any unreacted chemicals and then it was dried in oven at 70 °C. In this method, the carboxyl groups of CM-βCD directly reacted with the surface OH groups on the magnetite to form Fe-carboxylate [29].

Adsorption Experiments

The adsorption experiments of As(III) by CM-βCD modified magnetic nanoparticles (CM-βCD-

MNPs) were investigated in aqueous solutions. For each treatment, 30 mg CM-βCD-MNPs was added into 5mL of arsenic solution. The solutions of As(III) were obtained by dissolving arsenic trioxide (As_2O_3) into distilled water. The influence of experimental factors such as pH value, contact time, adsorbent dosage and initial arsenic concentration on the removal efficiency of As(III) was investigated. When equilibrium was achieved, magnetic nanoparticles were separated magnetically from arsenic solution using a magnet. The concentration values of As(III) were measured using the inductively coupled plasma spectroscopy (ICP) method. For comparison, the adsorption of As(III) by the naked Fe_3O_4 nanoparticles (unmodified MNPs) was also investigated.

The adsorption isotherm experiments were performed using different initial As(III) concentrations. The adsorption capacity (q_e, mg/g) and removal efficiency (E) of the adsorbent were calculated using the following equations:

$$q_e = \frac{(C_0 - C_e)V}{m} \qquad (1)$$

$$E(\%) = \frac{(C_0 - C_e)}{C_0} \times 100 \qquad (2)$$

Where C_0 and C_e are initial and equilibrium concentration (mg/L) of As(III) solution, respectively; V is the volume of the As(III) solution; and m is the weight of the β-CD modified magnetic nanoparticles.

RESULTS AND DISCUSSIONS

One of the most important steps in magnetic nanoparticle preparation is their characterization. It means the investigation of their size, shape and size distribution and also, their magnetic properties. For this purpose different methods were used and the results were shown following.

Characterizations of β-cyclodextrin modified Fe_3O_4 nanoparticles
Transmission Electron Microscopy (TEM)

Transmission electron microscopy (TEM) was utilized to investigate the size and morphology of magnetic nanoparticle samples. As shown in Fig. 1, the sample includes small nanoparticles that are relatively uniform in size and shape. The average size of the particles is about 10 nm. This reveals that

the binding process did not significantly result in changes in the size of the particles but it prevents agglomeration of nanoparticles which leads to lower distribution of modified nanoparticles rather than unmodified nanoparticles.

Dynamic Light Scattering (DLS)

In Fig. 2 the size distribution graphs for the naked and β-CD modified Fe_3O_4 nanoparticles obtained by DLS method are presented. The average diameters of the naked and β-CD modified magnetic nanoparticle were equal to 13.7 and 18.5 nm, respectively. It can be seen for Fig. 2 that, they are low-dispersed nanoparticles. As it evidence, the nanoparticles size observed by DLS is larger than those obtained by TEM because, β-CD coating increases the hydrodynamic diameter of nanoparticles [25].

FTIR analysis

The CM-β-CD binding on the surface of magnetic nanoparticles was confirmed by FTIR spectroscopy. FTIR spectra of naked MNPs, CM-β-CD-MNPs, and pure CM-β-CD in the range of 350–4000 cm^{-1} wavenumbers are shown in Fig. 3.

In the spectra of the Fe_3O_4 nanoparticles, the absorption peak at 585 is the characteristic of Fe-O-Fe bond in Fe_3O_4. The broad band around 3426 is due to –OH stretching vibrations. The spectrum of CM-β-CD shows the characteristic peaks at 948, 1031 and 1190. The peak at 948 is due to the R-1,4-bond skeleton vibration of β-CD, and the peaks at 1031 and 1190 corresponded to the asymmetric glycosidic (C–O–C) vibrations and coupled υ(C–C/C–O) stretch vibration [21]. All of these characteristic peaks in the spectrum of CM-β-CD (900–1200) can also be seen in the spectrum of CM-β-CD-MNPs with slight differences. The characteristic peak appeared at 1606 is due to bands of -COOM groups (M represents metal ions), indicates that the -COOH groups of CM-β-CD reacted with the surface of Fe_3O_4 particles [30]. These findings indicate that the binding of CM-β-CD on the surface of Fe_3O_4 nanoparticles were done successfully.

UV-Vis spectra

The UV–visible absorption spectra of the naked and β-CD modified nanoparticles are illustrated in Fig. 4. As can be seen in the absorption spectra

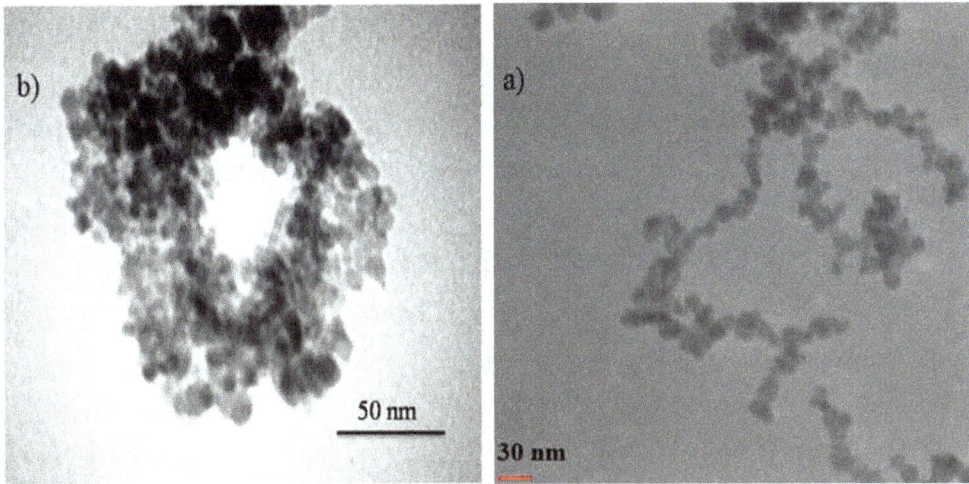

Fig. 1. TEM images of the (a) naked and (b) CM-βCD modified Fe$_3$O$_4$ nanoparticles

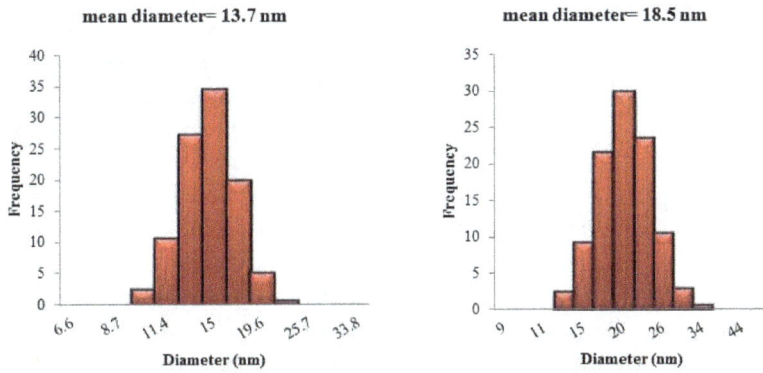

Fig. 2. DLS plots of the (a) naked and (b) β–CD modified Fe$_3$O$_4$ nanoparticles

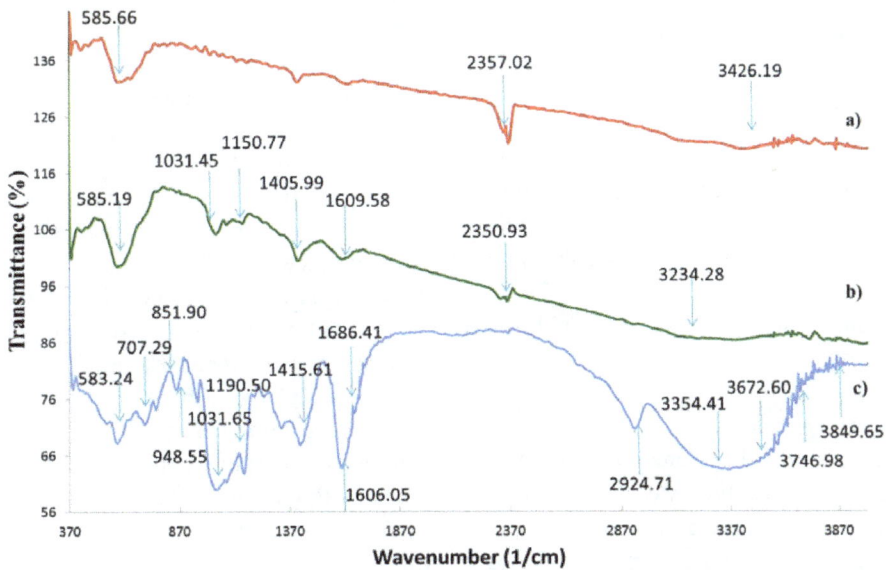

Fig. 3. FTIR spectra of (a) naked and (b) β–CD modified Fe$_3$O$_4$ nanoparticles and (C) pure CM-βCD

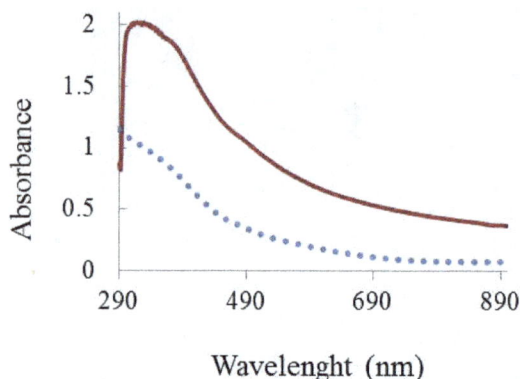

Fig. 4. UV–visible spectra of the naked (Dot line) and β-CD modified (Solid line) Fe₃O₄ nanoparticles

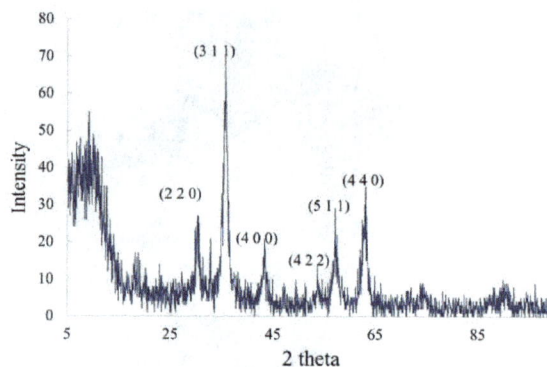

Fig. 5. XRD pattern of the β-CD modified Fe₃O₄ nanoparticles synthesized

reported in Fig. 4, after β-CD coating on the MNPs the UV-Vis spectrum of these nanoparticles was changed. This phenomenon can be also used as another confirmation for binding of β-CD on the Fe_3O_4 nanoparticles' surface.

XRD analysis

Fig.5 shows the XRD pattern of the CM-β-CD modified Fe_3O_4 nanoparticles. There are six characteristic peaks for Fe_3O_4 at 2θ values of 30.3°, 35.6°, 43.25°, 53.65°, 57.15°, and 62.85° marked by their indices (2 2 0), (3 1 1), (4 0 0), (4 2 2), (5 1 1), and (4 4 0), respectively. These observations confirm the presence of inverse cubic spinel structure of the resultant nanoparticles [29]. It is also clear that the coating did not lead any phase change of Fe_3O_4 nanoparticles. The average size of the CM-β-CD modified Fe_3O_4 can be calculated from Scherrer equation:

$$d = 0.9\lambda/\beta cos\theta \qquad (3)$$

Where d is particle diameter, λ is X-ray wavelength, β is the peak width of half-maximum, and θ is Bragg's diffraction angle in degree. The obtained size of CM-β-CD modified MNPs using strongest peak (3 1 1) at 2θ=35.6° and λ=1.54 A° was about 8.8 nm, which is in consistent with the results of TEM.

Vibrating sample magnetometer (VSM) analysis

In order to examine the magnetic properties of prepared nanoparticles, they were analyzed by vibrating sample magnetometer (VSM) at room temperature. The magnetization hysteresis loops of naked and CM-β-CD-MNPs were shown in Fig.

6. The saturated magnetizations for naked MNPs and CM-β-CD-MNPs are 54 and 48 emu. This demonstrates that all samples are superparamagnetic which can responsive to an applied magnetic field and retain no permanent magnetization after removing the applied magnetic field. The saturated magnetization of the Fe_3O_4 nanoparticles decreased after coating with CM-β-CD (Fig. 6). This is mainly attributed to the existence of non-magnetic materials on the surface of nanoparticles. Similar results are reported by Badruddoza et. al. [29]. Materials with superparamagnetic properties can be easily separated from the solution with the aid of an external magnetic field within several minutes.

Adsorption of arsenic onto β-CD modified MNP

The adsorption process can be influenced by some experimental parameters. The effects of these parameters should be investigated and the optimal conditions should be used for real samples analysis.

Effects of initial pH on arsenic adsorption

One of the most important parameters in arsenic removal process is the pH value of the sample. Since the pH variation can influence on the ionic state of the surface functional groups of nanoparticles, it can be effective on the adsorption process. On the other hand, the arsenic chemistry in water also highly depends on pH values. Therefore, arsenic adsorption ability of CM-β-CD-MNPs was investigated at room temperature by varying pH values in the range of 8 to 11 (Fig. 7). The removal efficiency of arsenic improved by increasing pH values and then remains almost constant at pH 9 -11 because of the pH-dependency of As (III).

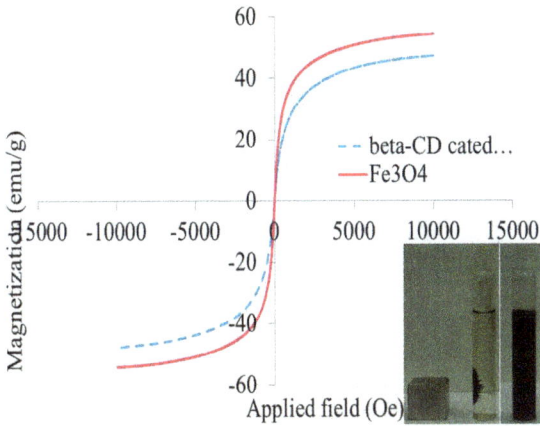

Fig. 6. Magnetic hysteresis curves of naked (solid line) and β-CD modified (dot line) Fe$_3$O$_4$ nanoparticles

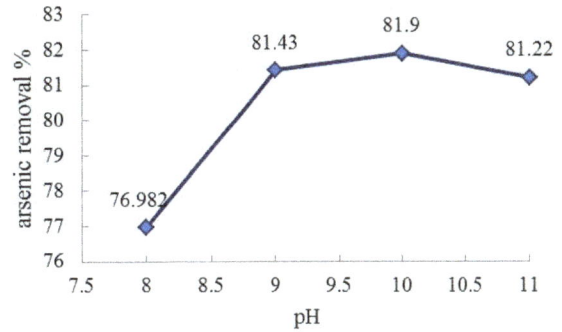

Fig. 7. The effect of pH on the arsenic removal efficiency by the β-CD modified Fe$_3$O$_4$ nanoparticle The removal studies were carried out by applying 6mg/ml β-CD-modified Fe$_3$O$_4$ on the As(III) solution with 10 ppm concentration, during 10 min contact time at room temperature)

However, pH 9 was selected as the optimal value since no significant improvement in arsenic removal was observed between 9 to 11 pH values.

Effect of Magnetic Nanoparticles Dosage

The effect of adsorbent dosage on arsenite adsorption capacity and removal efficiency was investigated and the results were shown in Fig. 8. It is evident that the removal efficiencies of arsenite increase with increasing adsorbent dosage, while the amount of adsorption capacity (q$_e$) decreases. The increase in the removal efficiency is due to an increase in the adsorbent amount which provides more adsorbent surface for the solute to be adsorbed.

Effect of Initial Concentration

As mentioned above, CM-βCD-MNPs provided good adsorption capacity for As(III) at pH 9 and 6 mg/ml adsorbent amount by applying 30 minutes contact time. The adsorption isotherm of arsenic was obtained by changing the initial concentration of As(III) at values ranging from 1 to 100 ppm. The removal efficiency of As(III) decreases with an increase in initial As(III) concentration (Fig. 9). This behavior was due to the fact that the total available adsorption sites for a fixed amount of adsorbent are limited, which leads to a decrease in removal percentage corresponding to an increased initial adsorbate concentration.

Adsorption Kinetics

Another important parameter in the arsenic wastewater treatment process is equilibrium contact time. The effect of contact time of CM-β-CD-MNPs with arsenic solution on the removal efficiencies was shown in Fig. 10. More than 75% of the arsenite was adsorbed during the first 5 min. There is no significant change from 30 to 45 min, and then equilibrium reached during only 90 min. Therefore, 30 min was considered as the optimal contact time for adsorption. The fast adsorption in the beginning of operation is due to the greater concentration gradient and more available sites for adsorption; however, as the process continues, the rate of As(III) adsorption becomes slower. This phenomenon may be attributed to decreasing of the binding sites on the surface of nanoparticles. Such behavior is common in adsorption processes and has been reported [22].

In order to investigate adsorption kinetics mechanism, pseudo-first order and pseudo-second order models were applied to fit the experimental data [20]. The correlation coefficient R^2 was used to express the uniformity between the experimental and model-predicted data. The equation used at pseudo-first order kinetic model is:

$$ln(q_e - q_t) = ln\, q_e - k_1 t \qquad (4)$$

where q$_e$ and q$_t$ are the amounts of arsenic adsorbed by each unit of CM-β-CD-MNPs at equilibrium state and time t, respectively (mg/g), k$_1$ is the pseudo-first order rate constant for the adsorption process and can be obtained from the plots of against t. The corresponding kinetic parameters are shown in Table. 1. Low value of correlation coefficient (R^2) and large difference between q$_{e,cal}$ and q$_{e,exp}$ indicated poor fit of the

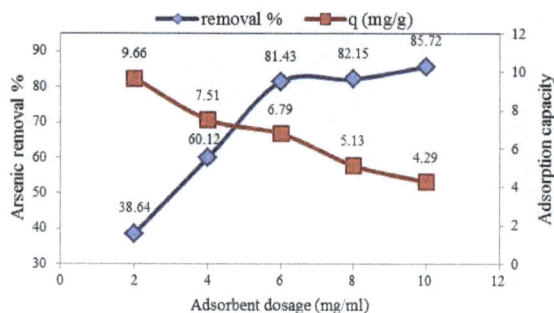

Fig. 8. The effect of β-CD-modified Fe₃O₄ nanoparticle adsorbent dosage on arsenic efficiencies and adsorption capacity

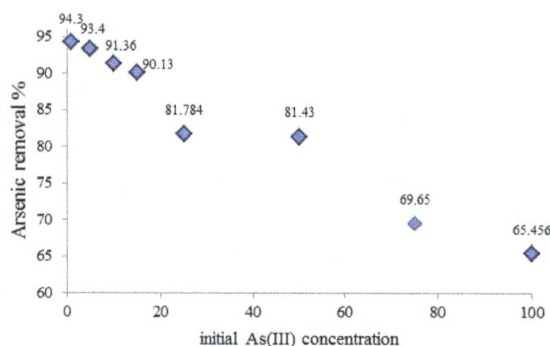

Fig. 10. The effect of contact time on the arsenic removal efficiency by the β-CD modified Fe₃O₄ nanoparticle. The initial concentration of arsenic was 50 ppm here

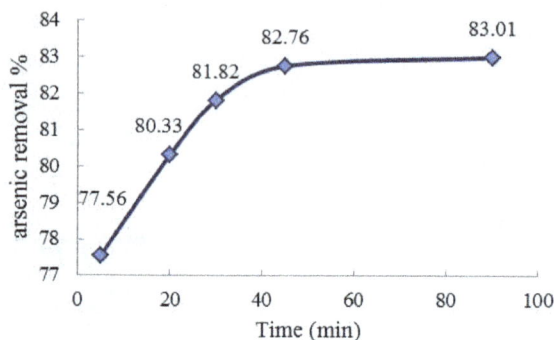

Fig. 9. The effect of initial As(III) concentration on arsenic efficiencies

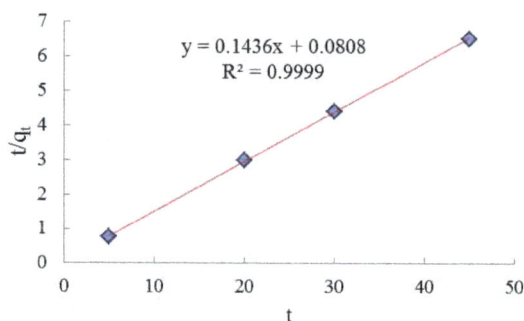

Fig. 11. plot of t/q_t against t that can show pseudo second-order kinetics for adsorption of As(III)

experimental data to pseudo-first order model. The equation of pseudo-second order kinetic model is:

$$\frac{t}{q_t} = \frac{1}{k_2 q_e^2} + \frac{t}{q_e} \qquad (5)$$

Where k_2 is the pseudo-second order rate constant for the adsorption process and can be calculated from the intercept of plot of t/q_t against t and shown in Fig. 11. Table 1 summarizes adsorption capacities determined by this model. A linear relationship ($R^2 = 0.9992$) obtained between t/q_t and t indicates that pseudo-second order model is better to describe the adsorption process than the other. The equilibrium adsorption capacity ($q_{e,cal}$) by CM- βCD-MNP was 6.99 mg/g, which is consistent with the experimental data. Also, it was found that the rate of the arsenic adsorption to be controlled by the chemisorption process [29].

Adsorption Isotherms

In order to examine the adsorption mechanism

and calculate the adsorption capacity of CM-βCD-MNPs, adsorption isotherms were investigated at room temperature by changing the initial concentration of As(III). The adsorption isotherms are important in describing the adsorption mechanism and yield certain constant values which express the surface properties and affinity of the adsorbent [30]. The equilibrium isotherms for the adsorption of arsenic ions by CM-βCD-MNPs at pH 9 are shown in Fig. 12. Generally, Langmuir and Freundlich adsorption isotherm models were used to describe the equilibrium data at a constant temperature. Langmuir model which is used widely assumes that adsorption takes place at specific homogeneous sites within the adsorbent and is applied to monolayer adsorption processes. The linear form of the Langmuir isotherm can be expressed as:

$$\frac{C_m}{q_e} = \frac{C_e}{q_m} + \frac{1}{q_m K_L} \qquad (6)$$

The Freundlich model is based on a multilayer adsorption. In this model the adsorption energy

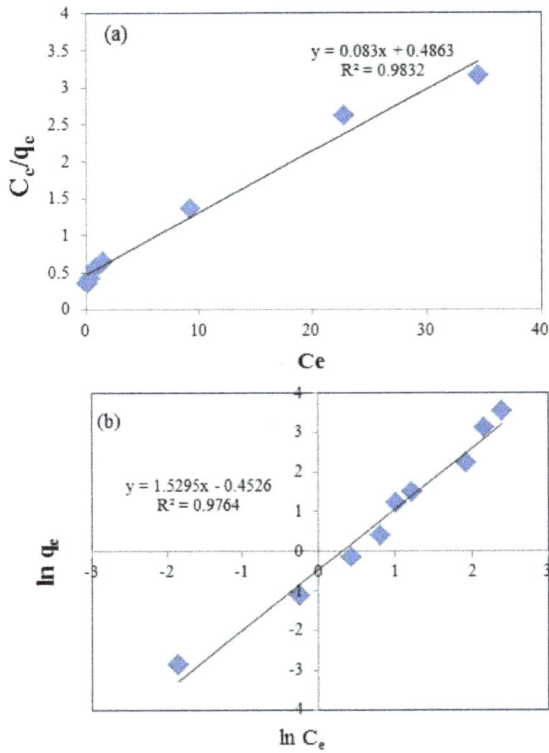

Fig. 12 The (a) Langmuir and (b) Freundlich isotherm plots for As(III) adsorption by CM-βCD-MNPs at pH 9

Table. 1. Adsorption kinetic parameters of As(III) onto CM-βCD-MNPs.

	Parameters	
	$K_1(min^{-1})$	0.077
Pseudo-first-order	$Q_{e,cal}(mg/g)$	0.834
	R^2	0.964
	$Q_{e,exp}(mg/g)$	6.917
Pseudo-second-order	$K_2,g/(mg\ min)$	0.256
	$Q_{e,cal}(mg/g)$	6.993
	R^2	0.999

Table. 2. Adsorption isotherm parameters for As(III) adsorption on β-CD modified magnetic nanoparticles at pH 9.

Isotherm models	Parameters	
	q_m	12.330
Langmuir	k_L	0.171
	R^2	0.983
	n	0.654
Freundlich	K_F	1.572
	R^2	0.976

decreases with the surface coverage. It is an empirical equation used to describe heterogeneous adsorption systems which can be represented as follows:

$$Ln\ q_e = \frac{1}{n} Ln\ C_e + Ln\ K_F \qquad (7)$$

Where q_e is the equilibrium adsorbate loading on the adsorbent in mg/g, C_e is the solute equilibrium concentration in mg/L, q_m the maximum capacity of adsorbent (mg/g), and K and $1/n$ are Freundlich constants related to adsorption capacity and intensity of adsorption, respectively. The values of q_m and K_l were determined from the slope and intercept of the linear plots of C_e/q_e versus C_e (Fig. 12a) and the values of K and 1/n were determined from the slope and intercept of the linear plot of ln q_e versus ln C_e (Fig. 12b). The isotherm parameters are shown in Table 2. The higher value of correlation coefficients indicates that the equilibrium data can be fitted better to Langmuir equation. Also, the maximum adsorption capacity obtained from the Langmuir isotherm was 12.33 mg/g. This result is comparable with the adsorption capacity of the other similar adsorbents reported in literature [25,

32, 33]. The result of this study indicated that the magnetic nanoparticles could be one of the best adsorbents for the removal of As(III) from aqueous solutions.

Effect of β-CD Coating on Arsenic Adsorption

To investigate the effect of coating on adsorption of trivalent arsenic, the adsorption capacity of CM-βCD modified and naked MNPs were compared. In Table 3, it can be seen that, the removal efficiency of As(III) by CM-βCD-MNPs are higher than the naked ones. Hereby the use of β-CD for coated magnetic nanoparticles can effectively enhance the arsenic removal. The reason is that the coating can prevent MNPs from aggregating in adsorption process, and this increases the effective adsorption area. Also β-CD can form inclusion complexes with arsenic oxyanions in its hydrophobic cavity through host-guest interactions. Hence, the adsorption capacity of CM-βCD-MNPs was enhanced through inclusion complex formation.

Desorption Experiments

Desorption experiments were conducted to evaluate the possibility of regeneration of the adsorbent. Desorption of As(III) from the surface of CM-βCD-MNPs was conducted using 0.1 M

Table. 3. Comparison of removal efficiency of As(III) by the (a) naked and (b) β-CD modified Fe₃O₄ nanoparticles

Adsorbent	Concentration (ppm)	% Removal
Unmodified Fe₃O₄ nanoparticles	10	74.60
β-CD modified Fe₃O₄ nanoparticles	10	91.36

Table. 4. Results of desorption and recovery of β-CD modified magnetic nanoparticles

	Time	% Adsorption	% Desorption	% Recovery
H_2SO_4	3 h	85.01	37.91	81.09
NaOH	3 h	82.80	33.71	68.13
Hot water (60 °C)	30 min	81.78	3.1	-

Table. 5. Removal efficiency of As(III) in real samples by CM-βCD-MNPs

Real sample	Initial As(III) (ppm)	Spiked As(III) (ppm)	% Removal
Gavandak	0.578	-	99.13
Ghochagh	0.186	10 ppm	92.97
Babagorgor	0.133	10 ppm	91.93
Industrial water	0.005	10 ppm	94.57

NaOH and 0.1 M H_2SO_4 solutions as eluent. The corresponding results are shown in Table 4. It can be seen that the eluents could desorb As(III) from the surface of nanoparticles lower than 40%. For examining the recovery of one time used nanoparticles, they were used again in another adsorption cycle using fresh arsenic solutions. The results in the last column of Table 4 indicate that they can be efficient in arsenic removal again. Desorption of As(III) from the surface of CM-βCD-MNPs was also carried out using heating at 60 °C in water. Since desorption efficiency of As(III) by heating was very low, it was indicated that arsenic adsorption process is chemisorption. Therefore, it can be hypothesized that β-CD modified magnetic nanoparticles remove As(III) through inclusion complex formation.

Real Samples

The ability of CM-βCD-MNPs in the adsorption of arsenic from real samples was investigated. Several experiments with CM-βCD-MNPs were conducted to remove arsenic from three surface water samples and one industrial water sample. The natural water samples were obtained from the villages of Kurdistan province, and industrial sample was collected from petroleum industrial waste water. To determine As(III) concentration in real samples, inductively coupled plasma mass spectrometry (ICP-Ms) was utilized. Table 5 lists the initial concentration and the removal efficiencies of As(III) after treatment with CM-βCD-MNP. In the observed samples, the obtained removals of As(III) were above 98%, and they were rarely affected by the commonly coexisted ions in real water samples. The arsenic concentration reached below WHO's drinking water standards through adsorption process using CM-βCD-MNPs.

CONCLUSIONS

In this study, β-cyclodextrin coated magnetic nanoparticles as novel adsorbents were prepared and characterized to remove As(III) from aqueous solution. These magnetic nano-adsorbents were fabricated by co-precipitation method which have the magnetic properties of Fe_3O_4 nanoparticles and adsorption properties of CM-β-CD. The TEM analysis indicated that the CM-β-CD modified Fe_3O_4 nanoparticles were monodisperse with a mean diameter of ~10 nm. These magnetic nano-adsorbents can effectively be used to remove As(III) from aqueous solution because the modification of Fe_3O_4 by CM-β-CD enhances the arsenic adsorption. Various factors affecting the uptake behavior such as contact time, pH, amount of CM-β-CD-MNPs, and initial concentration of As(III) were evaluated. The equilibrium data for CM-β-CD–MNPs fitted well the Langmuir isotherm model. The maximum monolayer adsorption capacity of CM-β-CD-MNPs was 12.33 mg/g at room temperature. Results of this work suggest that the CM-β-CD coated magnetic nanoparticles can have wide applications in the removal of hazardous species from water samples, and can be used instead of conventional adsorbents.

ACKNOWLEDGMENT

We acknowledge the nanotechnology research Institute of Shiraz University and the ministry of science and technology.

CONFLICT OF INTEREST

The authors declare that there are no conflicts of interest regarding the publication of this manuscript.

REFERENCES

1. M.I. Litter, M.E. Morgada, J.Bundschuh, Possible treatments for arsenic removal in Latin American waters for human consumption, *Environmental Pollution*, 158 (2010) 1105–1118.
2. K. Lizama A., T. D. Fletcher, G. Sun, Removal processes for arsenic in constructed wetlands, *Chemosphere*, 84 (2011) 1032–1043.
3. M.A. Malana, R.B. Qureshi, M.N. Ashiq, Adsorption studies of arsenic on nano aluminium doped manganese copper ferrite polymer (MA, VA, AA) composite: Kinetics and mechanism, *Chemical Engineering Journal*, 172 (2011) 721-727.
4. Sh. Zhang, H. Niu, Y. Cai, X. Zhao, Y. Shi, Arsenite and arsenate adsorption on coprecipitated bimetal oxide magnetic nanomaterials: $MnFe_2O_4$ and $CoFe_2O_4$, *Chemical Engineering Journal*, 158 (2010) 599-607.
5. A.H. Barati, A. Maleki, M. Alasvand, Multi-trace elements level in drinking water and the prevalence of multi-chronic arsenical poisoning in residents in the west area of Iran, *Science of the Total Environment*, 408 (2010) 1523-1529.
6. A. Gupta, M. Yunus, N. Sankararamakrishnan, Zerovalent iron encapsulated chitosan nanospheres-A novel adsorbent for the removal of total inorganic arsenic from aqueous systems, *Chemosphere*, 86 (2012) 150-155.
7. Y.Y. Park, T. Tran, Y. H. Lee, Y.I. Nam, G. Senanayake, M.J. Kim, Selective removal of arsenic(V) from a molybdate plant liquor by precipitation of magnesium arsenate, *Hydrometallurgy*, 104 (2010) 290-297.
8. D. Qu, J. Wang, D. Hou, Zh. Luan, B. Fan, Ch. Zhao, Experimental study of arsenic removal by direct contact membrane distillation, *Journal of Hazardous Materials*, 163 (2009) 874-879.
9. M. Sen, A. Manna, P. Pal, Removal of arsenic from contaminated groundwater by membrane-integrated hybrid treatment system, *Journal of Membrane Science*, 354 (2010) 108-113.
10. A. Figoli, A. Cassano, A. Criscuoli, M.S.I. Mozumder, M.T. Uddin, M.A. Islam, E. Drioli, Influence of operating parameters on the arsenic removal by nanofiltration, *Water Research*, 44 (2010) 97-104.
11. H. Saitua, R. Gil, A.P. Padilla, Experimental investigation on arsenic removal with a nanofiltration pilot plant from naturally contaminated groundwater, *Desalination*, 274

12. M.B. Baskan, A. Pala, A statistical experiment design approach for arsenic removal by coagulation process using aluminum sulfate, *Desalination*, 254 (2010) 42-48.
13. B. Pakzadeh, J.R. Batista, Surface complexation modeling of the removal of arsenic from ion-exchange waste brines with ferric chloride, *Journal of Hazardous Materials*, 188 (2011) 399-407.
14. D. Teclu, G. Tivchev, M. Laing, M. Wallis, Bioremoval of arsenic species from contaminated waters by sulphate-reducing bacteria, *Water Research*, 42 (2008) 4885-4893.
15. P.K. Srivastava, A. Vaish, S. Dwivedi, D. Chakrabarty, N. Singh, R.D. Tripathi, Biological removal of arsenic pollution by soil fungi, *Science of the Total Environment*, 409 (2011) 2430-2442.
16. L. Zhou, J. Xu, X. Liang, Zh. Liu, Adsorption of platinum (IV) and palladium(II) from aqueous solution by magnetic cross-linking chitosan nanoparticles modified with ethylenediamine, *Journal of Hazardous Materials*, 182 (2010) 518–524.
17. M. Velicu, H. Fu, R.P.S. Suri, K. Woods, Use of adsorption process to remove organic mercury thimerosal from industrial process wastewater, *Journal of Hazardous Materials*, 148 (2007) 599–605.
18. M. Abdollahi, S. Zeinali, S.M. Nasirimoghaddam, S. Sabbaghi, Effective removal of As (III) from drinking water samples by chitosan-coated magnetic nanoparticles, *Desalination and Water Treatment*, 56 (2015) 2092-2104.
19. V.M. Boddu, K. Abburi, J.L. Talbott, E.D. Smith, R. Haasch, Removal of arsenic (III) and arsenic (V) from aqueous medium using chitosan-coated biosorbent, *Water Research*, 42 (2008) 633-642.
20. A. Gupta, V.S. Chauhan, N. Sankararamakrishnan, Preparation and evaluation of iron–chitosan composites for removal of As(III) and As(V) from arsenic contaminated real life groundwater, *Water Research*, 43 (2009) 3962-3870.
21. D. Mohan, C.U. Pittman Jr, Arsenic removal from water/wastewater using adsorbents—A critical review, Journal of Hazardous Materials. 142 (2007) 1-53.
22. D.D. Gang, B. Deng, L.Sh. Lin, As(III) removal using an iron-impregnated chitosan sorbent, *Journal of Hazardous Materials*, 482 (2010) 156-161.
23. A.Z.M. Badruddoza, A.S.H. Tay, P.Y. Tan, K. Hidajat, M.S. Uddin, Carboxymethyl-β-cyclodextrin conjugated magnetic nanoparticles as nano-adsorbents for removal of copper ions: Synthesis and adsorption studies, *Journal of Hazardous Materials*, 185 (2011) 1177-1186.
24. S.M. Nasirimoghaddam, S. Zeinali, S. Sabbaghi, Journal of Industrial and Engineering Chemistry,
25. L. Feng, M. Cao, X. Ma, Y. Zhu, Ch. Hu, Superparamagnetic high-surface-area Fe3O4 nanoparticles as adsorbents for arsenic removal, *Journal of Hazardous Materials*, 217-218 (2012) 439-446.
26. E.M. M.D. Valle, Cyclodextrins and their uses: a review, *Process Biochemistry*, (2003).
27. Sh. J. Cathum, A. Boudreau, A. Obenauf, A. Dumouchel, C.E. Brown, M.Punt, Treatment of Mixed Contamination in Water Using Cyclodextrin-Based Materials, (2006).
28. A. Khodabakhshi, M. M. Amin, M. Mozaffari, Synthesis of magnetite nanoparticles and evaluation of its efficiency for arsenic removal from simulated industrial wastwater, *Iranian Journal of Environmental Health Science & Engineering*, 8 (2011) 189-200.
29. A. Z. M. Badruddoza, G.S.S. Hazel, K. Hidajat, M.S. Uddin, "Synthesis of carboxymethyl-β-cyclodextrin conjugated

magnetic nano-adsorbent for removal of methylene blue."*Colloids and Surfaces A: Physicochemical and Engineering Aspects* 367 (2010): 85-95.

30. M. Namdeo, S.K. Bajpai, Chitosan–magnetite nanocomposites (CMNs) as magnetic carrier particles for removal of Fe(III) from aqueous solutions, *Colloids and Surfaces A: Physicochem. Eng. Aspects.*, 320 (2008) 161-168.

31. Q. Peng, Y. Liu, G. Zeng, W. Xu, Ch. Yanga, J. Zhang, Biosorption of copper(II) by immobilizing Saccharomyces cerevisiae on the surface of chitosan-coated magnetic nanoparticles from aqueous solution, *Journal of Hazardous Materials,* 177 (2010) 676-682.

32. S. Luther, N. Borgfeld, J. Kim, J.G. Parsons, Removal of arsenic from aqueous solution: A study of the effects of pH and interfering ions using iron oxide nanomaterials, *Microchemical Journal,* 101 (2010) 30-36.

33. Y. Tian, M. Wu, X. Lin, P. Huang, Y. Huang, Synthesis of magnetic wheat straw for arsenic adsorption, *Journal of Hazardous Materials,* 193 (2011) 10-16.

22

The Potential of ZnO Nanoparticles to Reduce Water Consuming in Iranian Heavy Oil Reservoir

Masoumeh Tajmiri*, Mohammad Reza Ehsani

Department of Chemical Engineering, Faculty of Chemical Engineering, Isfahan University of Technology, Isfahan, Iran

ABSTRACT

Water is critically important, because its supply is under stress. In oil fields, the ratio-of-water-to-oil (WCUT%) can be 95% or higher. Managing this produced water is a great challenge whereas the best opportunity to reduce costs, improve profitability and preserve the natural environment. The oil industry is looking for more effective ways to reduce water consuming and improve the recovery rates. Nano materials are an obvious place to look. This study provides new insights into ZnO nanoparticles effects on residual oil saturation (SOR) and WCUT% through steam assisted gravity drainage (SAGD) process by experimental work. Laboratory tests were conducted in two experiments through the use of 2 dimensional scaled SAGD cell from an Iranian heavy oil reservoir. In the first experiment, the SAGD cell was saturated with heavy oil and in the second one, the cell was flooded with nanoparticles before saturation with oil. The amount of recoveries were monitored during 12 hours. Results show that the ultimate oil recoveries increase from 52.43% to 87.93% by adding ZnO nanoparticles, respectively. The experimental results provide the nanoparticles ability to reduce produced water and minimize fresh water use can contribute to water conservation.

KEYWORDS: *Nanoparticles; Steam Assisted Gravity Drainage; ZnO*

INTRODUCTION

The oil and gas industry is undergoing a series of dramatic shifts with one common outcome, extracting hydrocarbons is harder than ever before. Production from the world's largest conventional fields is in decline while national oil companies continue to control the majority of the world's oil reserves. Simultaneously, global demand for oil and gas continues to grow, fueled in large part by emerging economies.

As a result, producers have resorted to new techniques to bypass declining and inaccessible legacy sources of oil and gas. The last five years

have seen a dramatic increase in production from unconventional sources. These source, shale, oil sands and deep water offshore, represented 47 percent of capital spending in oil industry in 2016. Producers are using more to get less, more labor, more energy, more time and more water which all lead to higher costs for both producers and consumers. From the water used to flood declining conventional and offshore wells to the water injected to fracture underground shale to the steam required for oil sands extraction. Water is the most important input to the oil and gas industry. By 2030, if current trends continue,

* Corresponding Author Email: maral.tajmiri@gmail.com

Table 1. Experimental conditions and purposes of two steps

Step No of runs	Process	Purpose	Model width × height (cm²)	length (cm)	Pressure difference (KPas)
Step I	Usual SAGD	Effect of well spacing	20 × 20	15	20
Step II	Usual SAGD	Effect of ZnO nanoparticles	20 × 20	15	20

Fig. 1. The schematic of physical model;
1- The main body with inside dimensions 20×20×4 cm; 2- Upper cap with dimensions of 31.5×25.8×11 cm; 3- Lower cap with dimensions of 31.5×25.8×11 cm; 4- Steam injector well in 15 cm from producer; 5- Steam injector well in 10 cm from producerp; 6- Oil producer well in the bottom of model; 7- A total of 9 thermocouple (T1-T9) probes installed to the model. The thermocouples connected to the data acquisition instrument, which recorded and displayed the temperature profile within the model during the experiment; 8- The basis for rotating the model in different angels.

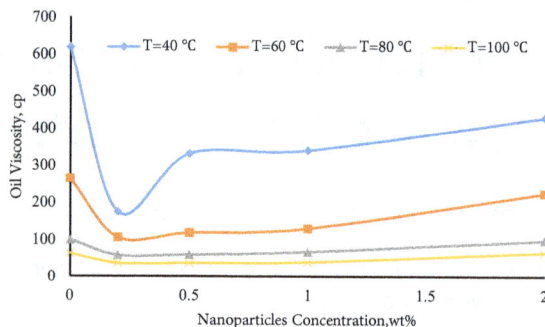

Fig. 2. The effect of different concentrations of ZnO nanoparticles on heavy oil viscosity.

global water requirements are expected to exceed supplies by forty percent. This trend is all the more relevant in oil and gas production, as many of the world's largest reserves reside in the most water-starved regions. Oil and gas producers should be concerned with water not only as a proactive step to be more efficient, but also as a defensive step against declining water supplies. From chemical and mechanical conformance tools to custom water treatment, water challenges with processes and technologies that reduce unwanted water production and treat produced water for disposal or reuse while satisfying a broad range of reservoir management and environmental objectives are solved. Globally, oil wells produce about 220 million BWPD (barrels of water per day) roughly three barrels of water for every barrel of oil. In older fields, the WCUT% can be 95% or higher. Managing this produced water is a great challenge for operators. Rising prices for energy coupled with the increasing environmental

awareness of consumers are responsible for a flood of products on the market that promise certain advantages for environmental and climate protection. Nanotechnology exhibits special physical and chemical properties that makes it interesting for novel and environmentally friendly products. In the chemical industry sector, nano materials are applied based on their special catalytic properties in order to boost energy and resource efficiency and nano materials can replace environmentally problematic chemicals in certain fields of application. High hopes are being placed in nanotechnology optimized products and processes for energy production and storage. These are currently in the development phase and are slated to contribute significantly to climate protection and solving our energy problems in the future [1]. The capability of nanotechnology usage in enhanced oil recovery processes is investigated. The rock pores may contain trapped oil, gas and water. Nanoparticles can be used to recover more residual oil. It is showed that nanotechnology affects on several parameters such as oil viscosity reduction [2]. High surface-to-volume ratio of nano fluids leads to improve thermal properties. The surface-to-volume ratio of nanoparticles may be 1000 times greater than of micro particles. It is tested a variety of particle sizes and types to find those best suited for plugging the rock pores, which turn out to be elastic nanoparticles made of polymer

Table 2. Properties of the oil used

Name	Oil density (API)	Dead oil viscosity (at 7.158 °C) (cp)	Compressibility (psi^{-1})	Thermal expansion factor (from 4 to 12.421°C) (°C^{-1})
KM	13	16000	4.67×10^{6}	4.00×10^{4}

Table 3. Specifications of ZnO nanoparticles used

Particle type	Formula	Form	Purity	Absolute particle size	Specific surface area	Appearance
Zinc oxide	ZnO	Nano powder	99%	<50 nm	>60 m^{2}g^{-1}	White powder

threads that retract into coils. Nanoparticles in solid form such as silica are less effective [3]. The effect of nano sized metal on decreasing viscosity through thermal process is reported. The results provide a good understanding of viscosity mechanism [4]. Silica nanoparticles fluid is flooded in water-wet sandstone and investigated the hydrophilic or hydrophobic monolayer role of them in the pore spaces. It is found that adsorption of SNPs can help to alter reservoir wettability [5]. Wettability can be changed depending on nanoparticles type through altering the chemical interactions in interfacial tension [6]. The oil recovery is increased about 4-5% compared to brine in the core flooding procedure by nano fluids. Hence, the potential of nanoparticles enhanced oil recovery is clear. The experimental results show that the IFT between water phase and oil phase can be reduced by nano fluids and the wettability of solid surface alters to more water wet and releasing oil drops by increasing capillary pressure is completely obvious [7]. Nano-size metal particles is used to reduce viscosity of heavy oil/bitumen through steam injection techniques. The experiments are a good proof of viscosity reduction by adding metal particles. The optimal concentration of metal particles are critical factors to affect on viscosity reduction [8]. The objectives of this study are to clarify the potential of ZnO nanoparticles on increasing oil recovery, decreasing the residual oil saturation (SOR) and WCUT%. By considering of environmental issues and characteristics of ZnO nanoparticles, the event of viscosity reduction by adding nanoparticles causes to use lower energy to supply steam in thermal process which helps to save more energy and water.

MATERIALS AND METHODS

A 2D sand pack model was designed for studying SAGD experiment. A single SAGD process well pair structure was deployed for steam injection and oil production. The vertical spacing between well pairs, l, set up to 15 cm. Two experiments were conducted which the major experimental condition sand purposes are listed in Table 1.

Scaled Reservoir Model

For investigating SAGD process, a rectangular physical model (Sand pack) designed for this study which was made of stain steel 316 with dimensions $20 \times 20 \times 4$ cm^3. As shown in Fig. 1, the physical model was consisted of different parts.

Materials

The crude oil used in this study was taken from an Iranian heavy oil field. Heavy oil properties are given in Table 2. The water phase for all experiments as base fluid was distilled water with viscosity of 1cp at ambient condition. To study the effect of nanoparticle on SAGD process, zinc oxide was selected. ZnO nanoparticles specifications are listed in Table 3.

Selection of the best concentration of nanoparticles is one of the challenging issues. Fig. 2 shows the effect of different concentrations of ZnO nanoparticles on heavy oil viscosity at different temperatures. It seems that by adding nanoparticles, viscosity reduction happens more. Another feature of Fig. 2 is that by decreasing nano concentration, viscosity decreases more. It is presumed that the main reason of this viscosity reduction can be related to the catalytic characteristics of nanoparticles on breaking the carbon and sulfur bonds in a chemical reaction. The most viscosity reduction happens at the lowest concentration of nanoparticles (0.2-0.5%wt).

Experimental Procedure

Prior to all experiments, the model was assembled. The thermocouples were placed back into the model and the pressure test was conducted. Usually the model was left pressurized with gas for

Fig. 3. Oil saturating set up; a) put the packed model inside the oven, b) fill the accumulator (600 ml) with heavy oil, c) connect the accumulator to injection pump and packed model, d) pump 2PV of heavy oil to the model to make sure it saturated with heavy oil completely.

Fig. 4. Flooding the packed model with nanoparticles; a) fill the accumulator 600 ml with homogeneous ZnO nanoparticles solution, b) connect the accumulator to injection pump and packed model, c) pump 2PV of nanoparticle and flood the model.

24h to make sure that there was no pressure leak. In the second step, the physical model was packed with sand from the reservoir. During packing, the model was vibrated and held at several different angles to make sure no gap would be left behind and a homogenous packing was created. The packing and shaking process typically took 24h. Table 4 shows the measured porosity, absolute permeability and pore volume of packed model for all experiments. The third step was to evacuate the model to remove air from the pore space. Finally, the model was connected to a vacuum pump and evacuated for 16h. The model was disconnected from the vacuum pump and kept on vacuum for couple of hours to make sure that it held the vacuum. If high vacuum was maintained, it was ready for saturating. Fig. 3 shows the packed model saturated with oil. The oil saturation of model took around 72h. After passing these steps, the packed model was ready for SAGD process.

For doing SAGD test with ZnO nanoparticles, the model preparation was achieved in a step-wise manner as follows; a) clean the physical model, b) vacuum the pore space of packed model, c) flood the ZnO nanoparticles to packed model about 72h, d) dry the packed model about 72h, e) saturate the model with heavy oil about 72h. After finishing the previous test, the packed model cleaned with toluene during 42 days and dried in oven for 7 days. Then the packed model flooded with ZnO nanoparticles. Water was selected as base fluid. By using ultrasonic device UP200S (Hielscher Ultrasonic, 200W, 24 kHz),

homogeneous nanoparticles solution was made. Maintaining the homogeneous solution during the experiment was one of the most important challenging issues. ZnO Nanoparticles tend to be disposed after around 7h. To avoid this problem and assure that the nanoparticles affected on the packed model directly, nanoparticles flooding procedure applied. First, the model was vacuumed and as shown in Fig. 4, flooding procedure was applied. Hence, the cell was ready to saturate with heavy oil for 72h the same as previous test.

Fig. 5 shows a schematic of the displacement apparatus used in this study. This apparatus included a water pump, steam generator, steam accumulator, 2-D scaled sand pack model, production control mechanism and the data acquisition system. Water was injected using nitrogen pressure via water accumulator cylinder. This positive pressure pump could inject at pressure range between 150 and up to 4000 psi. Steam generator heated water with an electrical element to vaporize water and injected the steam by nitrogen pressure pump with constant pressure. A temperature controller was used to inject the hot fluid at constant temperatures. The physical model was built from stainless steel with operating pressure up to 10000 psi and operating temperatures up to 500 °C. The physical model and measurement tools placed in a thermostatic oven as shown by dotted line in Fig. 5. In each experiment, the oven temperature fixed at 75°C and the system was allowed to reach thermal equilibrium. For steam injection, pressure was set

Table 4. Properties of packed model used

Absolute permeability (D)	Porosity (%)	horizonral permeability / vertical permeability	Initial oil saturation	Initial water saturation
4.87	34	1	100%	0%

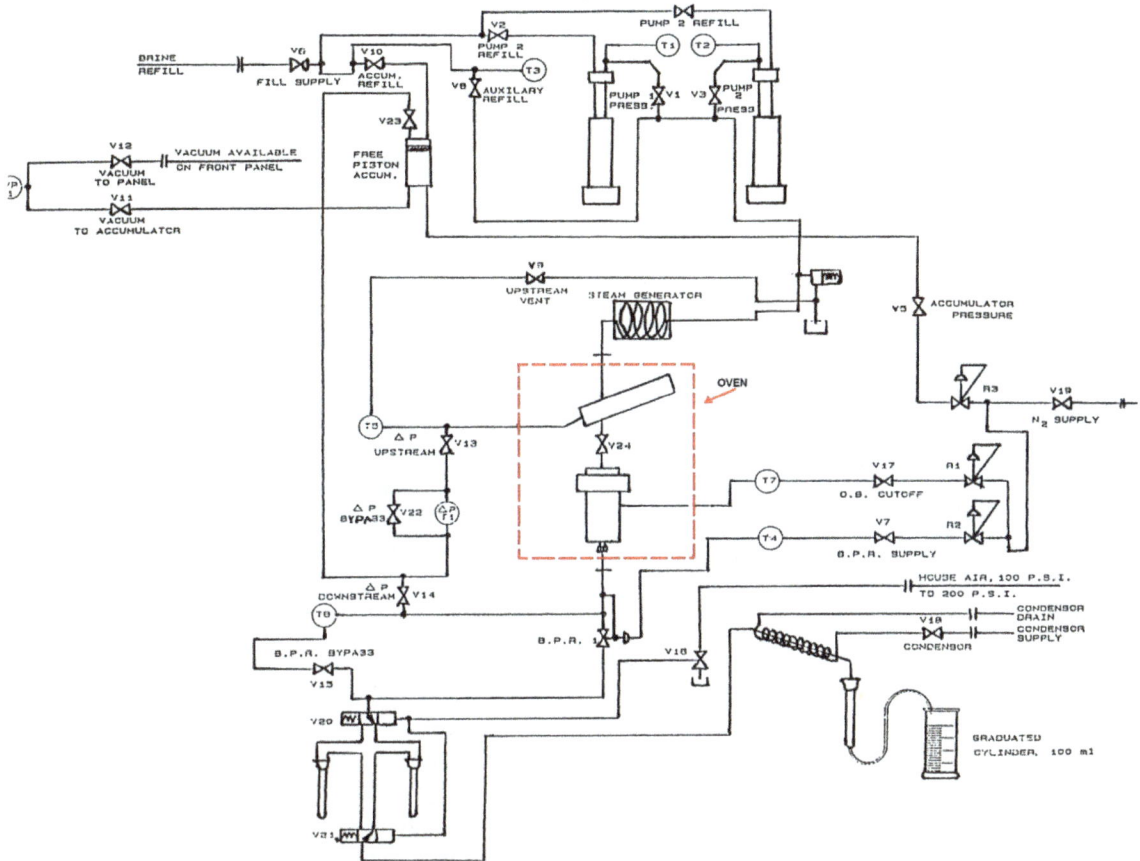

Fig. 5. Schematic representation of experimental apparatus.

in accumulator. When the system became steady state, oven and steam generation temperature and pressure injection were set and water became steam in vent tube. Once the oil production decreased below 1%, steam injection stopped. Mixture of heavy oil and produced condensate water collected in a sample bottle every 15 minutes. All oil production and condensed water measured in calibrated graduated cylindrical tubes. The oil and emulsion heated at 60 °C in the oven for 24h to break emulsion into oil and water and lately used centrifuged at 6000 RPM speed to separate the water and oil completely and the amount of separated oil was measured. The fluid cooled in a condenser before collection.

RESULTS AND DISCUSSION

The first experiment used the configuration with vertical distance between the horizontal well-pair of 15 cm. The oil recovery is shown in Fig. 6. When the injection well is at =15 cm from producer, it takes much time to invade heavy oil to start production. Despite fast creating steam chamber, improving the steam chamber domain is slow. Passing the time, more area of heavy oil heats by steam and gets mobile. As oil produces, steam chamber is able to touch and transfer heat to wider area of oil. Meanwhile, steam chamber effect is sensible and oil production increases. The steam breakthrough happens after 53 minutes. The ultimate oil recovery is 52.43% after 690 minutes. Obviously, locating

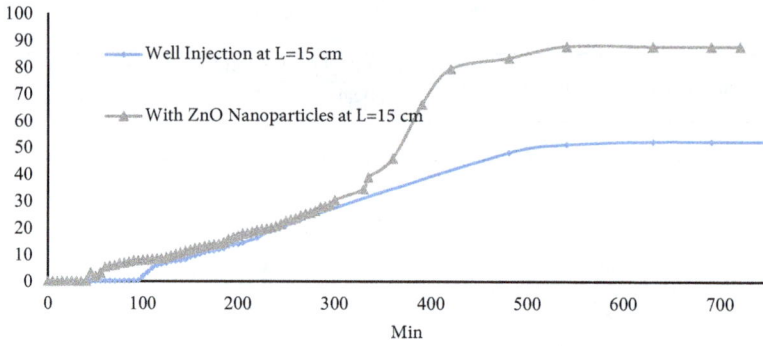

Fig. 6. Oil recovery for two sets of experimental conditions.

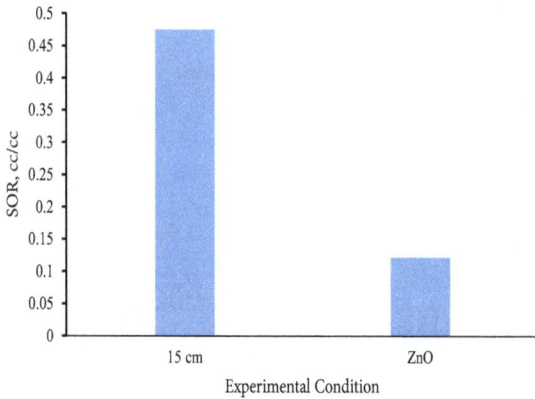

Fig. 7. SOR for well distance at well distance at 15 cm and with ZnO nanoparticles.

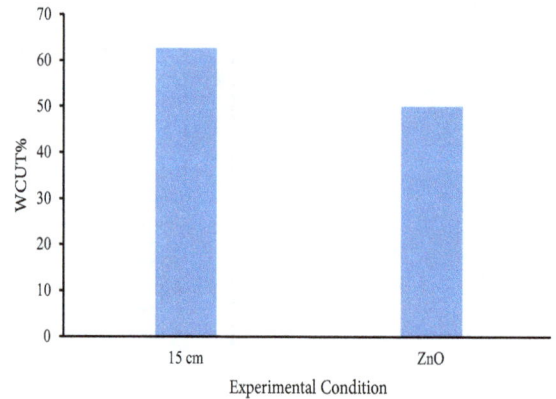

Fig. 8. WCUT% for two steps of experimental conditions; well distance at 15 cm and with ZnO nanoparticles.

well injection at 15cm causes more area of heavy oil is invaded by steam chamber heat therefore, experiment's time is longer and oil production increases more. Considering the findings from previous experiments in second experiment, the oil recovery is shown in Fig. 6. The condition of steam chamber formation is almost the same as previous experiment. After 62 minutes, the steam breakthrough happens. The role of nanoparticles becomes important once steam adds to the system. Nanoparticles have very small particle sizes which penetrate into pore volume of porous media and stick on the packed model surface. As shown in Fig. 2, by increasing the heavy oil temperature and ZnO nanoparticles, viscosity reduces dramatically. This feature of ZnO nanoparticles together with steam injection helps to intensify heavy oil viscosity reduction. It is clear from Fig. 6, the effect of ZnO nanoparticles on decreasing oil viscosity is sensible which helps to raise oil recovery compared to previous experiment. Another interesting finding is that ZnO nanoparticles causes the experiment

time shorter with high oil recovery. The ultimate oil recovery is 87.93% of OOIP at 630 minutes of experiment thanks to ZnO nanoparticles.

SOR for two experiments is shown in Fig. 7. It can be seen that, SOR for the first experiment (l=10 cm) is roughly 1.3 higher than second experiment (l=15 cm). Larger well spacing has considerable effect on SOR reduction. The important role of ZnO nanoparticles is clear for SOR reduction. As shown, SOR for second experiment is 3.89 higher than SOR of experiment with ZnO nanoparticles.

The comparison of WCUT% for two experiments also confirm that ZnO nanoparticles have potential to decrease the water content about 1.25 times than the second experiment and 1.6 times compared to the first experiment. The results of water cut for each experiment are shown in Fig. 8.

Due to increasing demand on fresh water sources, saving energy through improved efficiency and conservation by adding nanoparticles has a central role to play in reconciling the goals of economic development, energy security and environmental protection. Producers will need to understand their water demands, not just to improve the yield of their wells, but also to address the growing public concern over fracking. In particular fresh water resources which may already be fully allocated and wastewater storage and disposal can become hot-button issues with local communities. As shown in Fig. 8 the use of nanoparticles will be increasingly used to prevent wastewater fouling. As a result, oil companies are fast at work making their activities less water intensive. Hence, nanotechnology can help oil industry in this issue.

CONCLUSIONS

Oil and gas producers need to understand their water consumption, the requirements and limitations of the areas in which they operate and the potential of nanotechnology investment to protect and enhance their profitability into the future. This study shows the capability of ZnO nanoparticles to reduce viscosity and increases heavy oil recovery whereas WCUT% is decreased through SAGD process. The catalytic chemical reaction of nanoparticles on breaking the bonds between carbon and sulfur is assumed that the main reason of viscosity reduction. The most viscosity reduction happens at the lowest concentration of nanoparticles (0.2-0.5%wt). Based on ZnO nanoparticles characteristics on viscosity reduction, the final oil recovery is considerably increased hence, the SOR and WCUT% is definitely decreased.

CONFLICT OF INTEREST

The authors declare that there are no conflicts of interest regarding the publication of this manuscript.

REFERENCES

1. Shah, R.D., 2009. Application of Nanoparticle Saturated Injectant Gases for EOR of Heavy Oils. International Student Paper Contest at the SPE Annual Technical Conference and Exhibition, Louisiana, USA, SPE 129539- STU.
2. Fletcher, A and j. Davis, 2010. How EOR Can Be Transformed by Nanotechnology. Presented at the SPE Improved Oil Recovery Symposium, Tulsa, Oklahoma, 24-28 April, SPE 129531.
3. Kong, X and M.M. Ohadi, 2010. Application of Micro and Nano Technologies in the Oil and Gas Industry an Over view of the Recent Progress. Presented at International Petroleum Exhibition & Conference held, Abu Dhabi, Emirate, SPE 138241.
4. Hamedi Shokrlu, Y and T. Babadagli, 2010. Effect of Nano Sized Metals on Viscosity Reduction of Heavy Oil/ Bitumen thermal Applications. Presented at the Canadian Unconventional Resources & International Petroleum Conference, Calgary, Alberta, Canada, CSUG/SPE 137540.
5. Hendraningrat, L and L. Shidongm, 2010. A Glass Micro Model Experimental Study of Hydrophilic Nanoparticles Retention for EOR Project. Paper read at SPE Russian Oil and Gas Exploration and Production Technical Conference and Exhibition, 16-18 October, Moscow, Russia, SPE 159161.
6. Oglo, N.A and M.O. Onyekonwu, 2010. Enhanced Oil Recovery Using of Nano Particles. Presented at the SPE Saudi Arabia Section Technical Symposium, Al-Khobar, Saudi Arabia, 8-11 April, SPE 160847.
7. Hendraningrat, L and L. Shidongm, 2013. Improved Oil Recovery by Hydrophilic Silica Nanoparticles Suspension. 2-Phase Flow Experimental Studies. International Petroleum Technology Conference, 26-28 March, Beijing, China, IPTC 16707.
8. Babadagli, T and Y. Hamedi Shokrlu, 2014. Viscosity Reduction of Heavy Oil/Bitumen Using Micro- and Nano-Metal Particles during Aqueous and Non-aqueous Thermal Applications. Journal of Petroleum Science and Engineering, 119: 210-220.

Permissions

List of Contributors

Taher Yousefi, Meisam Torab-Mostaedi, Amir Charkhi and Abolfazl Aghaei
Nuclear Fuel Cycle Research School, Nuclear Science and Technology Research Institute, Tehran, Iran

Zahra Sarteep
Fouman Faculty of Engineering, College of Engineering, University of Tehran, Tehran, Iran

Azadeh Ebrahimian Pirbazari
Fouman Faculty of Engineering, College of Engineering, University of Tehran, Tehran, Iran Caspian Faculty of Engineering, College of Engineering, University of Tehran, Tehran, Iran

Mohammad Ali Aroon
Caspian Faculty of Engineering, College of Engineering, University of Tehran, Tehran, Iran

Ouahiba Bechiri and Mostefa Abbessi
Laboratory of Environmental Engineering, Department of Process Engineering, Faculty of Engineering, University of Annaba, Annaba, Algeria

André Savall and Karine Groenen Serrano
Laboratory of Chemical Engineering, University of Toulouse, CNRS, INP, 118 route de Narbonne 31062 Toulouse cedex, France

Ines Bouaziz
Laboratory of Chemical Engineering, University of Toulouse, CNRS, INP, 118 route de Narbonne 31062 Toulouse cedex, France Laboratory of Electrochemistry and Environment, National Engineering School, Sfax University, BP 1173, 3038 Sfax, Tunisia

Morched Hamza and Ridha Abdelhedi
Laboratory of Electrochemistry and Environment, National Engineering School, Sfax University, BP 1173, 3038 Sfax, Tunisia

Jafar Azamat
Department of Chemical Engineering, Ahar Branch, Islamic Azad University, Ahar, Iran

Leila Mahdavian
Department of Chemistry, Doroud Branch, Islamic Azad University, Doroud, Iran

Zahra Hassanzadeh Siahpoosh and Majid Soleimani
Department of Chemistry, Imam Khomeini International University (IKIU), Qazvin, Iran

Bal Chandra Yadav and Ravindra Kumar
Department of Applied Physics, School of Physical Sciences, Babasaheb Bhimrao Ambedkar University, Lucknow-226025, U.P., India

Ritesh Kumar
Department of Physics, University of Lucknow, Lucknow-226007, U.P., India

Subhasis Chaudhuri and Panchanan Pramanik
Department of Chemistry, IIT Kharagpur, W.B., India

Elham Shokri and Reza Yegani
Faculty of Chemical Engineering, Sahand University of Technology, Tabriz, Iran

Soheill Azadikhah Marian
Department of Chemical Engineering, Faculty of Engineering, University of Azad, North Branch, Tehran, Iran

Morteza Asghari and Zahra Amini
Separation Processes Research Group (SPRG), Department of Engineering, University of Kashan, Kashan, Iran

Mohsen Moghimi, Mohammad Ghorbanpour and Samaneh Lotfiman
Chemical Engineering Department, University of Mohaghegh Ardabili, Ardabil, Iran

Siroos Shojaei
Department of Chemistry, University of Sistan and Baluchestan, Zahedan, Iran

Somaye Khammarnia
Department of Chemistry, Payam-e-noor University, Zahedan, Iran

Saeed Shojaei
Department of Desertification, University of Yazd, Iran

Mojtaba Sasani
Analytical Chemistry, University of Sistan and Baluchestan, Zahedan, Iran

Hanieh Karimnezhad
Polymer Research Center, Department of Chemical Engineering, Razi University, Kermanshah, Iran

Farzane Arsiya and Mohammad Hossein Sayadi
Department of Environmental Sciences, School of Natural Resources and Environment, University of Birjand, Birjand, Iran

Sara Sobhani
Department of Chemistry, College of Science, University of Birjand, Birjand, Iran

Marjan Tanzifi, Mohsen Mansouri, Maryam Heidarzadeh and Kobra Gheibi
Department of Chemical Engineering, Faculty of Engineering, University of Ilam, Ilam, Iran

Maryam Adimi and Maziyar Mohammad Pour
Department of Chemical Engineering, Farahan Branch, Islamic Azad University, Farahan, Iran

Hassan Fathinejad Jirandehi
Young researchers and Elite Club, Farahan Branch, Islamic Azad University, Farahan, Iran

Eman Serag and Ahmed El-Nemr
Marine Pollution Department, Environmental Division, National Institute of Oceanography and Fisheries, Kayet Bey, El-Anfoushy, Alexandria, Egypt

Azza El-Maghraby
Fabrication Technology Department, Advanced Technology and New Materials Institute, City for Scientific Research and Technology Application, Alexandria, Egypt

Marjan Tanzifi and Kianoush Karimipour
Department of Chemical Engineering, Faculty of Engineering, University of Ilam, Ilam, Iran

Marzieh Kolbadi Nezhad
School of Chemical Gas and Petroleum Engineering, Semnan University, Semnan, Iran

Shahryar Nazarpour Laghani
Caspian Faculty of Engineering, College of Engineering, University of Tehran, Rezvanshahr, Iran

Azadeh Ebrahimian Pirbazari
Caspian Faculty of Engineering, College of Engineering, University of Tehran, Rezvanshahr, Iran
Fouman Faculty of Engineering, College of Engineering, University of Tehran, Fouman, Iran

Saeed Tizro and Hadi Baseri
School of Chemistry, Damghan University, Damghan, Iran

Sedigheh Zeinali and Maryam Abdollahi
Department of Nanochemical Engineering, Faculty of Advanced Technologies, Shiraz University, Shiraz, Iran

Samad Sabbaghi
Department of Nanochemical Engineering, Faculty of Advanced Technologies, Shiraz University, Shiraz, Iran
Nanotechnology Research Institute, Shiraz University, Shiraz, Iran

Masoumeh Tajmiri and Mohammad Reza Ehsani
Department of Chemical Engineering, Faculty of Chemical Engineering, Isfahan University of Technology, Isfahan, Iran

Index